Pretending and imagination in animals and children

It is well known that children's activities are full of pretending and imagination, but it is less appreciated that animals can also show similar activities. This is the first book to focus on comparing and contrasting children's and animals' pretenses and imaginative activities. In this book, overviews of recent research present conflicting interpretations of children's understanding of the psychology of pretense, and describe sociocultural factors that influence children's pretenses. Studies of nonhuman primates provide examples of their pretenses and other simulative activities, explore their representational and imaginative capacities and compare their skills with children's. Although the psychological requirements for pretending are controversial, evidence presented in this volume suggests that great apes and even monkeys may share capacities for imagination with children, and that children's early pretenses may be less psychological than they appear.

ROBERT W. MITCHELL is Professor of Psychology at Eastern Kentucky University. He is currently interested in exploring the significance of kinesthetic-visual matching in human and animal behavior and psychological understanding, and is writing a history of scientific attitudes toward using anthropomorphism to understand animals. Professor Mitchell's previous books include *Self-awareness in animals and humans* (1994, ISBN 0521441080), edited with S. T. Parker and M. L. Boccia, and *The mentalities of gorillas and orangutans* (1999, ISBN 0521580277), edited with S. T. Parker and H. L. Miles.

Pretending and imagination in animals and children

Edited by
ROBERT W. MITCHELL
Eastern Kentucky University

PUBLISHED BY THE PRESS SYNDICATE OF THE UNIVERSITY OF CAMBRIDGE
The Pitt Building, Trumpington Street, Cambridge, United Kingdom

CAMBRIDGE UNIVERSITY PRESS
The Edinburgh Building, Cambridge CB2 2RU, UK
40 West 20th Street, New York, NY 10011–4211, USA
477 Williamstown Road, Port Melbourne, VIC 3207, Australia
Ruiz de Alarcón 13, 28014 Madrid, Spain
Dock House, The Waterfront, Cape Town 8001, South Africa

http://www.cambridge.org

© Cambridge University Press 2002

This book is in copyright. Subject to statutory exception
and to the provisions of relevant collective licensing agreements,
no reproduction of any part may take place without
the written permission of Cambridge University Press.

First published 2002

Typeface Lexicon (*The Enschedé Font Foundry*) 9/13 pt   *System* QuarkXPress™   [SE]

A catalogue record for this book is available from the British Library

Library of Congress Cataloguing in Publication data

Pretending and imagination in animals and children / edited by Robert W. Mitchell.
    p. cm.
  Includes bibliographical references (p.   ) and indexes.
  ISBN 0 521 77030 0
  1. Imagination in children.   2. Psychology, Comparative.   I. Mitchell, Robert W., 1958–

BF723.I5 P74 2002
156′.33–dc21   2001037368

ISBN 0 521 77030 0 hardback

Transferred to digital printing 2003

This book is dedicated
to the memory of
Ina Č. Užgiris

Contents

List of contributors x
Foreword by Sue Taylor Parker xiv
Preface and acknowledgments xvii

I Historical, developmental, and comparative overviews

1 Imaginative animals, pretending children 3
   ROBERT W. MITCHELL

2 A history of pretense in animals and children 23
   ROBERT W. MITCHELL

3 Pretending as representation: a developmental and comparative view 43
   LORRAINE McCUNE AND JOANNE AGAYOFF

II Pretense and imagination in children

4 Language in pretense during the second year: what it can tell us about "pretending" in pretense and the "know-how" about the mind 59
   EDY VENEZIANO

5 A longitudinal and cross-sectional study of the emergence of the symbolic function in children between 15 and 19 months of age: pretend play, object permanence understanding, and self-recognition 73
   PIERRE-MARIE BAUDONNIÈRE, SYLVIE MARGULES, SOUMEYA BELKHENCHIR, GWÉNNAELLE CARN, FLORENCE PÈPE, AND VÉRONIQUE WARKENTIN

6 Caregiver–child social pretend play: what transpires? 91
   ROBERT D. KAVANAUGH

7   Just through the looking glass: children's understanding of
    pretense 102
    ANGELINE LILLARD

8   Young children's understanding of pretense and other fictional
    mental states 115
    JACQUELINE D. WOOLLEY

9   Pretend play, metarepresentation and theory of mind 129
    PETER K. SMITH

10 Replica toys, stories, and a functional theory of mind 142
    GRETA G. FEIN, LYNN D. DARLING, AND LOIS A. GROTH

11 Young children's animal-role pretend play 154
    OLIN EUGENE MYERS, JR.

12 Imaginary companions and elaborate fantasy in childhood:
    discontinuity with nonhuman animals 167
    MARJORIE TAYLOR AND STEPHANIE M. CARLSON

III Pretense and imagination in primates

13 Pretending in monkeys 183
    ANNE ZELLER

14 Pretending primates: play and simulation in the evolution of
    primate societies 196
    PETER C. REYNOLDS

15 Representational capacities for pretense with scale models and
    photographs in chimpanzees (*Pan troglodytes*) 210
    SARAH T. BOYSEN AND VALERIE A. KUHLMEIER

16 Pretending in free-ranging rehabilitant orangutans 229
    ANNE E. RUSSON

17 Seeing with the mind's eye: eye-covering play in orangutans and
    Japanese macaques 241
    ANNE E. RUSSON, PAUL L. VASEY, AND CAROLE
    GAUTHIER

18 Possible precursors of pretend play in nonpretend actions of captive
    gorillas (*Gorilla gorilla*) 255
    JUAN CARLOS GÓMEZ AND BEATRIZ MARTÍN-ANDRADE

19 Pretending culture: social and cognitive features of pretense in apes
    and humans 269
    WARREN P. ROBERTS AND MARK A. KRAUSE

20 Empathy in a bonobo 280
    ELLEN J. INGMANSON

21 Pretend play in a signing gorilla 285
MARILYN L. MATEVIA, FRANCINE G. P. PATTERSON, AND
WILLIAM A. HILLIX

IV Prospects

22 Exploring pretense in animals and children 307
ROBERT W. MITCHELL

References 317
Author Index 353
Subject Index 362

Contributors

JOANNE AGAYOFF
Department of Educational Psychology, Graduate School of Education, Rutgers – The State University of New Jersey, New Brunswick, NJ 08903, USA

PIERRE-MARIE BAUDONNIÈRE
Neurosciences Cognitives et Imagerie Cerebrale, UPR CNRS 640, Equipe de Psychobiologie du Developpement, Hôpital de la Salpétrière, Paris Cedex 13, France

SOUMEYA BELKHENCHIR
Neurosciences Cognitives et Imagerie Cerebrale, UPR CNRS 640, Equipe de Psychobiologie du Developpement, Hôpital de la Salpétrière, Paris Cedex 13, France

SARAH T. BOYSEN
Comparative Cognition Project, Department of Psychology, The Ohio State University, Columbus, OH 43210, USA

STEPHANIE M. CARLSON
Department of Psychology, University of Washington, Box 351525, Seattle, WA 98195, USA

GWÉNNAELLE CARN
Neurosciences Cognitives et Imagerie Cerebrale, UPR CNRS 640, Equipe de Psychobiologie du Developpement, Hôpital de la Salpétrière, Paris Cedex 13, France

LYNN D. DARLING
Department of Human Development, 3242 Benjamin Building, University of Maryland, College Park, MD 20742, USA

GRETA G. FEIN
Department of Human Development, 3242 Benjamin Building, University of Maryland, College Park, MD 20742, USA

CAROLE GAUTHIER
Departement d'Anthropologie, l'Université de Montréal, C.P. 6128, Succursale Centre-ville, Montréal, PQ H3C 3J7, Canada

JUAN CARLOS GÓMEZ
School of Psychology, University of St. Andrews, St. Andrews, UK

LOIS A. GROTH
Department of Human Development, 3242 Benjamin Building, University of Maryland, College Park, MD 20742, USA

WILLIAM A. HILLIX
The Gorilla Foundation, Woodside, CA 94062, USA

ELLEN J. INGMANSON
Dickinson College, Department of Anthropology, Carlisle, PA 17013, USA

ROBERT D. KAVANAUGH
Department of Psychology, Williams College, Williamstown, MA, USA

MARK A. KRAUSE
Department of Psychology, Mezes 330, The University of Texas, Austin, TX 78712, USA

VALERIE A. KUHLMEIER
Department of Psychology, Yale University, 2 Hillhouse Dr., New Haven, CT 06520, USA

ANGELINE LILLARD
Department of Psychology, P.O. Box 400400, University of Virginia, Charlottesville, VA 22903, USA

SYLVIE MARGULES
Neurosciences Cognitives et Imagerie Cerebrale, UPR CNRS 640, Equipe de Psychobiologie du Developpement, Hôpital de la Salpétrière, Paris Cedex 13, France

BEATRIZ MARTÍN-ANDRADE
Departamento de Psicología Evolutiva, Universidad Autónoma de Madrid, 28049 Madrid, Spain

MARILYN L. MATEVIA
The Gorilla Foundation, Woodside, CA 94062, USA

LORRAINE MCCUNE
Department of Educational Psychology, Graduate School of Education, Rutgers – The State University of New Jersey, New Brunswick, NJ 08903, USA

ROBERT W. MITCHELL
Department of Psychology, Eastern Kentucky University, Richmond, KY 40475, USA

OLIN EUGENE MYERS, JR.
Huxley College of Environmental Studies, Western Washington University, Bellingham, WA 98225, USA

SUE TAYLOR PARKER
Department of Anthropology, Sonoma State University, Rohnert Park, CA 94928, USA

FRANCINE G. P. PATTERSON
The Gorilla Foundation, Woodside, CA 94062, USA

FLORENCE PÈPE
Neurosciences Cognitives et Imagerie Cerebrale, UPR CNRS 640, Equipe de Psychobiologie du Developpement, Hôpital de la Salpétrière, Paris Cedex 13, France

PETER C. REYNOLDS
pcr@aya.yale.edu

WARREN P. ROBERTS
Assistant Director, Project Chantek, 3830 Woodcrest Circle NW, Cleveland, TN 37312, USA

ANNE E. RUSSON
Department of Psychology, Glendon College, York University, Toronto M4N 3M6, Canada

PETER K. SMITH
Department of Psychology, Goldsmiths College, New Cross, London SE14 6NW, UK

MARJORIE TAYLOR
1227 Department of Psychology, University of Oregon, Eugene, OR 97403, USA

PAUL L. VASEY
Assistant Professor, Department of Psychology & Neuroscience, University of Lethbridge, 4401 University Drive, Lethbridge, Alberta, T1K 3M4, Canada

EDY VENEZIANO
Laboratoire de Psychologie de l'Interaction, Université Nancy 2, Boulevard Albert 1er, 54015 Nancy Cedex, France

VÉRONIQUE WARKENTIN
Neurosciences Cognitives et Imagerie Cerebrale, UPR CNRS 640, Equipe de Psychobiologie du Developpement, Hôpital de la Salpétrière, Paris Cedex 13, France

JACQUELINE D. WOOLLEY
Department of Psychology, Mezes 330, The University of Texas, Austin, TX 78712, USA

ANNE ZELLER
Department of Anthropology and Class Studies, University of Waterloo, Waterloo, Ontario N2L 3G1, Canada

Foreword
by Sue Taylor Parker

The collection of articles herein focuses on the mysterious liminal region that lies between pre-symbolic and symbolic abilities. The transition between the two remains the least charted area, most intriguing of all developmental transformation in human childhood, and the least understood of all transformations in hominoid evolution. The mystery is deepened by disagreements over terminology. *Simulation, imitation, pretense, symbolic play, representation, meta-representation, theory of mind, intentionality,* and *imagination,* the very definition of these terms is contested territory. Authors of articles in this volume do not simplify the task because many of them disagree on these matters. Rather, their articles provide readers with a fascinating array of perspectives on these and related concepts. They also provide comparative data on a rich array of great ape species: humans, bonobos, chimpanzees, gorillas, and orangutans, plus some macaque monkeys.

Following in the wake of his earlier work on deception, self-awareness, and anthropomorphism, Robert Mitchell's new collection carries us further into contested twilight zones between infancy and childhood, and between other great ape and human minds. The juxtaposition between animals and children in the title is more than accidental since many of the same frameworks have been used to study and compare children of our species with those of our closest living relatives, the nonhuman great apes, an approach that has come to be known as comparative developmental evolutionary psychology (Parker, 1990).

One of the key debates in this volume, for example, revolves around Bateson's (1956) idea that the play face in monkeys constitutes metacommunicative, intentional deployment of gestures to convey such messages as "this is only play" or whether it is simply an evolved communicative display that directly reflects the animals motivational state. In other words, authors differ over whether or not "metacommunication" and perhaps pretence exists in monkeys. At one extreme, Reynolds argues that "evolutionary changes in primate social organization presuppose the cognitive functions of pretence and simulation." At the other extreme Gómez and Martín-Andrade argue that, even in gorillas, apparent examples of pretend play can better be explained by such precursors of true pretense as the ability to decontexualize and mentally represent actions.

Similar differences exist among human developmental psychologists regarding the emergence of pretence and symbolism. Many draw heavily on Piaget's stages of developmental of imitation and pretend play. Some, like Leslie (1988), argue that the earliest forms of pretend play with objects representing other objects entails meta-representation whereas others such as Perner (1991) argue for intermediate stages, a position taken by Veneziano and Fein *et al.* in this volume.

Before exploring all this, readers will find Mitchell's introductory chapter provides a useful roadmap to these and other contrasts in viewpoint of various contributors. Likewise, his chapter on the history of ideas about pretense in animals and humans provides a much-needed perspective on the incredible persistence of many of these issues. Finally, Mitchell's concluding chapter and that of Roberts and Krause, explicitly address implications of comparative studies of pretense for understanding the evolution of human culture and cognition.

Given the developmental and evolutionary proximity between pretense and early language, perhaps it is inevitable that interest in the developmental and evolutionary emergence of language lurks behind much of the work on pretense. Piaget (1945/1962) and others have long argued for the common origins of symbolic play, language, and drawing. In recent years, developmental psychologists have moved in two opposing directions in their modeling of language acquisition. On the one hand, Pinker (1994) and other neo-Chompskians have emphasized the early, virtually imperturbable unfolding of innate grammar. On the other hand, Nelson and other constructivists have emphasized the prolonged and contingent course of language acquisition: "The child does not immediately make a leap from prelinguistic to linguistic, or from sensorimotor to representational, or any of the other stages that have been proposed as explanations of developments between 1 and 3 years of age. The transition is long and composed of a complexity of developments in different parts of the social-linguistic–cognitive system" (Nelson, 1996, p. 120).

These two positions, in turn, project onto two different scenarios for the pace of language evolution. According to the first, language evolved recently and rapidly. According to the second, symbolic abilities began to evolve early in hominid evolution, indeed in ape evolution, but became fully linguistic only recently. Contributors to this volume generally fall more on the constructivist side.

Those who enjoy contending hypotheses and a rich pallet of color and flavor in their comparative studies, will surely find a feast within! Choose your partner and let the dance begin again!

Preface and acknowledgments

As the only comparative psychology graduate student at Clark University, Massachusetts, I was surrounded by graduate students studying human development assured in their belief that humans were distinctly different psychologically from other animals. When imaginative pretense was brought up as one of many "uniquely human" capacities (including language, intentional deception, imitation, and self-recognition), I mentioned the chimpanzee Viki's imaginary pulltoy (Hayes, 1951), to the general response that it was only one example compared to the myriad instances exhibited by children. The intellectually stimulating debates that followed my frequent disagreements led to this book, and fueled my desire to publish three other books, which I initiated prior to this one – *Deception: perspectives on human and nonhuman deceit* (Mitchell & Thompson, 1986), *Self-awareness in animals and humans* (Parker, Mitchell & Boccia, 1994), and *Anthropomorphism, anecdotes, and animals* (Mitchell, Thompson & Miles, 1997). I hope the skills I learned from my talented co-editors – all ardent educators – are evident in my first run as solo editor.

While I was at Clark, Ina Užgiris introduced me to Paul Guillaume's (1926/1971) *Imitation in children* in her course on "Imitation, internalization, and identification," and to Piaget's (1947/1972) *Psychology of intelligence* in her course on "Piaget's theory." Reading Guillaume and Piaget was a gift, but reading them with Ina's guidance was an extraordinary gift, for which I am grateful.

I would like to thank Sue Parker, Anne Russon, and Angeline Lillard for editorial assistance and support, Mark Spina, Shyamala Venkataraman, Cathy Clement, and Ron Mawby for intellectual stimulation amidst years of friendship, Eastern Kentucky University (EKU) for financial support, and my students at EKU for putting up with my frequent absences to work on this book. I also appreciate the patience of the authors in this volume, as well as my editor Tracey Sanderson, none of whom ever complained about the year-long delay in publication (precipitated in part by my grandmother's extended dying during the summer of 2000). I also wish to acknowledge the first-hand delights of pretending with my young friends Benjamin Mawby, Elizabeth Vincelli, and Elliot Mawby, and the continuing affirmation, love, and support of my partner Randy Huff.

Part I
Historical, developmental, and comparative overviews

1

Imaginative animals, pretending children

> Could one imagine a world in which there could be no pretence?
>
> (WITTGENSTEIN, 1949/1992, p. 37e)

This book is a delightful collection of scientific writings about pretending and imagination in animals and children. The impetus for the present volume derives not only from observations of animals' activities similar (perhaps identical) to pretense (e.g., Groos, 1898; Mitchell & Thompson, 1986; Mitchell, 1987, 1990, 1991a, 1993c, 1994a; Byrne & Whiten, 1990; Miles, 1991; Russon, 1996), but also from developing ideas about children's understanding of pretense (Harris & Kavanaugh, 1993; Lillard, 1993a,b, 2001a), philosophy of art (Walton, 1990), and the evolution of image-making (Davis, 1986), all of which concern organisms' understanding or creating reproductions of various sorts (Mitchell, 1994a). The continuing influences of Piaget (1945/1962), Guillaume (1926/1971), Bateson (1955/1972, 1956), Vygotsky (1930–1966/1978), and Leslie (1987) are also apparent. The purpose of the book is primarily to present and examine evidence for the existence and nature of pretense in animals and children, and secondarily to examine various aspects of why or how. Evidence of pretense in animals may eventually allow us to provide a "psychologically and evolutionarily plausible account of 'fictive acts of perceiving'" (Davis, 1986, p. 211; see Mitchell, 1994a; Reynolds, PIAC14).[1] A full appreciation of what needs to be incorporated into such an account is provided by reading chapters in this volume, in which the topics range from relatively simple simulative actions to complex fantasizing about nonexistent objects and agents.

Answering questions about animal pretense requires us to look to behavior not only of various nonhuman species, but also of those prototypical pretenders – human children. From an early age, children act "as if" things are the case, and this "acting as if" seems present, if not always ubiquitous, in extant human cultures (Millar, 1968; Schwartzman, 1978; Roopnarine *et al.*, 1994; Kavanaugh, PIAC6; Smith, PIAC9). Because children are the prototypical pretenders, comparisons between children's and animals' pretenses are essential, and understanding children's pretense is necessary to get some idea as to the phenomenon itself and its scope. Children clearly outrank nonhuman animals as pretenders, but this does not mean that animals do not or cannot pretend. As scientists studying autistic children's pretense have learnt, failure to do something in some circumstances does not mean an inability to do it in others (Lewis & Boucher, 1988). Children's pretense builds on precursors that are themselves not pretense (Piaget, 1945/1962; Fein & Moorin, 1985), so that even nonpretending animals may exhibit precursors (Gómez & Martín-Andrade, PIAC18). In addition, children's pretense may not be as complex as many believe. Children may understand pretense as simply a unique type of action – "acting as if" (Fein & Moorin, 1985; Harris & Kavanaugh, 1993; Lillard, 1993a,b, 1998b; PIAC7; Jarrold *et al.*, 1994; Smith, PIAC9). Although this view is controversial (see Woolley, PIAC8; Taylor & Carlson, PIAC12), the fact that it is even plausible opens a window to explore pretense in animals.

Pretense and imaginative activities

Pretense or make-believe is a mental activity involving imagination that is intentionally projected onto something (Goldman, 1998; Lillard, PIAC7). More elaborately, make-believe is "the use of . . . props in imaginative activities" (Walton, 1990, p. 67), where props are "objects of imaginings" (p. 25). Props include pretenders themselves, who are simultaneously also imaginers, imagining about objects (including themselves) that they are something else. Pretense in play is called "symbolic play," but pretending also occurs outside play, and need not be "playful." Autistic children can enact pretend scenarios with little apparent pleasure (Wulff, 1985) and, in some cases of trauma, children compulsively (and unhelpfully) repeat their experiences in grim pretense (Terr, 1990; Gordon, 1993; Smith, PIAC9), apparently gaining new understanding about the trauma, rather than catharsis (Coates & Moore, 1997).[2]

Imagination, so central to pretense, is a tricky concept (Walton, 1990). Minimally, imagination requires that an organism has an idea which it seeks to examine in its actions or mind.[3] Most children's early pretense seems to be imaginative in that they are acting out ideas (schemas), based on conventional experiences (e.g., eating, sleeping) or variations on these, which they are just coming to understand (Sully, 1896; Vygotsky, 1930–1966/1978; Fein & Apfel, 1979a). Imagination can also imply innovation, as when unconventional ideas are put into play. Such innovative imagination is present in some nonpretend actions, as when animals engage in "what if" scenarios, such as eye-closing games (Gómez and Martín-Andrade, PIAC18; Russon, Vasey & Gauthier, PIAC17). Imagination is present by definition in the imaginary companions of linguistically skilled children (Taylor & Carlson, PIAC12), referring here to the invented and fictional nature of the companions.

Although external manifestations are not essential for pretense (Lillard, PIAC7), outsiders need them to discern pretense. Consequently, observational definitions *require* external manifestation. Pretense often depends on imitation of activities out of context or otherwise different from the original activities (Bates *et al.*, 1979; Bretherton, 1984). Although some require that the imitation be of another's actions (Mitchell, 1987, 1990), others acknowledge that more rudimentary pretense can involve imitation of an individual's own actions (Groos, 1898; Piaget, 1945/1962; Bretherton, 1984; McCune & Agayoff, PIAC3). Children's earliest pretenses appear to be imitations of their own activities (self-pretenses), but this priority is questionable (Bretherton, 1984) as some research suggests a simultaneous occurrence of imitation of self and others in the earliest pretenses (Guillaume, 1926/1971; Lowe, 1975; Fein & Apfel, 1979a). Whether self-pretense occurs in animal play requires exploration (Gómez & Martín-Andrade, PIAC18); though it seems common in deception and teasing (Mitchell, 1994a), authors disagree as to whether these manipulations are pretense (Russon, PIAC16) or not (McCune & Agayoff, PIAC3).

Similarity and simulation

Recognizing pretense requires knowledge of the relationship between the pretend "copy" and its "model." Discerning this model (or that there is a model) often relies upon knowledge of regularities (norms) of behavior, whether the behavior is bodily action (see Gómez & Martín-Andrade, PIAC18), language (Veneziano, PIAC4), or other cultural phenomena (Fein, Darling & Groth, PIAC10). In some pretenses the model is purely imaginary, such that there is no externally observable model (Taylor & Carlson, PIAC12), and in others the model is used as only a springboard for inventive and imaginary characterization (see discussion in Harris & Kavanaugh, 1993). Language users can talk to clarify what their pretenses are about, and in the middle of their second year children discover the need for clarification (Veneziano, *PIAC4*). With nonlinguistic organisms, pretense can be detected only if an observer can recognize an intended similarity between the pretend copy and the model (Fein & Moorin, 1985; Mitchell, 1987; McCune & Agayoff, *PIAC3*).

Concern with similarity between pretenses and their models is prominent in Bateson's (1955/1972; 1956) notion of metacommunication in play and Grice's (1982) ideas about the evolution of non-natural meaning (Mitchell, 1991a; *PIAC2*). In Bateson's conception, organisms recreate activities in such a way that simulation (something's being designed to resemble something else; Mitchell, 1991a, 1994a) is obvious as such; in Grice's conception, they produce intentional simulation for recognition as such. Grice imagines organisms producing simulations (of actions with "natural" meaning) for or as communication (thereby creating "non-natural" meaning), suggestive not only of representation, but also of understanding other minds (Mitchell, 1987, 1990, 1991a, 1994a). Metacommunication (intentional or not), implicit in orangutans' acting "nicer" than usual or being otherwise different from normal, directed Russon (*PIAC16*) to the pretend nature of these actions. Similarly, metacommunication is present in exaggerated actions by children and chimpanzees (McCune & Agayoff, *PIAC3*). Non-natural meaning occurred in metacommunicative pretenses by a chimpanzee who "offered his leg in an exaggerated way to his partner ... and then feigned effortful attempts to run away" to instigate playchase with a playpartner who engaged in leg-pulling (Tomasello *et al.*, 1985, p. 181), by chimpanzee mothers who slowly reenacted their own nut-cracking techniques for their infants (Boesch, 1991a), and by a young bonobo who "made twisting motions toward containers ... he needed help in opening ... , or made hitting motions toward nuts he wanted others to crack for him" (Savage-Rumbaugh, 1986, p. 386).

Bateson's ideas are expanded by Reynolds (*PIAC14*), who argues that primate social behaviors are based on simulation of innate behavior patterns, which are "redeployed in a symbolic manner." This simulative propensity is taken to the extreme in synchromimesis, in which individuals produce highly similar behaviors simultaneously – a phenomenon common in humans (Hatfield, Cacioppo & Rapson, 1994), but also present in other socially sophisticated animals such as dolphins (Fellner & Bauer, 1999). Like

Reynolds, Zeller (*PIAC13*) maintains that macaques engage in self-simulation and recognize the effects that repetitions of their actions have on other monkeys (particularly in deception), although she acknowledges that they provide few examples of either imitation of others or pretense.

Similarity seems to be an essential aspect of pretense for children. In now classic experiments, Lillard told children that a doll Moe is hopping like a rabbit but the doll does not know what rabbits are and is not trying to act like a rabbit. When she asked whether the hopping doll is pretending to be a rabbit or not, most children from 3 to 6 years of age claimed that he is (Lillard, 1993b). These and other data (Harris & Kavanaugh, 1993; Lillard, *PIAC7*) suggest that children for some time are unaware of mental aspects of pretense and see pretense as a form of action: "acting as if." By contrast, other studies suggest that even young children are aware of some mentalistic aspects of pretense (see discussion in Woolley, *PIAC8*; Taylor & Carlson, *PIAC12*). To my mind (and Lillard's) these studies instead suggest that children distinguish pretense from reality – a distinction which (suggest McCune and Agayoff, *PIAC3*) might not be of much concern to animals (however much distinguishing between other animals' feigned and natural actions is).

Although children show greater success on tasks similar to Lillard's "Moe task" which rely less on language and which depict mentalistic and action components more deliberately (Woolley, *PIAC8*), these changes alone cannot explain children's typical failure on the Moe task because children often succeed on Lillard's task when they themselves are the purported "pretenders." I replicated aspects of Lillard's studies but replaced Moe with the child him/herself or another person. Children (aged 4.5–6.5 years) generally succeeded at recognizing that they themselves were not pretending (even after watching their movements in a mirror) when their actions were described as looking like those of a (real or fictional) animal (e.g., a cat reaching for a ball). However, as with Moe, these children generally failed to recognize that another person was not pretending when that person's actions were described as looking like those of an animal (Mitchell, 2000). In fact, these children's attributions of pretense were directly related to their attributions of similarity between the action produced and the purportedly similar action. Children who agreed (or disagreed) that an action looked like a purported pretend action tended to agree (or disagree) that the action was pretense; the self/other difference in pretense attribution resulted because children tended to agree more that another's actions looked like an animal's actions, than that their own actions did (Mitchell & Neal, 1999; Mitchell, 2000). Variations on Lillard's studies (Woolley, *PIAC8*) often show greater success by young children when the doll's actions are similar to two potential models (only one of which the doll knows about), and children are forced to choose which model the doll is pretending about, thereby precluding the use of similarity to detect pretense.

Young children's sensitivity to similarity is present not only in their attributions of pretense, but also in their self-simulations, imitations of others, and even the objects with which they pretend. Children initially (prior to age 3) prefer to pretend with objects that prototypically resemble real objects more than with ones that do not (El'konin, 1969; Elder & Pederson, 1978; Jackowitz & Watson, 1980; Fein, 1981; Pederson, Rook-Green & Elder, 1981; Fein & Moorin, 1985). Learning and culture influence the recognition of resemblance (see Davis, 1986; Walton, 1990; Mitchell, 1994a; Noble & Davidson, 1996), and sometimes even strong similarities are not enough for young children (or apes) to recognize resemblance outside pretense (DeLoache, 1991; Boysen & Kuhlmeier, *PIAC15*). Objects that look like other things to players stimulate them to inquisitively try out using these objects as if they were these things (Fein & Apfel, 1979a; Musatti & Mayer, 1987) – a phenomenon Lorenz (1950/1971) detected in animals' object play. While noticing similarities is widespread among animals (Guthrie, 1993; Fagot, 1999), creating resemblances is not (Mitchell, 1991a).

The fact that creating and using resemblances is common in human experience suggests that it may have had an important place in human evolution. Indeed, in his discussion of the origins of human image-making, Davis (1986) suggests that accidental recognition of resemblance and subsequent attempts to recreate or develop the resemblance induced (some) hominids to arrive at the idea that a mark (e.g., a curved slash in a cave wall) could represent something else (e.g., a horse) (see also Dowson, 1998). (A similar recognition sometimes occurs for children in their scribblings; Luquet, 1927.) This recognition of resemblance between divergent things indicates *seeing* or *experiencing* something *as* something else. Recognizing and recreating resemblance can occur with any medium – including bodily actions, gestures, and sounds (Davis, 1986) – and has considerable consequence, in that it allows organisms to *experience* something *as* something else – a doll as a baby, a stick as a horse, another's bodily actions or gestures as one's own – which is essential for pretense (Mitchell, 1994a).

Resemblances acted out bodily or affirmed for things can turn bodies and things into objects of imagining, props for games of make-believe (Walton, 1990). To know that something is a prop for an organism, that something is experienced as something else, requires some knowledge of the possibilities of what the prop could be experienced as. Specifically, we expect that organisms must have experience of real things in order to represent them in pretense. Consequently, we need to know behavioral patterns of individuals intimately before we can begin to interpret what their actions are simulating (Köhler, 1925/1976; Davis, 1986; Mitchell, 1986, 1987, 1994a). This knowledge is exactly what several authors provide given their intense involvement in the lives of their subjects (e.g., Goodall, 1973; Miles, 1986; Veneziano, PIAC4; Zeller, PIAC13; Russon, PIAC16; Gómez & Martín-Andrade, PIAC18; Matevia et al., PIAC21).

Detecting the similarity between animals' and young children's actions and their models seems relatively easy given that the possibilities in their experiences appear limited and saliently related to recurrent experiences (Fein & Moorin, 1985; Mitchell, 1994a; Miles, Mitchell & Harper, 1996), but more is needed for evidence of pretense – the similarity must occur out of context, that is, be "decontextualized." When animals repeat their own behaviors, such as repetitively rolling down a hill, they engage in self-imitation, but self-pretense requires that they reproduce their own behavior in appearance only – usually in a new context or for a new purpose. This is why deception, in which actions are used in new contexts, seems particularly relevant to pretense, and why its usefulness might have created an evolutionary context from which skills at pretense could develop (Mitchell, 1994a; Russon, PIAC16). Repeating others' behaviors, by contrast with self-simulation, seems more like pretense, in that the reenactment of the behavior is immediately decontextualized. The development from recreating one's own actions to recreating another's exhibits "decentering," a movement away from oneself as the center of action (Fenson, 1984; Musatti & Mayer, 1987; Lyytinen, 1991). Recreating one's own or another's behavior in a new context may be only a preliminary stage in the development of pretense, a precursor ("re-presentation" – Bates *et al.*, 1979) necessary for eventual representation (Piaget, 1945/1962; Gómez & Martín-Andrade, PIAC18).

McCune & Agayoff (PIAC3) acknowledge pretense only when activities "evoke" other activities experienced by the pretender, rather than recreate these activities for "practical goals" (as in deception). Similarly, Harris & Kavanaugh (1993, p. 72) distinguish children's indication of something via iconic representation from deception. In their view, "during pretense, a signifier (lying down, closing their eyes momentarily) does not signify the real act of sleeping. Even though it may be inspired by and [be] a partially accurate reproduction of the real act of sleeping, it can still be a piece of make-believe in that it stands for a fictional act of sleeping." By contrast, in deception, as when a child pretends to be asleep in hopes of seeing Santa Claus, the child is representing real sleep: "the child is pretending to be asleep in order to convey to Santa Claus that he or she is really asleep." In effect, deceiving organisms must produce an appearance closely matched to the modeled activity (avoiding metacommunication), whereas pretending organisms need only produce an aspect of the modeled activity, without regard to exact correspondence, to evoke the idea of it (which is inherently metacommunicative) (Mitchell, 1986, 1987, 1991a). Still, both pretense and deception evince abilities for simulation suggestive of an understanding of their underlying fictionality (Reddy, 1991; Mitchell, 1994a, 1996). The usefulness of pretense in deceit may constrain nonlinguistic organisms to recreate more exact, rather than more imaginative or evocative, replications, for more effective deception.

The complex deceptions among wild apes suggest that human enculturation is not essential for the expression of complex cognition in apes (Russon, PIAC16, citing Whiten & Byrne, 1991). However, even those rare cases of deception by apes which strongly suggest "higher" cognition (Whiten & Byrne, 1988; Byrne & Whiten, 1990) offer ambiguous evidence of it (Mitchell, 1988, 1993c, 1997c). I suggest that, instead of complex metarepresentational abilities, knowledge structures concerning action–reaction regularities in their experience ("scripts") explain almost all deceptions by apes, whether human-reared or wild, as well as nonlinguistic deceptions by young children (Mitchell, 1999a). Indeed, highly complex deceptions show integration among various scripts in much the same way that complex pretenses do (Nicolich, 1977; Fenson, 1984; McCune & Agayoff, PIAC3). As Russon (PIAC16) shows, little exposure to action–reaction sequences is necessary for orangutans to recognize which of their own actions they must repeat to deceive. For apes and young children, sometimes one experience is enough to recreate it, indicating rapid script development (Mitchell, 1999a). Adult human deceptions also depend upon scripts, but their more elaborate deceits require extensive use of props and numerous steps in their planning, a pretending perhaps beyond that of apes (Mitchell, 1996, 1999a). Even by 2.5 years of age children, enabled by their elaborate linguistic skills, begin to have deceptions "too complex to be merely behavioural routines" (Newton, Reddy & Bull, 2000, p. 313). Likewise, in pretend play development, children at this age and older appear to be examining more and more elaborate possibilities inherent in scripts and invented plans (Fein & Moorin, 1985; McCune & Agayoff, *PIAC3*).

Deceiving animals clearly recognize the usefulness of enacting a script in a never-experienced-before context, which implies attentiveness to the possibilities of their current situation (Russon, *PIAC16*). Such "decontextualization" suggests that their behavior is "detached" from its typical real-life supports (Fein & Moorin, 1985). Still, deceiving animals (and young children) may initially learn the action–reaction regularities embodied in scripts without attention to other contextual specifics, such that any contextual novelty may be in the eye of the observer, not the deceiver (Miles, 1986; Mitchell, 1986). In effect, the deceptive act may be experienced as identical to the "real" act for the deceiver, who does not think that the deceptive act *represents* the real act. Similarly, young children may be able to find a hidden object in a room from its location in a photograph of the room (because the photograph is "identical" with the room), but have trouble finding the hidden object when shown its location using a miniature replica of the object in a scale model *representing* the room (DeLoache, 1987; see Boysen & Kuhlmeier, *PIAC15*). Actions used to deceive, and photographs used to show an object's location, can be viewed as "the same thing," not as representations. (Indeed, children and apes sometimes attempt to listen for sounds from photographs of sound-producing objects; see Mitchell, *PIAC2*.) Describing deceptions (or pretenses) as symbolic or complexly representational requires knowing that animals recognize that their actions stand for other actions, not only recreate them in hopes of their having similar consequences. For such knowledge, one needs to look at the development of an organism's simulations (Fein & Moorin, 1985; Davis, 1986; Mitchell, 1986, 1993c, 1997c).

Development: symbols and language

> A child must have developed far before it can pretend, must have learned a lot before it can simulate.
>
> (WITTGENSTEIN, 1949/1992, p. 42e)

In Piaget's (1945/1962) view, symbolic pretense derives initially from imitation of self which extends to others over the course of sensorimotor development. McCune & Agayoff (*PIAC3*) elaborate this view for the development of pretense with objects (props). In this developmental scheme, prior to actual pretense children simply reproduce actions they know well, showing that they understand regular (normative) uses of objects. (This is necessary not only for the child to be able to pretend, but for the researcher to know that the child understands realistic uses of the objects.) Next, they show similar uses of these objects, but with some part missing – a spoon contains no food, a bottle no liquid. In these early pretenses, their own body can be a prop, as when they pretend to sleep, eat, or drink. The intent seems to be to evoke these activities, suggesting a focus on enjoying their fictionality. Children then branch out, performing toward dolls or people actions which they or others normally performed toward themselves, as well as enacting activities of others. By the end of this developmental sequence, children show planning in their pretense, indicating the "interiorized imitation" (mental images) to which Piaget believed sensorimotor development led. This analysis combines sometimes distinguishable changes in children's understandings of roles, actions, and objects (Bretherton, 1984) in a useful framework for examining and comparing pretenses by animals. (Note that this developmental scheme focuses on children's optimal performance during ontogeny, not the typical type of pretense they enact – which tends to be less sequential in ontogeny; Hoppe-Graff, 1993.) The more developed pretenses by apes are shown in captivity, and these employ representational skills used for more practical purposes in the wild (McCune & Agayoff, PIAC3).

As Guillaume (1926/1971) and Piaget (1945/1962) noted, pretense is related to other symbolic activities. Guillaume posited that the development of symbolic bodily imitation led from visual matches experienced repetitively and contiguously between one's own and others' actions in relation to objects, to an ability to match between the kinesthetic experiences of one's own body and visual images similar to one's own body – a "kinesthetic-visual matching" present in symbolic activities such as mirror-self-recognition, planning, and pretending to be another (see Mitchell, 1993a,b, 1997a,b; Reynolds, PIAC14). Piaget acknowledged the developmental coordination among these activities as representing the development of symbolic or representational capacities that also include language and understanding object permanence. Using both cross-sectional and longitudinal approaches with children, Baudonnière *et al.* (PIAC5) examined the developmental coordination of object-permanence understanding, self-recognition, and pretending (specifically, pretending to eat imaginary food after an adult modeled and verbally described her actions and then asked the child to partake). In their study, understanding object permanence (typically present in pretense, which requires "the ability to represent an absent object"; Lowe, 1975, p. 33) preceded self-recognition and pretense. By contrast, these latter two skills developed virtually simultaneously, indicating a common ability which, to these authors, is secondary representation. I suggest more particularly that it is kinesthetic–visual matching (Mitchell, 1993a, 1997a). Sign-using apes also show both self-recognition and pretense, and sometimes use mirrors imaginatively to transform their body images, a form of "dress up" common among children (Hayes, 1951; Miles, 1994; Miles *et al.*, 1996; Roberts & Krause, PIAC19).

Development: cultural scaffolding and scripts

Describing global developments in pretend play, Fein & Apfel (1979a, p. 99) wrote: "On the one hand, there is a tendency toward expansion and elaboration which makes anything possible, and, on the other, there is the beginning of a consensus which narrows and reduces the possibilities which are realized." Oddly, the imaginative component of early pretense becomes culturally canalized toward conventionality. This canalization occurs through toys – apparently quite ancient cultural devices (King, 1979) – but also through language. Indeed, language and toys apparently enable older autistic children (who rarely pretend spontaneously) to produce significantly imaginative pretense with object substitution in somewhat structured contexts (Lewis & Boucher, 1988; Boucher & Lewis, 1990; Jarrold *et al.*, 1993; Wolfberg, 1999). Language, that totally conventional medium, frees pretense from convention by creating possibilities undreamed of through imitation of self or others (Fein *et al.*, PIAC10), but simultaneously constrains most pretenses to be about things that can be articulated for others to participate in and enjoy (Veneziano, PIAC4). In fact, young children apparently find pleasure in fitting their pretend actions to others' commands (Rheingold, Cook & Kolowitz, 1987). Although speech predominates in children's pretense at 26 months (Fenson, 1984), it is not essential for young children's pretense (Fein & Moorin, 1985; Casby, 1997), which at 20 months occurs largely without speech (Fenson, 1984). It remains unknown whether pretend play assists in developing language or the reverse (see Lillard, PIAC7; Smith, PIAC9), if both share a common basis in symbolic development (Piaget, 1945/1962; Werner & Kaplan, 1963; Bates *et al.*, 1979), or if they are unrelated (Folven,

Bonvillian & Orlansky, 1984/1985). While sign-using and speech-comprehending great apes appear more competent in pretense than captive and wild apes (Hayes, 1951; Savage-Rumbaugh & McDonald, 1988; Miles, 1991; Jensvold & Fouts, 1993; Miles *et al.*, 1996; Matevia *et al.*, PIAC21), this greater skill may have to do with the greater wealth of things to pretend about in the typically middle-class environments in which these apes are raised.

Language in pretense begins as confirmation of obvious aspects of children's actions, but by 18–23 months of age children also use language more elaborately to describe aspects of the pretense which might not be clear to an observer (Veneziano, *PIAC4*). This pragmatic elaboration of speech co-occurs with other uses, such as references to the past and justifications for their actions, suggesting a change in the understanding of communicative functions of language (a view which seemingly overturns Leslie's (1987) concerns that language used in pretense should be ultimately confusing to a child, in that it equates real and pretend objects). The gorilla Koko's sign-use in pretense bears resemblance to children's earliest language, in that she largely describes what is obvious, but in a few instances she makes clear what might not be so to her playmates (Matevia *et al.*, *PIAC21*).

Language in pretense develops still further in storytelling, where children become more articulate in defining story parameters such as context and internal states. The "narrative thought" developed in storytelling becomes an important "tool" of culture (Fein *et al.*, *PIAC10*), in that sequential events can be coherently organized (scripted) for understanding. As Kavanaugh (*PIAC6*) notes, human mothers (and siblings) not only "enhance and elaborate" infants' "fleeting and sporadic" early attempts to pretend, but initiate, demonstrate, and otherwise "create a context" conducive to infant pretend play *until* such limited pretending takes hold (prior to which infants can be largely unresponsive) (see also Užgiris, 1981, 1999). Indeed, one might argue that a great deal of parents' initial involvement with infants is based on pretend that the infant is a culturally competent agent (Kaye, 1980). Such pretense is sometimes applied to human-raised animals such as dogs, apes, and parrots, and, as with children, can lead to linguistic skill and psychological achievements (Tomasello & Call, 1997; Mitchell, 1997d, 2001; Miles, 1999; Roberts & Krause, *PIAC19*; McCune & Agayoff, *PIAC3*). Story-elaboration shown by sign-using gorilla Koko (Matevia *et al.*, *PIAC21*) reminds one of early stories by children which, like their pretenses, are heavily scaffolded both culturally and socially (Peterson & McCabe, 1996).

Sociocultural influences in wild apes (Whiten *et al.*, 1999; Russon, *PIAC16*) evince continuity in this domain between animals and humans (contra Taylor & Carlson, *PIAC12*). Still, unlike some human mothers (Snow, Dubber & de Blauw, 1982; Hoppe-Graff, 1993; First, 1994), ape mothers show little scaffolding of bodily imitation or other forms of self-other matching in games with infants (Mitchell, 1993a; Parker, 1993). Scaffolding, whether by adults or siblings, seems to require free time, where the scaffolding agents and children are not overwhelmed by subsistence activities (McCune & Agayoff, *PIAC3*; Roberts & Krause, *PIAC19*). The presence of nonreproductive grandmothers to act as babysitters for collections of children in early human evolution may have freed time for symbolic development in collaborative activities among peers (Roberts & Krause, *PIAC19*).

Cultural beliefs influence the occurrence of scaffolding of children's learning about pretend as well as the themes of pretense (Bretherton, 1984; Taylor & Carlson, *PIAC12*). Adult encouragement leads children in normative directions, and pretending children in part act upon or against norms and other salient aspects of their culture (Kavanaugh, *PIAC6*). As Karl Bühler (1930, pp. 93–4) somewhat exaggeratedly noted, "Everything that happens in the family, and later on, everything the child hears about, is repeated in play." More specifically, children "represent in pretend play life's challenges, not life's commonplaces" (Fein & Apfel, 1979a, p. 97). Infants typically start pretending by imitating their own and others' regularly experienced actions, which leads to the conclusion that much early pretense is acting out scripts (Bretherton, 1984; Smith, *PIAC9*). As previously noted, early object pretense requires that children understand some cultural regularities (norms) about the objects of their pretense (Lowe, 1975; McCune & Agayoff, *PIAC3*). Pretense can also involve plans ("the mechanisms that underlie scripts"; Schank & Abelson, 1977, p. 70), such that players can create scenarios using an internal model without any specific external model.

The development of pretense shows increasing integration among schemas (Bretherton, 1984; McCune & Agayoff, *PIAC3*), but initial pretenses can be remarkably simple. Children can pretend with only an idea, as when Guillaume's (1926/1971, p. 117) 11-month-old daughter "pretends to put a marble in her mouth immediately after she has been forbidden to do so [and] smacks her lips defiantly." Sign-using gorilla Koko sometimes pretends by "almost" signing the correct sign in contexts in which she clearly knows the correct sign (Matevia *et al.*, *PIAC21*), and other apes similarly "almost" reenact routinized activities in pretense and deceit (Quiatt, 1984; Mitchell, 1999a,b). Enacting normative or regularly experienced behaviors toward inappropriate objects (acting as if eating inedible food) by orangutans can serve to defuse tensions and signal friendly relations (Russon, PIAC16), perhaps because such "script violation" is so interesting to onlookers (see Goodall, 1973, on chimpanzee "leaf grooming").

Toys and animals, scripts and schemas

If we could be certain that apes treat lifeless objects as dolls, this act would be in the foremost rank of illusion plays.

(GROOS, 1898, p. 302)

Children's pretenses commonly involve toys, often miniature props which resemble and thereby indicate real-life objects. What young children do with these toys often borrows extensively from real-world (scripted) uses of their referents (Lowe, 1975). Toys seem a largely human phenomenon intended to canalize children's pretense. Toys attach to scripts which inform children about what to pretend: empty glasses suggest drinking, empty spoons eating, dolls infants, and toy alligators attacking. Chimpanzees and 4-year-old children can also recognize when miniature objects represent specific objects which are visually similar but larger (Boysen & Kuhlmeier, PIAC15).

Some toys do not resemble what they represent in pretense, as when children use corncobs or sticks as dolls (Hall, 1914). Other toys may not resemble anything known to pretenders, but are props nonetheless. For example, both children and the gorilla Koko enjoy pretending with alligator toys, where the alligator acts as a nasty character (Fein et al., PIAC10; Matevia et al., PIAC21); yet these pretenders likely have little or no experience with alligators. These pretenders presumably learn that alligators are unpleasant interactors through language, and enjoy imagining scenarios in which they are prominent.

Children's projections of animism and personality to objects, effectively making them dolls, is mirrored by some human-enculturated apes (Gardner & Gardner, 1969; Miles, 1991; Matevia et al., PIAC21). Deciding whether dolls exist in nonenculturated animals is difficult. One macaque represented an infant with a coconut shell, moving the shell to the same bodily positions that the mother moved her infant, but otherwise made no response to the shell (Breuggeman, 1973). Descriptions of a captive bonobo treating a dead squirrel as a baby (Savage-Rumbaugh, Shanker & Taylor, 1998), and of wild chimpanzees treating logs in ways similar to the ways mothers treat infants, in one case even making a nest for one (Wrangham & Peterson, 1996; Matsuzawa, 1997), are more suggestive of dolls, but similar descriptions for captive gorillas unfamiliar with maternal behavior toward infants suggest that instinctive caregiving activities might be enacted without representing infants (Gómez & Martín-Andrade, PIAC18). Similarly, role behavior developed in play may lead to knowing how to participate in particular roles in adulthood without the play necessarily representing these adult roles (Parker, 1999). We must be careful not to make the same mistake children do in Lillard's Moe task of assuming that because an activity looks like another one (which it is not) that pretense is involved; we need to know if the organism knows about the purported model and is trying to create its appearance. Gómez & Martín-Andrade (PIAC18) assert that many seeming pretenses by their captive gorillas might be better described as using one object *instead of* another, rather than *as if* or *to represent* another. Still, though rudimentary, some instances of gorilla reenactment, such as repeatedly putting a spoon into an experimenter's mouth (similar to how the experimenter feeds them) would be classified as pretense if enacted by infants (Fein & Apfel, 1979a; McCune & Agayoff, PIAC3). Perhaps some early object substitution pretenses by children are also "using instead."

Some pretenses derive from knowledge structures (schemas) other than scripts. Person or animal schemas are applied to pretenders themselves, to others, to objects, and to nothing (as in imaginary companions), and these schemas can be incorporated into scripts. Myers (PIAC11) describes how fertile many animal schemas are for children. Children match their bodily actions and thereby represent activities of a diverse array of animals (some known only through drawings and speech). By contrast, if animals act like another being, it is almost always a conspecific (Zeller, PIAC13; Reynolds, PIAC14); acting as (or as if) a nonconspecific is rare, described only for captive dolphins and apes (Hayes, 1951; Tayler & Saayman, 1973; Miles, 1990; Matevia *et al.*, PIAC21).

Schemas of animals and humans are so well-developed in children that they use them to create imaginary companions (ICs) with thoughts and feelings (Taylor & Carlson, PIAC12; see Mead, 1934/1974; Guthrie, 1993). Sometimes these ICs are dolls or stuffed animals, but usually they are purely imaginary. Although many American children perceive these ICs as pretense (yet, paradoxically, remind others that they are "only pretend"), people in many cultures attach diverse levels of reality to the existence of imaginary beings – resulting in, from a different vantage point, "believed-in imaginings" (de Rivera & Sarbin, 1998). For example, as a Catholic child, I believed in the existence of a guardian angel who watched over me and ensured my safety – a kind of ghostly version of myself. Such guardian angels are culturally constructed ICs intentionally created for children, similar to but more persistently salient than Santa Claus. Knowledge of such culturally acceptable ICs may pave the way for children to create their own unique ICs, as might playing with dolls. Some children's ICs show marked creativity, presumably in part because language allows for variation from normative experiences. Still, ICs typically have human or animal attributes, suggesting a constraint on imaginings. Even unique ICs seem composites of aspects of real creatures – who happen to be invisible!

Although ICs seem uniquely human (Taylor & Carlson, PIAC12), animals sometimes show behaviors suggestive of ICs. For example, a long-tailed macaque named Rodrique with which I interacted, was caged alone and would sometimes turn away from me and threaten a nonexistent being, acting with repeated threat as if the nonexistent other in turn threatened him. Another example is Mills' (1921, p. 71) description of a pet beaver engaging in play with "imaginary playfellows. He raced or wrestled with them and occasionally simply annihilated an imaginary enemy." Although such activities might suggest vacuum activities or stereotypy induced by confinement (Lorenz, 1932/1970), the presence of imaginary others is more clearly implicated in the pretend displays of wild chimpanzees imitating more dominant conspecifics (Goodall, 1973), the aggressive displays seemingly directed toward the sky during storms (Guthrie, 1993), and the fear-inducing creatures (monsters, cats, and dogs) which an orangutan and a bonobo invented (and designated via signs) to engage others in their pretend fear (Miles, 1986; Savage-Rumbaugh & Lewin, 1994). The sign-using orangutan Chantek "dart[ed] about during play and look[ed] over his shoulder as if he were being chased[,] ... signed to his toys and offered them food and drink, ... and appeared to express sympathy for his toys by rescuing them from caregiver[s'] apparent threats" (Miles, 1991, p. 16). The gorilla Koko acted similarly, and provided names, emotions, and desires in repeated scripted activities with her stuffed animals, suggesting similarities to ICs (Matevia et al., PIAC21).

Not all pretense indicates experientially derived scripts, in that pretenders can act in culturally unspecified ways (Moore, 1964). For example, a child might turn an empty toy cup over a doll's head and say "pouring coffee over her" (Lowe, 1975, p. 36), perhaps never having seen anyone pour coffee over someone's head. Still, such pretenses seem combinations of various schemas children know. Given that pretending relies on knowledge of reality (Leslie, 1987; Harris, 1998; 2000), where does imagination come into play? One possibility is that imagination is the novel combination of schemas which might then be put into action (Piaget, 1945/1962), as in Lowe's example. Similarly, imagination seems present in primate eye-closing games (Russon *et al.*, *PIAC17*), in which the frivolous (closing eyes while moving) is combined with the mundane (moving from one place to another) for fun or exploration (Mitchell, 1990). Another possibility, suggested by Smith (*PIAC9*), is that imagination is filling in and reasoning about the details of a world that might be, accepting what's taken as true about that world (see also Harris & Kavanaugh, 1993). As Wittgenstein (1921/1974, p. 41) noted, "A proposition constructs a world with the help of a logical scaffolding, so that one can actually see from the proposition how everything stands logically *if* it is true. One can *draw inferences* from a false proposition" (see also Harris, 1991). How immersed children are with their pretend inferences varies. For example, when dealing with imaginary creatures "hidden" in boxes, "skeptical" children tend to deny these creatures' existence, whereas "credulous" children wonder about them and open the boxes to look for them, the latter pattern being (paradoxically) more typical of older (6-year-old) than younger (5-year-old) children (Johnson & Harris, 1994; Bourchier & Davis, 2000).

Theory of mind

Children's attributions of mental states to dolls are, ironically, often used by researchers to evaluate children's understanding of other minds (Fein *et al.*, *PIAC10*). In effect, researchers apparently accept the pretense that dolls are animate beings with knowledge, feelings, and thoughts, and ask children to apply these ideas to the doll, and children readily comply. Apparently children feel no need here to remind researchers that these mental states are only pretend. Children's attributions concerning pretense to dolls are quite similar to their attributions to humans. When children were asked questions about another (real) person similar to those used in Lillard's Moe task, their answers were just like those with dolls (Mitchell, 2000).

Both Veneziano (*PIAC4*) and Fein *et al.* (*PIAC10*) posit an intermediate state in psychological development between understanding action, on the one hand, and understanding consciousness or having a theory of mind on the other. This intermediate state is variously called "know-how about the mind" (in contrast with knowing that the mind is such and such), "functional theory of mind" (in contrast to a contemplative theory of mind), or understanding of "internal state behavior" (rather than of internal states and behavior separately). In this view, children (like adults) view others as psychological beings whose actions are infused with mentality, but (unlike adults) are not completely clear that consciousness and action can be separable. This idea fits nicely with the evidence from other chapters which indicates that children understand that pretense differs from reality, and imbue people's actions with intentions and other psychological states (Woolley, PIAC8; Taylor & Carlson, PIAC12) even though children are unclear about the particular relations among actions, situations, and mental states (Hall, Frank & Ellison, 1995; e.g., they may not always view knowledge as required for pretense – Lillard, PIAC7; Smith, PIAC9). Consequently, older children may make faulty distinctions between the mental and the physical when the distinctions are difficult (e.g., describing pretense as needing only a body and not a mind; see discussion in Lillard, PIAC7; Woolley, PIAC8). For young children as well as for some animals, behavior and mental states may not be distinguishable or distinguished (Mitchell, 2000). The same seems true at times even for scientists' knowledge of other animals: As Yerkes & Yerkes (1929, p. 365) wrote about studying apes, "we shall not endeavor to distinguish sharply between phenomena of experience and those of behavior, but instead shall assume that each is implied when such expressions as imaginal processes, memory and imagination, are used" (see discussion in Mitchell & Hamm, 1997). However much they "comprehend the effects of their own actions and act intentionally" (Zeller, PIAC13), animals such as monkeys may not develop a functional theory of mind (Mitchell, 2000, 2002a; cf. 1991a). Other animals such as apes seem, like young children, to develop to the level of a functional theory of mind in their sympathy and role-play (Parker & Milbrath, 1994; Russon, PIAC16; Ingmanson, PIAC20), but unlike children they may never get beyond it. At present, it is too early to tell what animals are capable of (Mitchell, 1993c, 1997c,d; Byrne, 1995; Whiten, 1996).

Pretense itself seems to point to a disparity between internal experience and external behavior. As Wittgenstein (1949/1992, p. 42e) noted, "The possibility of pretense seems to create a difficulty. For it seems to devalue the outer evidence, i.e., to annul the evidence [of internal experience]. One wants to say: Either he is in pain, or he is experiencing feigning. Everything on the outside can express either one." Recognizing this disparity seems to require extensive knowledge about the relations between pretend actions and real actions, a knowledge which likely develops (luckily) only *after* children perceive behavior and internal states as simultaneous occurrents. In keeping with this idea, Strawson (1959/1963) proposed that ascribing experiences to oneself requires knowing how to ascribe experiences to anyone – that it is impossible to ascribe experiences to oneself if one has no idea how to ascribe experiences in general. This does not mean, however, that one's ascriptions will immediately be accurate. "If there is any learning in [understanding people's minds], it seems more necessary to learn to check and refine our projections and attributions than to learn to make them . . . The trick is to make them at the appropriate level of generality, and unfortunately *that* trick is not part of our initial, constitutional dispositional equipment" (Rorty, 1995, p. 215). I suspect that ascriptions of experience to self and other start to develop simultaneously when children begin to develop kinesthetic–visual matching skills such as imitating others and recognizing themselves in mirrors (Guillaume, 1926/1971), after which they learn more elaborate criteria for mental events (Mitchell, 1997a, 2000). Children's generous ascriptions of psychological capacities to people (including themselves), animals, dolls and imaginary companions suggests that we humans have at quite an early age a rudimentary but useful means to begin our understanding of minds. Perhaps experience at pretending to be another leads to differentiation of behavior and experience because the pretender understands the world from multiple vantage points (Bretherton, 1984). How much pretense and imagination are involved in our psychological attributions to animals and people – and in animals to each other (Ingmanson, PIAC20) – is as yet unclear (Harris, 1992; Mitchell, 1993c, 1994a; Currie, 1995; Quiatt, 1997).

Language seems important in the development of our early mental appreciation, and understanding and/or producing language influences great apes to develop advanced pretend skills suggesting a know-how about the mind (McCune & Agayoff, PIAC3). Their nonlinguistic conspecifics also appear to share similar skills (Parker & McKinney, 1999; McCune & Agayoff, PIAC3; Ingmanson, PIAC20), but exhibit them less frequently. Although developing language increasingly allows for articulation of pretend intentions and imaginings (Veneziano, PIAC4; Taylor & Carlson, PIAC12), even with language pretend play can be difficult to interpret (Fein & Moorin, 1985). Hopefully reflection on the diverse and even contradictory approaches and assumptions of the authors in this volume will lead to greater understanding of our own pretenses and those of other animals.

Endnotes

1. Throughout this volume, chapters herein will be cross-referenced by the abbreviation *PIAC* and the chapter number.
2. Catharsis may work for dogs: one I observed engaged solely in repetitive games with her master of runaway and self-keepaway followed by happy submission to her chasing master, potentially a recreation of her real-life experiences with a previous owner who often beat her, revised to have a happy ending.
3. Imagination in this sense is believed to be potentially attributable to bees in their use of mental maps (Gould & Gould, 1988).

2

A history of pretense in animals and children

Interest in animal behavior suggestive of pretense and imagination followed directly from Darwinian ideas about the evolutionary relationship between humans and other animals. This chapter presents a history of ideas about, and evidence of, animal and childhood pretense. After describing the scientific climate about animal pretense prior to Groos' (1898) *The play of animals*, I present Groos' ideas and the scientific response. I then articulate ideas about pretense from three approaches: developmental psychology; its evolutionary offshoot in ape–human comparison; and ethology.

Evolutionary implications

Given the desire which Darwin's (1859/1902; 1871/1896) theory of natural selection produced by the late 1800s to see humanity in animals' activities, it is not surprising that, from the fact of simulation, a knowledgeable "pretense" was at times inferred. Both animals' and children's play exhibit "simulated actions in place of real actions" (Spencer, 1878, p. 630), a "dramatizing" of unsatisfied instincts or adult activities (Spencer, 1878, pp. 630–1), or "imitations of purposive voluntary acts" (Wundt, 1884/1907, p. 357).

Much as it does today, pretending included obvious simulation of the organism's own functional behavior ("self-simulations" such as feigning sleep, fighting, or retrieval; teasing; deception), responding to nothing as if something (imaginary objects), and treating one object as if another (object substitution). Feigning (Hornaday, 1879; Romanes, 1889/1975) and "pretending" (Darwin, 1871/1896, p. 69; Wundt, 1884/1907, pp. 358) described the inhibited fighting and biting of playfighting ants, dogs,

[23]

and orangutans, as well as instincts enacted out of context (James, 1890). Feigning in teasing seemed deliberately deceptive, as when a dog or ape offered something and then retracted it, for fun or *schadenfreude* (Darwin, 1871/1896; Lindsay, 1880), and other deceptions suggested intentional pretense (Watson, 1870; Romanes, 1881/1906; Jones, 1889).

Object substitution by pets also indicated pretense: "dogs and cats resemble children in their play of 'pretending' that inanimate objects are alive, and this betokens a comparatively high level of the imaginative faculty" (Romanes, 1889/1975, p. 56). Some animals showed a "still higher level" of pretending that imaginary objects are present, as when one dog showed a "dramatic faculty" in (purportedly) chasing imaginary pigs whenever his mistress uttered "Pigs" (p. 56; see also Groos, 1898), and another pretended to catch a nonexistent fly, precisely imitating his normal fly-catching (Romanes, 1881/1906).

Yet, by contrast with human pretense, animals were deficient. Although animals' playful actions are clearly "imitations" (pretense) because "the end pursued is only a fictitious end," they are without imagination: "against the countless varieties of the play of children . . . stands the single form of mock fighting among the animals . . . Animal play never shows any inventiveness, any regular and orderly working out of some general idea. And only [this implies] the expression of really imaginative activity" (Wundt, 1884/1907, pp. 357–8). (Girls' maternal play similarly shows a "remarkable monotony" (Stern, 1914/1924, p. 318).) By contrast with animals, children pretend to be someone else: "passing out of one's ordinary self and assuming a foreign existence is confined to the child-player; the cat or the dog, though able, as Mr. Darwin and others have shown, to go through a kind of make-believe game, remaining always within the limits of his ordinary self" (Sully, 1896, p. 38).

Groos on animal and childhood play

Consolidating these ideas, Groos (1898) conceptualized play as the pleasurable acting out of instincts before an animal has any serious need of them, thereby explaining much seemingly intentional pretense in young animals as unintentional simulation of adult activity (cp. James, 1890). Play does not originate in imitation of others, as young animals have no model and thus no notion of assuming a role or mock activity. Rather, play activities execute instincts without an appropriate aim or serious occasion, providing practice and exercise for later serious use.

The difficulty for Groos' view was play among mature animals. These animals, capable of enacting instinctive actions in appropriate circumstances, need no practice, yet still play. Groos (1898, pp. 145, 294) reasoned that they must be pretending: "the adult animal, though already well acquainted with real fighting, still knows how to keep within the bounds of play, and must therefore be consciously playing a rôle, making believe . . . the full-grown dog romps with his master and the make-believe is fully developed and conscious, for his bite is intentionally only a mumbling, his growl pure hypocrisy" (see also Baldwin, 1898/1904). Pretense includes "the conscious repetition of our own previous acts" (Groos, 1901, p. 386).

Groos (1898, p. 301; also Sully, 1896) explained adult make-believe with the notion of divided consciousness (dissociation) or "conscious self-deception." Initially pretend involves "complete . . . absorption and self-forgetfulness": the "child actually approaches the hypnotic state when she says that the pillow is a lady on the sofa, and chats with her" (Groos, 1901, pp. 387–8; see also Baldwin, 1894/1903; Sully, 1896; cf. Breuer & Freud, 1895/1955; Sarbin, 1954). Later, in "Making believe . . . the illusion is perfect, while . . . there is full knowledge that it is an illusion" (Groos, 1901, p. 386). Such conscious self-deception (or "self-illusion") can occur only when play activities have been "so frequently repeated that the animal recognises their pleasurable quality"; only then "can we assume that even an intelligent creature begins consciously to play a part" (Groos, 1898, p. 297).

Groos (1898, pp. 301, 302) vacillated about conscious self-deception in animals. He attributed it as "the strongest possibility" to "most" animal play; but also expressed uncertainty because, unlike human children, "speech is wanting to these creatures." As evidence of conscious self-deception, Groos proposed sham activity and using a sham object, as well as deception: "When we see deception used so effectively to serve practical ends . . . it can hardly be doubted that there is in all probability more consciousness of shamming in play than we have any means of demonstrating" (1898, p. 299).

Response
Baldwin (1901/1960, pp. 37, 513) supported Groos' views, but extended the notion of make-believe to include acts creating intentional resemblance to influence another (deception), acts of unintentional resemblance (feigning), and "consciousness of unreality attaching to certain mental

constructions, notably those of play and art." All of these are forms of "semblance," a "general term for those mental contents which simulate the real, but do not prove to be it. Under this heading we have the class of facts covered by the imaginative faculty when working under claim to cognitive validity, whether or not its claim be recognized as artificial and false by the subject. It has been called an 'inner imitation' of reality, in that it is – especially when its false claim is detected – a simulation of the marks of reality" (Baldwin, 1901/1960, p. 513). In this view, imagination creates mental contents which simulate reality and are (sometimes) demarcated as simulations.

Others were less happy with Groos' ideas. Against conscious self-deception by animals, Morgan (1900, p. 270) invoked his canon: "we should not interpret animal behaviour as the outcome of higher mental processes, if it can be fairly explained as due to the operation of those which stand lower in the psychological scale of development."

> Surely the difference of behaviour in [Groos'] examples [of deception and role-playing by animals], is sufficiently explained as the outcome of diverse situations, without having recourse to anything so psychologically complex as the conscious self-illusion of make-believe – interesting and important as this is in the psychology of children. To suppose that a monkey who nurses a bit of blanket has any ideas about its being a make-believe baby is *not* to interpret the behaviour of animals in accordance with the canon.
>
> (MORGAN, 1900, p. 281)

Hall (1928, p. 73), who interpreted play as recapitulation of ancestral patterns, portrayed Groos' practice theory as "partial, superficial, and perverse," and noted that girls' doll play was not practice for motherhood.

Soon after Groos' positing of animal pretense, interest waned. Pretending was attributed to animals such as bears and gorillas, but usually infrequently and superficially (Mills, 1919/1976; Cunningham, 1921; see Mitchell, 1999b). Complex examples among apes include feigned indifference, teasing with objects, and noiselessly creeping up on unsuspecting individuals to frighten them (Köhler, 1925/1976; Nissen & Crawford, 1936). (Children and dogs show similar teasing patterns; Groos, 1901; Lorenz, 1953/1977.) Köhler also noted some odd games of apparent solitary pretense in the "extraordinary antics" of an intelligent chimpanzee: "Often, as he squatted on the ground, he would take hold of one of his own legs with both arms, stroke it, rock it to and fro, and generally treat it as some pleasant, but wholly exterior, object. Or he would stretch out either one or both legs on the ground, limp and motionless, and shuffle along on his powerful hands" (pp. 312–13). Such games may have represented stereotypies induced by captivity, but suggest a concern with "what if," an imaginary or symbolic component (Piaget, 1945/1962, p. 99), perhaps similar to that present in the games of primates and mongooses who move about with covered eyes (Rensch, 1972; Russon, Vasey & Gauthier, PIAC17; Gómez & Martín-Andrade, PIAC18).

Children's pretense

Surveys of children (and adults recollecting their childhood), anecdotes, and diaries of children's development documented pretense in children's use of dolls and toys, their personification and animism, their drawings and art, their daydreams and general fantasy life (Sully, 1896; Preyer, 1889, 1890; Claparède, 1911/1975; Hall, 1914; Stern, 1914/1924; see Valentine, 1942/1950). This evidence of pretense raised questions about its psychological interpretation. Whereas Stern (1914/1924) agreed with Groos that children develop from complete involvement in pretense to recognition of it as make-believe, Karl Bühler (1930, pp. 93–4) believed that this recognition was there initially: "However firmly the child may insist that everyone else should join in keeping up the pretence of treating the piece of wood like a child, or of calling the chair a coach, the little player would nevertheless get a big shock if the doll she is trying to calm down really began to cry, or if any toy really carried out its imaginary motions." But early pretenses may require little of either involvement or recognition; rather, infants may simply be enacting poorly understood ideas when objects or activities stimulate associations (Preyer, 1889). Sully (1896, p. 37) suggested a similar but more active interpretation of children's pretenses – a child initially does not "knowingly act a part [or exhibit] a fully conscious process of imitative acting . . . [Rather] when at play he is possessed by an idea, and is working this out into visible action."

Whatever their level of complexity, the ideas enacted in pretense derive from reality, which is imitated or imaginatively recreated. The child's early imitations develop into the more complicated reenactments of pretense (Preyer, 1890). Pretense is the child's "first great period of apprenticeship" (Taylor, 1898, p. 163) in its reproducing reality. "The intensity of the realising power of imagination in play is seen . . . in the stickling for fidelity to the original in all playful reproduction, whether of scenes observed in everyday life or of what has been narrated" (Sully, 1896, pp. 47–8). Children's imagination is, however, selective. Typically, boys act agonistically, girls maternally (James, 1890; Stern, 1914/1924; K. Bühler, 1930), a distinction relatively consistently observed (Pintler, Phillips & Sears, 1946; Liss, 1983; Hutt *et al.*, 1989; cf. Rottnek, 1999).

Imagination itself derives from perceptions, as in visual images. When images come exclusively from reality (as in memory), imagination is "reproductive," but when images are transformed, imagination is "productive" or "creative" (James, 1890; Arlitt, 1928; Yerkes & Yerkes, 1929). Reproductive imagination appears first, and accounts for reality-based pretense. Confusions between reproductive and creative imagination result in young children's unintentional deceptions (Stern & Stern, 1909/1999; Arlitt, 1928). Creative imagination accounts for pretenses that combine elements of experiences in new ways (Claparède, 1911/1975; Stern, 1914/1924), including daydreams, imaginary companions, "planning a trip" and "attempting to recreate the experience of another" by children (Arlitt, 1928, p. 157). Also included are more mundane examples by apes and children – dreams, problem solving, planning about object use, and inventive play (Baldwin, 1909; Yerkes & Yerkes, 1929).

Whereas children's pretend actions were based on reality, the objects they acted upon often were not – "the little child shows striking unconcern as to the character of the outer object with which the fantasy-percept is associated"; this "disregard of detail" derives from the "sketchy nature of the percept" (Stern, 1914/1924, p. 268). In their animation of dolls and other objects, children used real-world knowledge and "a strong vitalising or personifying element" toward things only imagined to be real (Sully, 1896, p. 31). Almost anything could be personified, not only dolls but flowers, blocks, ears of corn, peanuts, sticks, and apples (Hall, 1914). Children attributed psychological experiences to such objects, fed and doctored them, put them to sleep, disciplined them, had funerals for them, and generally treated them as if human (Sully, 1896; Hall, 1914; Stern, 1914/1924; Freud, 1920/1972; Valentine, 1942/1950; see Fein, Darling & Groth, *PIAC10*). Both boys and girls played with dolls (Sully, 1896; Hall, 1914), but typically in different ways, girls showing more nurturance, boys less extended interest and more movement (Stern, 1914/1924; cf. Formanek-Brunell, 1993). Stern (1914/1924) believed – contrary to doll manufacturers (Formanek-Brunell, 1993) – that children prefer to use unconventional rather than realistic objects as dolls, developing toward an "unconcern" with the realism of toys which "reaches its highest point when imagination dispenses entirely with a tangible starting-point and brings into being something very like hallucinations" (p. 270).

Children's animism extended to nothing, as when fear of the dark derived from "the imaginative filling of the dark with the forms of alarming animals" (Sully, 1896, p. 212). Thus, actual "imagination-stimuli" (objects) are unnecessary for imagination: "When the child begins its 'confabulation,' or when, lying in its cot in the evening darkness before sleep comes, it holds forth in a monologue blending, in wild confusion, relatives and toys, the day's events and wishes for the future, then all imagination-stimuli have disappeared, and we see the play of the inner perceptual power in its purest form." (Stern, 1914/1924, p. 270). One wonders what Stern would have thought about the evening monologues of African grey parrot Alex (Pepperberg, Brese & Harris, 1991), so similar to those of children (Weir, 1962).

Imitation, symbols, language, and development

In the early age of the individual and of the race what we enlightened persons call fancy has a good deal to do with the first crude attempts at understanding things.

(SULLY, 1896, p. 28)

Children's development showed interrelations between early pretense and imitation. Guillaume (1926/1971, p. 116), studying his own children's development, placed pretense firmly as a form of symbolic imitation which begins at the end of the first year. Initial examples (at 11–15 months) included imitating both their own actions and those of others. Guillaume posited that the human infant's recognition of the similarity between its own actions and those of others (which would lead to flexible bodily imitation) derives from an infant's initially attempting to recreate the objective effects of another's actions on objects, and only later trying to reproduce the movements that led to the effects. Through imitation of effects, in which the visual (objective) stimuli of the model become a cue for the production of the same act by the child, gradually the kinesthetic (subjective) feeling of the child's act is associated with the model's act; the ability to imitate others' bodily actions results. For Guillaume, this kinesthetic-visual matching promotes not only pretending to be another, but also mirror-self-recognition, sympathy, and knowledge of other minds (see Mitchell, 1997b). Similarly, Piaget (1945/1962, pp. 44, 56) recognized the significance of "translating the visual into the kinesthetic" for these representational activities, and Wallon (1965/1984) believed that such "body modeling" is the initial evocative representation, "the first step toward fiction" (Nadel, 1994, p. 186).

The consequence of body modeling in pretense is that players experience different roles, thereby developing appreciation of multiple attitudes and building a self through personal responses to these actions (Mead, 1934/1974). In such pretending, "children are practicing acquiring and mastering new forms of social relations which are transferred to ... real relationships [which themselves] exert an organizing and controlling influence on the make-believe relationships" (El'konin, 1969, p. 176). Children's impulse for role-play is so strong that they invent imaginary companions (Mead, 1934/1974; see Taylor & Carlson, PIAC12).

The developmental relations between imitation and pretend (symbolic) play described by Guillaume (1926/1971) were further elaborated by Piaget (1945/1962; see McCune & Agayoff, PIAC3). For Piaget, pretense or make-believe first appears with the "union" of the "application of ... schema to inadequate objects and evocation for pleasure" (p. 97), the latter usually evinced by laughing or smiling. Self-imitation initiates pretense. Piaget described his children repeating their own actions, using inappropriate or nonexistent objects and assimilating the available stimuli to their actions. Such self-imitation might seem simple, but it involves an active avoidance of reality toward the objects the child pretends with – for example, treating a nonpillow as a pillow upon which to sleep. Pretense (which largely assimilates to known schemas) incorporates imitation (which largely accommodates schemas), such that the two processes work together – pretense about an object providing what is signified (i.e., a nonpillow signifies the pillow), and imitation being the signifier.

Piaget's (1945/1962; 1947/1972) ideas about pretense and imitation are part of his larger theory about the development of intelligence. Pretense and deferred imitation occur at the end of the sensorimotor period, during the first year and a half, when internal representation is achieved in the sixth (final) stage. During the fourth stage, children apply schemas they already know to new situations, such that their acts are presymbolic, creating the groundwork for symbolization in both ritualized playful self-imitations and imitation of others' acts which children already enact themselves. Such behavioral simulation is "enactive representation" (Bruner, 1966, p. 12), reduplicating action to re-instantiate something. For example, Piaget's 7-month-old son reproduced his own act of swinging a cigarette box after the box had dropped from his hand. The fifth stage initiates symbolism. Here children produce novel behaviors often accidentally produced in their own behavior, as well as novel behaviors of others in imitation. This incorporation indicates the development of internal representations, in that children abstract schemas from their original context and thereby "evoke them symbolically," being "almost the symbol in action" (p. 95). Yet Piaget does not grant children awareness of make-believe until the sixth stage (at about 18 months), because at the fifth stage "the child confines himself to reproducing the schemas as they stand, without applying them symbolically to new objects" (p. 95). At the sixth stage, children apply schemas to inadequate stimuli and delight in doing so, thereby indicating their awareness of the discrepancy between signified and signifier. Consequently, "external imitation becomes internal or 'deferred' imitation" or "interiorized imitation" (pp. 95, 75) – i.e., imitation produces mental imagery and representation (a view presented earlier by Baldwin, 1901/1960; and Werner – see Werner & Kaplan, 1963, p. 91).

Vygotsky (1933/1978), by contrast with Guillaume and Piaget, argued that early pretense is not symbolic imitation, but instead paves the way for symbolic understanding (cp. Sully, 1896). In pretense, a child creates an imaginary situation in which "things lose their determining force" such that (Vygotsky, 1933/1978, p. 97):

> *the child begins to act independently of what he sees* ... Action in an imaginary situation teaches the child to guide her behavior not only by immediate perception of objects or by the situation immediately affecting her but also by the meaning of this situation ... [In play] an object (for example, a stick) becomes a pivot for severing the meaning of horse from a real horse.

Symbolism is not present in young children because their imagination is not completely free, in that not all objects can represent something: "Any stick can be a horse but, for example, a postcard cannot be a horse ... [B]ecause of the lack of free substitution, the child's activity is play and not symbolism" (Vygotsky, 1933/1978, p. 98). Applying imagination to objects requires that children use rules of behavior, such that pretense is acting out those rules: "The child imagines himself to be the mother and the doll to be the child, so he must obey the rules of maternal behavior" (Vygotsky, 1933/1978, p. 94). Vygotsky notes Sully's (1896) description of two sisters who paradoxically pretend to be sisters as a revealing illustration of acting out roles according to rules, in this case of how ideal sisters act:

"the essential attribute of play is a rule that has become a desire ... *play gives a child a new form of desires*. It teaches her to desire by relating her desires to a fictitious 'I,' to her role in the game and its rules" (Vygotsky, 1933/1978, pp. 99–100; cp. Wittgenstein, 1949/1992; Oughourlian, 1991). For Vygostsky (1933/1978, p. 93), "Imagination is a new psychological process for the child; it is not present in the consciousness of the very young child, is totally absent in animals, and represents a specifically human form of conscious activity."

Whereas Piaget assumed that children's pleasure in acting toward new objects using old routines indicates their awareness of pretense, psychologists following Vygotsky argued that object substitutions by 18-montholds are not pretense, but simply acting out normative routines (much like "scripts"; Bretherton, 1984) but with inappropriate objects, an indication of their "weakness of critical thinking" (El'konin, 1969, p. 255; cp. Preyer, 1889). A child's action of putting toy objects "to sleep" on a pillow is "determined by the pillow, which is associated with putting something to bed, and not by the toy objects which are being put to bed on the pillow"; in "eating" from an empty spoon, "drinking" from an empty cup, the child (apparently mirthlessly) simply repeats normal acts (Repina, 1964/1971, p. 257). The "logic of the actions performed by a child ... at play exactly reproduces the logic of the real actions of adults" (El'konin, 1969, p. 177), and young children strongly resist the use of arbitrary (nonprototypical) objects in play. Symbolic pretense is relegated to the third year because only then does the child name imaginary objects (cf. Veneziano, *PIAC4*). Other psychologists similarly emphasized the importance of language in guiding pretense, consequently discerning a difference between apes and children (K. Bühler, 1930; C. Bühler, 1935; Repina, 1964/1971).

The relation between symbolic play and language was further articulated by Werner & Kaplan (1963). They viewed representation (whether linguistic or not) as deriving from experienced correspondence between signifier and signified, present in self-imitation and imitation of others. From such imitation, children show an increasing *distancing* between "depictive expression and presented content" (Werner & Kaplan, 1963, p. 89) which eventually results in deferred imitation. This latter phenomenon "presupposes the liberation of gestural depiction from concretely presented content; in other words, for the gestural depiction to take place in the absence of concrete content implies that it must be activated and regulated by some kind of *internal model or 'schema.'*" (Werner & Kaplan, 1963, p. 91). The same distancing occurs in pretense (see also Drucker,

1994). As support for distancing, Werner and Kaplan presented (as had Guillaume) children's increasing ability to use fewer and fewer realistic media to represent, culminating in "empty gestures" – gestures without any objects – not present in nonhuman animals.

Ape–child comparisons

A baby chimp at play is a subject worthy of deep meditation.

(LORENZ, 1956, p. 637)

When compared directly to the "imitative play" or pretense of children, that of chimpanzees seemed impoverished (K. Bühler, 1930). When Kellogg & Kellogg (1933/1967) raised the chimpanzee Gua with their own son Donald, they observed little imitative play by Gua but lots by Donald. A similar contrast was evident between the chimpanzee Joni raised by Ladygina-Kots (1935/1982) and her own son Roody (raised years after Joni had gone). For the latter pair, the child's instinct for imitation of others superseded the ape's and he also showed "more daring" in his impersonations: "While Joni could only imitate separate actions, Roody appeared to be capable of reproducing a whole series of them. The boy liked to impersonate professional men and tradesmen ... mimiking ... voice, gesture and pitch of tone" and dressing up (p. 161). Joni's pretenses were self-simulations: in one deceptive hiding, Joni "sought shelter under a pillow trying to make-believe that he was behaving in exactly the right manner – while he was in reality steadily biting at the wood of the taboo table" (p. 139; cp. Sheak, 1917). Only Roody created drawings and responded animistically to toy animals and dolls. Perhaps "the only cases when the imagination of the ape does come into the picture is when he is found to play with some part of his own body or when he purposefully erects obstacles to be overcome in course of his playful activities" (p. 164). Yet, whereas the child's pretenses showed imagination and conscious differentiation between pretend and real, the chimpanzee was either too absorbed in and unaware of the pretense as such, or not pretending at all – views earlier applied to children's pretense.

From descriptions like these, Karl Bühler (1930, p. 93) could safely write that "The illusory games of the child are based on imitation to a far higher degree than are those of the chimpanzee" – or any other animal. Bühler (1930, p. 48) named the end of children's first year as the "chimpanzee-age" to denote similarities in both species' experimentation with tools, but noted that only children went on to pretend. Piaget (1945/1962), by contrast, credited apes with some representational capacity in pretense, but felt that "pretenses" by other animals "are merely patterns of behaviour begun but not carried through. Kittens which fight with their mother and bite without hurting her are not 'pretending' to fight, since they do not know what real fighting is"; still, perhaps these are "a preparation for representational symbols" (Piaget, 1945/1962, p. 101; see Gómez & Martín-Andrade, PIAC18). Overall, instances of animals' pretense described repeating or dramatizing their own actions, nothing like the rampant impersonations of others present in children's play.

Still, indications that apes might show complex pretense appeared. In Crawford's (1937, p. 57) study of cooperation between chimpanzees, one chimpanzee solicited another's help in pulling an object by touching her on the shoulder and then acting out what she needed: she "took her rope, braced herself ready to pull, and turned back to look at [the other chimp] as if expecting her to come up and help" (see discussion of communicative pretense in Keller, 1902/1965, p. 178; Grice, 1982; Bates *et al.*, 1983; Mitchell, 1987; 1990; 1994a). Another ape, the gorilla Toto, raised as a spoiled child by Maria Hoyt (1941), enjoyed hiding things and then pretending to look for them by going through "an elaborate routine of showing me that she didn't have them at all by calling attention to every place on her body where they might be except the one where they were" (Hoyt, 1941, p. 149). Once, when Hoyt (1941, pp. 211–12) discovered that a pearl was lost after Toto had stolen her necklace, Hoyt demanded the pearl back, but Toto "shook her head, showed me her empty hands and the soles of her feet and opened her mouth to prove to me that she didn't have it"; when Hoyt threatened to leave, Toto gave her the pearl after removing it from her mouth. "Then we both laughed and she stood on her head to show me how happy she was to have me with her again."

The belief that apes do not pretend to act like others or impersonate them could be maintained until the chimpanzee Viki was studied by Keith and Cathy Hayes in the late 1940s. They raised Viki as a human child, and were rewarded with a humanized ape. At about 16 months, Viki began to imitate her adoptive mother by "crudely copying my household routine – dusting, washing dishes, pushing the vacuum cleaner about ... one day she claimed the grater from my lemon-pie-making residue, helped herself to a lemon ... and grated it all over the living-room rug ... every tool we used, every little action, was apt to result in her attempts at duplication – hair brushing, fingernail filing, eyebrow tweezing, the use of a saw, a drill, a bottle opener, a pencil sharpener" (Hayes, 1951, pp. 181–2). After observing her mother dabbing at a skirt with a wet washcloth, the next day Viki did the same thing, having wet the washcloth herself. In playful imitation of her mother, "She dug at her fingernails with a nail file, patted a powder puff over her face with startling results, and insisted on being given a dab of lipstick, which she smoothed on, as I do, with a little finger" (Hayes, 1951, p. 87).

> Often she imitated our motions with no idea of what she was doing. For instance, every morning she ran outdoors to pick up the newspaper. Seated on the couch, she would first open the paper wide, and hold it at arm's length as if to scan the headlines. Then settling back she would turn the upper corners a hand's breadth, one by one, as though she were looking for the sports page or the comics. Occasionally she would grin and open a page part way as people do when some fascinating tidbit catches the eye.
>
> (p. 87; cp. PIAGET, 1945/1962, p. 122)

Viki's imitativeness was certainly supported by the Hayeses, who taught her to repeat their actions when they said "Do this," though Viki began imitating before they taught her this command. Viki also enjoyed using mirrors to observe and change her appearance (Hayes & Hayes, 1955; Hayes & Nissen, 1971).

Viki's favorite playthings were pull toys and picture books. Like some other human-raised chimpanzees and children (Stern, 1914/1924, p. 188; Kellogg & Kellogg, 1933/1967; Ladygina-Kots, 1935/1982, pp. 142–3; Savage-Rumbaugh & Lewin, 1994), she responded to pictures as if real – one photo (Hayes, 1951, opposite p. 201) shows Viki putting her ear to a picture of a watch. Perhaps her most famous instance of pretense was her invention of an imaginary pulltoy which she "pulled" through the house by an imaginary string (Hayes, 1951, pp. 80–4). When asked what she was doing "trailing the fingertips of one hand on the floor" Viki stopped short "with a look of guilt and embarrassment" and then "pretended to be very busy examining a knob" on the toilet. Over the course of a month or so, Viki continued to play with the imaginary pulltoy, even to the point of its getting "stuck" and having her adoptive mother "untangle" it. Upon her mother's creation of her own imaginary pulltoy, Viki was initially intrigued but then became frightened, leapt into her mother's arms, and never played with the imaginary pulltoy again. (Viki seems to satisfy Karl Bühler's (1930, pp. 93–4) expectation that pretenders would receive quite a shock if their imaginary objects suddenly became real.) Viki, with her imitation, pretense, and self-recognition, reminds one very much of Guillaume's (1926/1971) children.

Ethology

> [T]o see into the minds of animals one has to live with them.
> (SMYTHE, 1961, p. 182)

Groos' concerns about instincts in animal play were not empirically examined until ethologists began to describe their observations of developing captive and wild animals. Ethologists such as Lorenz observed many activities by animals that bore close similarity to functional adult activities, many resembling pretense (e.g., cats pouncing on nothing). Lorenz, like Groos, characterized play as acting out instinctive activities. However, Lorenz believed not that play is practice, but that much "play" is simply immature activity (see also James, 1890). By contrast, real play forms "have the common quality that they are fundamentally different from 'earnest'; at the same time, however, they show an unmistakable resemblance, indeed an imitation of a definite, earnest situation" (Lorenz, 1953/1977, p. 155), a view clearly indicating pretense (Thorpe, 1956/1963). For Lorenz (pp. 158–9),

> The relationship of all play to play-acting lies in the fact that the player 'pretends' to be obsessed with an emotion which he does not really feel ... the emotions pertaining to the earnest situation are [not] present in an attenuated form ... they are altogether missing ...

By contrast with play, simulative activities by "animals of restricted intelligence" had "central nervous processes ... little different from those in normal elicitation," as when rabbits experience real fear during escape "play" (Lorenz, 1935/1970, p. 165). More intelligent animals, however, experience similar activities less realistically, as play.

Observation supported the idea that playing animals pretend about their emotions (though not that emotions are "altogether missing"). For example, a pet badger recreated fear in chase play, and "invited a bush to play" when no other play partner was available (Eibl-Eibesfeldt, 1950/1978, p. 145; cp. similar invitation in a chimpanzee: Hayaki, 1985). The badger also played with "pseudo-prey," which he scent-marked and grabbed, and toward which he "endlessly repeated the killing shake" used to kill real prey (Eibl-Eibesfeldt, 1950/1978, p. 146). By contrast, a dog showed "more plasticity, more 'fantasy' and more expression" in play, in that he repeatedly threw his "prey" object, chased it, jumped on it, and "growled at his play objects ... although one could never assume that he was angry, because he vigorously wagged his tail when doing so" (pp. 147,

148), the wagging tail suggestive of children's evocative laughter during simulation.

Ethologists noted diverse "imitative" (simulative) activities, not all of which are clearly pretense. These simulative activities include:

- immature enaction of adult activities;
- vacuum activities (acting as if doing something toward nothing);
- displacement activities (engaging in an activity unrelated to the present activity (e.g., "pretense pecking" by fighting cocks; Leyhausen, 1952/1973, p. 61), also called "redirected activity");
- substitute activities (acting as if doing something toward an inappropriate object);
- play.

Motivation distinguished these simulative activities. Lorenz posited "response-specific energy" which, if unused in normal responses, would motivate instinctive actions toward inadequate or nonexistent stimuli: "the damming of response-specific energy alters the *perceptive field* of the animal or human subject so far that a normally inadequate object will be subjectively experienced as adequate, or – in the extreme case – experienced as a pure hallucination ... in empty space" (Lorenz, 1942/1970, p. 363). Such innate motivational energy explained, for example, birds which performed species-typical behavioral sequences for which they had no experience (Lorenz, 1937/1970; Hinde, 1958). Different simulative activities derive from distinct conflicts and convergences between response-specific drives of various intensities, or from interaction between response-specific drives and a nonspecific activity drive (Meyer-Holzapfel, 1956/1978). Play in particular occurs when the nonspecific activity drive is of weak or medium strength, and the response-specific drive is weak or absent. The "hydraulic model" upon which this explication depends, in which motivation-specific energy builds up and overflows, was subsequently discounted (Sevenster, 1977).

The description of animals as acting simulatively raised again Groos' question of animals' conceptions of their own activities. For example, Leyhausen (1965/1973, p. 232) argued that "it would be quite absurd to assume that a cat which in its daily existence had already killed and eaten thousands of prey animals could not recognize the difference between a mouse and a ball of paper; it knows that a ball of paper is not a mouse" (see also Ewer, 1968). Still, such knowledge may indicate only that animals learn to differentiate cues distinguishing play and nonplay situations

(Millar, 1968). Substitute object play may merely be "the inquisitive trying-out of species-specific behaviour patterns on novel objects possessing significant Gestalt-applicable features . . . [which stimulate] innate releasing mechanisms" (Lorenz, 1950/1971, p. 177).

Pretense as metacommunication

> Above all pretence has its own outward signs. How could we otherwise talk about pretence at all?
>
> (WITTGENSTEIN, 1949/1992, p. 42e)

Lorenz's implications of pretense in higher mammals' play influenced Bateson's (1955/1972) idea of metacommunication. Metacommunication occurs when organisms produce a "simulation of . . . an involuntary or automatic . . . sign for use as a message (for example, the child makes a noise *like* automatic crying in order to make its mother come)," where the simulated sign is recognizable as a simulation; the differences between the automatic sign and the simulation communicate the fact of simulation, thereby indicating metacommunication (Bateson, 1956, pp. 167–8). Because play is obviously simulative to human observers, Bateson suggested that, "[i]n so far as animals recognize that what they are doing is play" (p. 168), they metacommunicate. Bateson offered no evidence that animals are intentionally simulating actions in play. Rather, he offered an explicit account of ethologists' implicit notions: *if* animals knowingly simulate their own activities, and/or *if* the fact of their simulation is detected by other animals, then their actions evoke one sort of activity and simultaneously indicate a difference from that sort of activity. In playfighting, for example, "the unit actions or signals [are] similar to but not the same as those of combat," such that to players playfighting is categorized as "not combat" (Bateson, 1955/1972, pp. 301–2). The animals "classify the components of their messages" (Bateson, 1978, p. 8), and consequently use the same actions to communicate diverse meanings (see Mitchell, 1991a). In his scenario for the evolution of non-natural meaning, Grice (1982) presented a similar view, positing that humanlike communication developed from organisms saliently simulating their natural communicative signals (which would produce "non-natural" meaning). Bateson tied metacommunication to various other simulative phenomena, including teasing, threat, hysteria, hazing, and art, a view which has intriguing implications for primatology (Reynolds, PIAC14).

Given that young animals who playfight often have little knowledge of fighting, it is unlikely that playfighting initially implicates fighting.

Yet, as Groos' suggested, *adult* animals might recognize the simulation in play and thus metacommunicate. Indeed, older rhesus monkeys' playfighting was sometimes a "veiled threat" (Breuggeman, 1978, p. 185; Emory, Payne & Chance, 1979), and boys sometimes use the appearance of playfighting to hide fighting (Smith, 1997). Even young children obviously simulate their own emotions (Valentine, 1942/1950). Pretend cries by children, saliently (though unintentionally) metacommunicative, are sometimes used in attempted deceptions to fulfill their desires. Similarly, dogs produce deceptive barks (obviously false to people, but apparently not so to other dogs) to get other dogs to leave and thereby make available a comfortable location (Mitchell, 1993c; Rose Perrine, pers. comm., 1999).

Bateson's intuitive analysis of play as pretense has some objective basis, in that play is inconsistent with "real" activities of pursuit, flight, or object possession, whatever animals' intentions (Aldis, 1975; see also Reynolds, *PIAC14*). In playchase (unlike real chase), the fleeing animal initially approaches (rather than retreats from) the proposed chaser, seems to invite the chase, and waits to be chased; the chaser often engages in self-handicapping, slowing down to let the chasee get away; and players reverse roles. In playwrestling, animals self-handicap by playing less forcefully than they can. In some games, animals trying to supplant one another act as if an unimportant location has special value; and in object-keepaway, the object has no inherent value, but players act as if it does. Some of these contradictions may occur because one animal won't or can't play unless the other self-handicaps, or because animals repetitively enact variable versions of the same activity to develop their skill (Simpson, 1976; Fagen, 1981; Mitchell & Thompson, 1991; Biben & Suomi, 1993; Pereira & Preisser, 1998). However such contradictions arise, they seem to be the basis for intuitions that playing animals are pretending: not "really" doing what they appear to be doing.

Imitative pretense in primates and cetaceans

Piagetian analyses of nonhuman primate cognitive development began in earnest in the 1970s, providing evidence of up to sixth stage imitation and pretense by apes (chimpanzees and gorillas) (Jolly, 1972; Chevalier-Skolnikoff, 1977; Mignault, 1985; McCune & Agayoff, *PIAC3*; see Mitchell, 1994a, and Parker & McKinney, 1999, for more examples). Wild chimpanzees enacted self-imitation in communication and deception, and toward inappropriate objects, as when grooming leaves (Goodall, 1973;

Menzel, 1973; Plooij, 1978; see Russon, *PIAC16*, for similar behavior in orangutans). Chimpanzees also apparently pretended in (sometimes imitative) solitary aggressive displays directed toward no one, and imitated a mother's tool modification, tool use, and ant-fishing to, perhaps, "an imaginary nest" of ants (Goodall, 1973, 1986, p. 591). Yet another human-reared chimpanzee, Washoe, pretended and frequently imitated (much as had Viki).

> One day, during the 10th month of the project, she bathed one of her dolls in the way we usually bathed her. She filled her little bathtub with water, dunked the doll in the tub, then took it out and dried it with a towel. She has repeated the entire performance, or parts of it, many times since, sometimes also soaping the doll.
> (GARDNER & GARDNER, 1969, p. 666)

Washoe not only understood language, like Viki, but also communicated by sign-language, and sometimes imitated others' signs. Washoe also imitated her caregivers' actions toward objects, and was observed "signing spontaneously to herself, in front of a mirror, or in bed at nap time" (Bronowski & Bellugi, 1970, p. 671); when asked who was in the mirror, she signed "ME, WASHOE" (Desmond, 1979, p. 175). The developmental connections among imitation, symbol use, self-recognition and pretense initially suggested by Guillaume and seconded by Piaget and others (see Mitchell, 1997b) were again supported in this and other language projects: the gorilla Koko (Patterson & Cohn, 1994), the orangutan Chantek (Miles, 1990, 1991, 1994, 1999), and the bonobo Kanzi (Savage-Rumbaugh, 1986; Savage-Rumbaugh & McDonald, 1988) showed similar abilities. These apes also reenacted or simulated activities for communication, deception and enjoyment (Miles, 1986; Miles, Mitchell & Harper, 1996; Matevia *et al.*, *PIAC21*).

Captive dolphins also enacted sixth stage imitative pretenses (Tayler & Saayman, 1973). One male Indian Ocean bottlenose dolphin imitated the actions, sounds, and air discharge of a human cage cleaner, including putting a flipper on the glass while "cleaning" the glass with diverse objects in his mouth, much as the diver put his hand on the glass while cleaning with the other. Dolphins in this male's group also imitated various activities of cohabiting pinnipeds and visiting humans. One young dolphin who observed a human blow cigarette smoke through a window rushed to her mother, obtained milk, and returned to the window, spewing forth milk in what looked much like cigarette smoke.

Less imaginative pretense is apparent in an Atlantic Ocean bottlenose dolphin which teased a grouper in object-keepaway (Aldis, 1975).

Only one report suggested sixth stage pretense in macaques (Mitchell, 1987) – an instance of imitation of a mother's actions toward her infant and object substitution (Breuggeman, 1973). Few other reports indicated imitation of others (see Zeller, *PIAC13*). Macaques more frequently showed self-imitative pretense suggestive of fourth and fifth stage sensorimotor intelligence: feigned feeding; feigned ignoring; pretending a need to defecate; aggressive uses of play gestures; and teasing (Jolly, 1972; Bertrand, 1976; Parker, 1977; Chevalier-Skolnikoff, 1977; Breuggeman, 1978; see McCune & Agayoff, *PIAC3*; Zeller, *PIAC13*). (Note that most of these characterizations in relation to Piagetian categories are suggestive, in that information on their development is unavailable.)

The close similarity and developmental congruence among language, pretense (symbolic play), and deferred imitation in human development led researchers to argue that deferred imitation and pretense are identical, and that children's gestural "vocabularies" in action and pretense at a particular age are virtually the same as their spoken vocabularies (Bates *et al.*, 1979; also Lock, 1978), thereby positioning pretense as important in language development much as Piaget (1945/1962) and others (Werner & Kaplan, 1963) had predicted (cf. Folven, Bonvillian & Orlansky, 1984/1985). The idea of schemas, so prevalent in Piagetian theory, was further elaborated to distinguish types, including scripts: schemas about the sequence of elements that constitute an event (Bretherton, 1984).

Evolutionary models of animal play and human pretense provided some organization to the morass of details. One model posited that playfighting is practice of strategies (rather than behaviors) used in fighting, but denied that monkeys are aware of the similarities in strategies (Symons, 1978). Another model placed pretense as part of a suite of cognition-based actions. Building on the idea that tool use is an adaptation for extractive foraging, Parker & Gibson (1977; 1979) further articulated that intelligent tool use – based on delayed imitation of others – could lead to planning, pretense, teaching, and language, essentially predicting the evidence of chimpanzee mothers' teaching of tool use to their infants which was eventually observed by Boesch (1991a). In this view, the "ability to engage in make-believe games seems to be an adaptation for practicing domestic and extra-domestic subsistence tasks and associated sex-role differentiations" (Parker, 1984, p. 276). Although some evidence supports evolutionary explanations based on play as an adaptation for skill practice in many species, it is equivocal (Smith, 1982). Diverse adaptational explanations may explain play behaviors in diverse species (Fagen, 1981), leaving researchers a wide-open field for exploration (Smith, *PIAC9*).

Summary

Darwin's evolutionary ideas and anthropomorphic approach to animal psychology initiated the search for pretense in animals. Groos refined the search by delimiting pretense to simulations which organisms are aware of *as* simulations. He and other child psychologists questioned the extent of children's awareness and understanding of their pretenses as such. Developmental psychologists focused attention on ontogenetic changes in the nature of children's ideas in pretense, examining the onset of representation and symbolism in children's pretend play, self-recognition and imitation, and their incorporation of rule- and role-governed behavior. Researchers studying human-raised apes related the importance of apes' cultural involvement in the creation of their pretenses. Ethologists noted that the simulative quality of pretending is common in many activities of animals, which prompted the categorization and cataloguing of animals' simulations, promoted evolutionary explanations of animal pretense, and reiterated Groos' question about animals' understanding of their own simulations. Ethologists and animal behaviorists, particularly those studying primates and cetaceans, have continued to describe simulative activities suggestive of pretense. Hopefully continuing scientific exploration for evidence of pretense in diverse animal species will provide an answer to Groos' question about animals' knowledge of the simulative nature of their activities in play. As the disagreements through history and in this volume attest, the answer may depend upon one's point of view.

3

Pretending as representation: a developmental and comparative view

Humans enjoy perceptual experiences of the world from birth, while construing of an imaginal world only becomes possible with development. Considering adult experience, Sartre (1948/1966) distinguishes two aspects of experiencing a portrait of an absent friend. If the focus is on pictorial properties – color, brush stroke, shape – the experience is primarily perceptual. If the portrait serves to evoke thoughts of the friend who is its subject, then the portrait becomes a vehicle for imaginal reflection. There is little controversy regarding the existence of these differing human experiences. However, current discussions of the topic emphasize representation as storage, rather than as a form of psychological experience. To quote a definition, "Representation is defined most simply as stored information" (Mandler, 1998). Given this climate, perhaps it is not surprising that the basic distinction between "perception" and "representation" (more accurately termed "mental representation") is often not acknowledged in contemporary analyses (e.g., Leslie, 1987; Mandler, 1992, 1998) where infants' processing of perceived events and experience is described as "representation." Consequently it becomes problematic to discuss distinctions between earlier infant behaviors that seem based on perceptual, sensory and motor experiences, and later behaviors that suggest a basis in mental representation, for example, recall of previous experiences which now come to mind and affect present activities. Developmental changes in representational play can mark the transition from activities based on perceptual and sensorimotor experiences to those indicating imaginal contents (McCune, 1993).

Piaget (1937/1954; 1945/1962) suggested that the process of development from purely sensorimotor, perceptual reactions to the capacity to act on the basis of mentally represented experiences is a gradual one, based on infants' growing capacity to use their own bodies to mimic observed activities of others. Similarly, Werner & Kaplan (1963) attribute the development of the capacity for symbolic thought and language to the tendency of humans to make motions and sounds in response to the activities of objects and people in the environment, and a recent evolutionary theory (Donald, 1991) attributes the evolution of human language and culture to this same capacity, termed "mimesis" (see also Parker, 1985; Davidson & Noble, 1989; Kendon, 1991; Noble & Davidson, 1991; Mitchell, 1994a). All of these views assume that mimetic reactions become internalized and affect underlying neurological organization leading to developmental change, although none of them emphasize this assumption nor describe mechanisms for its effects.

Piaget (1945/1962) described a sequence of six levels of development tending toward imitative ability in human infants and culminating in imitative abilities that demonstrate the imitation is now based on an internal model of the observed behavior. The capacity to utilize such internal models of previous experience is considered to be the foundation of the capacity to engage in mental representation, and hence pretending. Imitation as a skill exhibits gradual development from simple repetition of one's own behavior, to repetition of familiar behaviors when stimulated by a model, to imitation of novel behaviors in the presence of a model, and finally to imitation based on mental representation. Two culminating capacities observable at Stage 6 are taken to demonstrate the existence of underlying mental representation: (1) immediate imitation of unfamiliar behaviors; and (2) the capacity to defer imitation over hours and days, producing the behavior in the absence of a model. From this representational form of imitation (Stage 6 in Piaget's imitation sequence) the continued development of mental representation throughout the second year of life in humans is then described in advancing levels of representational play, culminating in play exhibiting a hierarchical structure (Piaget, 1945/1962; Nicolich, 1977; McCune-Nicolich, 1981; McCune, 1995).

Comparative data demonstrate similarities and differences across several species of monkeys and apes in relation to the Piagetian imitation skills exhibited in human infants, modifying the Piagetian description as necessary to account for distinctions observed in other primates (Chevalier-Skolnikoff, 1977, 1989; Parker & McKinney, 1999). Any behavior following and matching that of a model finds its way into this catalogue. These rich descriptions allow consideration of how complex a given task is, whether its components are "novel" for the animal, etc. This approach allows careful designation of how similar behaviors across species may be, and where critical differences seem to lie. In the present chapter this comparative method will be used to consider instances of pretending in nonhuman primates in relation to the known sequence of representational play developments in humans.

With the decline among psychologists in the use of Piagetian descriptions of sequential developments there has been a loss of benchmarks for comparison across domains of development as well as across species. More recent discussions of imitation in humans and other animals have shifted the question from the original: "What capacity for imitation can be demonstrated?" to "What behaviors must have been learned by imitation?" (e.g., Byrne & Russon, 1998). Making this shift entails questions regarding, for example, whether behavior is novel in the species and individual repertoire, whether the performance is identical to the mimed behavior, or whether other mechanisms can account for the behavior. These are valuable questions, but different from more fundamental questions regarding similarity and differences in the capacity to produce behavior based on a model. Some reports of nonhuman primate activities described as imitation actually seem more akin to activities that would qualify as pretending if observed in human children (e.g., Mitchell, 1987; 1990) and will therefore be included in this chapter.

Instances described as "deception" are sometimes considered as simulations, similar to pretending. For example, a female chimpanzee, observed leading a male away from, rather than toward, food then turning to the food herself once the male was distracted, might be engaged in such deception. However, the female's sequence of actions may result from separate and shifting intentions for each act: first keeping the male from the food, then consuming it herself. This may be clever, perhaps planned behavior, but need be neither pretending nor deceiving. The animal's internal state over the course of the event cannot be known, so a "true" interpretation cannot be determined (see Searle, 1992; McCune, 1999).

Pretend play, completely separate from practical goals, is somewhat easier to interpret securely as representational. In this chapter we seek to determine the internal representational structure of behavior, taking a somewhat conservative stance. Using Piaget's original descriptions as a guide, McCune distinguished a series of five developmental steps in representational play, which can serve as a guide, analogous to the commonly used sensorimotor levels of the earlier phase, as a measure for examining the representational level of pretend behaviors. This sequence begins with the child's first expression of the social meanings of objects at about 10 months of age, and culminates in hierarchically organized pretend which uses present action to evoke absent realities (Nicolich, 1977; McCune-Nicolich, 1981; McCune, 1995). The chapter will describe this sequence and examine the theme of "play as representation" by describing the relationship of such play in humans to the development of language.

Next, we examine reported cases of representational behavior (including pretend play) in nonhuman primates in comparison with the representational developments universally observed in human children. Because of the ambiguity of intentional structure in communicative simulations, these will not be the subject of this report. See Mitchell (1994a) for a careful examination of the representational structure of these forms of communication. Study of cognition in nonhuman primates has emphasized the development of object permanence, tool use and imitation, without specific reference to mental representation. However, recent work (Parker & McKinney, 1999) emphasizes mental representation in relation to object permanence, imitation, and tool use. These skills are taken in the human literature to mark the transition from sensorimotor to representational intelligence, culminating at sensorimotor Stage 6 which marks the onset of mental representation (Piaget, 1937/1954; 1945/1962). These skills may not develop in synchrony in humans, and may not characterize different species to the same extent.

Representational play: a predictor of language development in humans

The earliest level of representational play (Level 1) merely mimes the expressive acts characteristic of use of particular objects with the object in hand, such as placing an empty cup to the lips. These are brief, isolated actions, without evidence that the child is aware of the relationship between the played acts and the routines of life they mimic. With differentiation come new forms of integration as the child uses sound effects and elaborations to link the played act (symbol) with the referenced real experience (Level 2). This level is considered the beginning of "pretend" because the addition of elaborations such as sound effects or exaggerated lip movements (e.g., drinking from an empty cup) suggests a mental comparison of the present pretend situation with real drinking experienced in the past. At Level 3 the child may place a cup to the doll's mouth, thus using a different motor pattern from that used in real self-drinking, or may play at activities not physically experienced by the child, such as play cleaning. Here motor patterns observed in others are adapted in pretend, indicating that the meaning has been learned by observation, and now internalized.

Play at Level 4 involves linking two related pretend activities together in sequence. The play sequences (e.g., feed mother, feed doll; place doll in car, roll car) may involve repetitions and do not necessarily follow the appropriate order of the portrayed events. Level 5, the most advanced level in this sequence, is defined by hierarchical structure such that an internal component or "plan" precedes and guides the pretend action. This is equivalent to the concept of "decoupling" (Leslie, 1987). This may involve verbal or nonverbal designation of an object as a substitute (e.g., "This banana will now function as a telephone"), and actions treating the real object "as if" it were the object symbolized. Alternatively, an internal plan may be inferred from search, preparatory actions, a verbal statement of intention, or treating a doll as if it could act independently, indicating that the doll now functions as a play partner, rather than merely as a recipient of the child's play action (e.g., place a cup in its hand rather than to its lips). See Table 3.1 for further examples and a comparison of the McCune categories (Nicolich, 1977; McCune-Nicolich, 1981; McCune, 1995) with Piaget's (1945/1962) original analyses. Children will, by 2 years of age, spend 30 minutes or more using toy replicas and other objects to create pretend scenes either alone or with a partner. During the preschool years this is a highly preferred activity resulting in elaborate scenarios for pairs or small groups of children.

This sequence of play developments demonstrates the growth of an internal representational ability that can, in conjunction with additional capacities (McCune et al., 1996; McCune & Vihman, 2001) and an appropriate social and linguistic environment, serve the development of reference and syntactic ability. Transitions in language acquisition are predictable from representational play development. First words, associated with specific events, may develop contemporaneously with single representational play acts (Level 2). Referential words as well as first word combinations occur only after combinatorial pretend play (Level 4) have been observed (McCune, 1993). First rule-based or syntactic combinations in language occur at Level 5, hierarchical pretend (McCune, 1995; McCune-Nicolich & Bruskin, 1982). In producing referential words the

Table 3.1. Sequence of representational play levels

Piaget (1945/1962)	McCune (1995) and Nicolich (1977)	
	Levels and criteria	Examples
Sensorimotor period		
Prior to Stage 6	Level 1. Pre-symbolic schemes: the child shows understanding of object use or meaning by brief recognitory gestures. • No pretending • Properties of the present object are the stimulus for action • Child appears serious rather than playful	The child picks up a comb, touches it to his hair, drops it The child picks up toy telephone receiver, puts it to his ear, sets it aside The child gives the toy mop a swish on the floor
Stage 6	Level 2. Self-pretend (auto-symbolic games): child pretends at self-related activities while showing by elaborations such as sound effects, affect, and gesture, an awareness of the pretend aspects of the behavior	The child simulates drinking from an empty toy baby bottle The child eats from an empty spoon The child closes her eyes, pretending to sleep
Symbolic Stage I	Level 3. Single representational play acts:	
Type I A (assimilative)	(A) Including other actors or receivers of action, such as doll or mother	Child feeds mother or doll (A) Child grooms mother or doll (A)
Type I B (accommodative; imitative)	(B) Pretending at activities of other people and objects such as dogs, trucks, trains, etc.	Child pretends to mop floor (B) Child pretends to read book (B) Child moves toy car with appropriate sounds of a vehicle (B)

Piaget does not distinguish single acts from simple multiple act combinations	Level 4. Combinatorial pretend: Single scheme combinations Multi-scheme combinations	Child combs own, then mother's hair (single scheme) Child stirs in pot, feeds doll, pours food into dish (multi-scheme)
Type II A Type II B Piaget distinguishes the assimilative case where the child identifies one object with another (A) from the accommodative or imitative case (B) where the child identifies her own body with some other object or person	Level 5. Hierarchical pretend An internal plan or designation is the basis for the pretend act. Evidence is of three types: (1) child engages in verbalization, search or other preparation; (2) one object is substituted for another with evidence that the child is aware of the multiple meanings expressed; (3) a doll is equated with a living being, treated as if it could act independently	Child picks up a toy screwdriver, says "toothbrush" and makes motions of toothbrushing Child picks up comb and doll, sets comb aside, removes doll's hat (preparation) then combs doll's hair Child places spoon by doll's hand
Type III A	Level 5. Hierarchical combinations Any combinations including an element qualifying as Level 5 are included here	Child picks up the doll, says "baby," then feeds the doll and covers it with a cloth Child puts play foods in a pot, stirs them. She dips the spoon in the pot, says "hot," blows on spoon then offers it to mother. She waits, says "more," and offers it again.

child demonstrates internal representation of both meaning and representational form (word or gestural sign); hence the co-occurrence with combinatorial representational play. Syntactic language exhibits a hierarchical structure first observable in children's simple ordered utterances, but also in their hierarchical representational play.

Representation and pretending in nonhuman primates

Representational play has been assessed in four chimpanzees over a period of 3 years 6 months, beginning when the animals were between 10 months and 2 years of age (Mignault, 1985). They had been raised by humans in an environment containing objects typically encountered by young children where they could observe the activities of their caregivers. During the study all of the animals exhibited increasing frequencies of pre-symbolic acts (Level 1) and exhibited a few instances of self-pretend (Level 2) or imitative pretend (Level 3) after these acts were modeled. In follow-up studies several months later the two oldest (age 5 years 6 months) both exhibited pretend grooming behavior, brushing a doll, and one exhibited a pretend combination (Level 4) by scraping an empty plate with a spoon handle and "eating."

Observations of the spontaneous imitation of formerly captive adult orangutans being rehabilitated to free-living showed that the animals tended to imitate favored human caregivers and high-ranking conspecifics, selecting actions within their competence (Russon & Galdikas, 1995). These included complex "real" activities imitated some time after they were observed, such as hanging a hammock, then swinging; and untying a canoe, then taking a ride. These actions were clearly learned by observation and enacted from memory, but may not qualify as "pretend" if the animal's goal was to participate in the real activity rather than evoke it representationally. However, "some cases were empty functional routines, often performed after delay (e.g., making motions of brushing teeth or siphoning fuel from an empty drum" (Russon & Galdikas, 1995, p. 13; see also Russon, *PIAC16*), suggesting representational play at Levels 2 and 3.

Hayes (1951) provided optimum conditions for representational development as defined by Werner & Kaplan (1963). The infant chimpanzee Viki was continually with her foster parents, living in a manner differing from that of a human child only in those areas demanded by species differences (e.g., her crib was roofed to prevent wandering by a motorically sophisticated infant). Viki began imitative caregiving and household routines at about 16 months of age (Level 3, comparable to humans), as well as putting on make-up and primping at the mirror (Level 4, play combination). At 18 months, Hayes' observations suggested that Viki developed a play routine involving an imaginary pull-toy which she held by an imaginary string, turning to "watch" it follow her. She also worked at guiding it around obstacles and untangling its string. If this indicated a general skill, rather than a unique activity, Viki would be credited with Level 5, hierarchical pretend. It is of interest that when Cathy Hayes attempted to join in Viki's play, pulling her own invisible pull-toy, Viki appeared startled and fearful and did not engage in this behavior again. This is in contrast with human children who are stimulated to more frequent performance by participation of a play partner (Slade, 1987a) and, by 3 years of age, develop the ability to coordinate play actions with a partner (Forys & McCune-Nicolich, 1984). Bonobo Kanzi (Savage-Rumbaugh & McDonald, 1988), chimpanzee Washoe (Gardner & Gardner, 1969), orangutan Chantek (Miles, 1990) and gorilla Koko (Patterson, 1978c; Tanner, 1985) have all been observed to enact sequences of human actions which would be credited as pretend play in humans, although none of them with the extent of elaboration shown by Viki. Examples of these sequences of human actions include, for example, eating imaginary food while spitting out the "bad parts" (Kanzi in Savage-Rumbaugh & McDonald, 1988), bathing a doll (Washoe in Gardner & Gardner, 1969), feeding toy animals (Chantek in Miles, 1990), and using a lobster claw as a cigarette (Koko in Tanner, 1985). It is notable that of these cross-fostering experiments, Viki was brought most completely into her caregivers' own lives (Hayes, 1951), in contrast with the other cases cited where the human "foster parents" maintained enriched human-like environments for the apes, and spent large blocks of time in their company, but did not actually live with them on a moment-to-moment basis.

Observations in the wild by Goodall (1986; Flaum, Goodall & Wells, 1965) describe infant chimpanzees enacting adult activities "out of context," which would qualify as pretend play if observed in human infants. These involve use of a short stick in a termite-fishing manner, constructing and climbing into bedding nests on the ground, and treating a stick as a "play partner" in a game of chase. All of these are easily dismissed by critics as "something other than pretend" and "something other than play," but they are significant in relation to skills demonstrated by cross-fostered animals. Representational ability is clearly available to chimpanzees.

Sensorimotor tasks, as noted, assess pre-representational skill, but culminate in Stage 6 tasks which are considered transitional to mental representation. Apes' performance on these tasks, in contrast with monkeys' performance, indicates representational ability, supporting the potential pretend play competence of apes. Stage 5 object permanence was confirmed in small samples of crab-eating macaques and cebus and a single gorilla and Japanese macaque (Natale, 1989). By controlling for learning of operant strategies that might improve success over trials, Natale & Antinucci (1989) found that only the gorilla exhibited Stage 6 object permanence.

Confirming Stage 6 representational tool use with Piagetian tasks requires the difficult judgment that the same problem is solved by "insight" rather than trial and error. Chevalier-Skolnikoff (1989) examined spontaneous tool use, categorizing the observations by stage and found that cebus monkeys and chimpanzees, but not spider monkeys, showed tool-use that could be characterized as Stage 6. This is in keeping with recent findings (Visalberghi, Fragaszy & Savage-Rumbaugh, 1995) that cebus and three species of ape (chimpanzee, bonobo, and orangutan) are capable of modifying sticks presented to them and successfully solving a tube-retrieval task. However, the apes, in contrast with the monkeys, showed a decline in errors across blocks of trials. This result suggests that the apes but not the monkeys used representation or remembered experience, rather than continued trial and error, in solving the task on subsequent trials. Limongelli, Boysen & Visalberghi (1995) modified the tool-retrieval task by introducing a trap into which the lure would fall if pushed from the "incorrect" end of the tube. Two of five chimpanzees solved this task and demonstrated in a follow-up experiment that their success was based on prediction of the trajectory of the lure. A single cebus monkey that solved this "trap" problem demonstrated rote responding on the control task (Visalberghi & Limongelli, 1994).

The capacity of apes to utilize the experience of video representation of real events as a guide to successful problem solving (Menzel, Premack & Woodruff, 1978; Menzel, Savage-Rumbaugh & Lawson, 1985) and to identify photographic solutions to problems presented by video (Premack & Woodruff, 1978) further substantiate their representational capacity.

Evidence also indicates representational skills are used in the wild. Chimpanzees are able to recall the locations of up to five stones appropriate for nut-cracking, and retrieve that nearest to the goal tree (Boesch & Boesch, 1984). This would require keeping the goal in mind (e.g., nut cracking) while retrieving the tool, an act similar in representational structure to Level 5 play, but a real rather than a pretend activity.

Byrne & Russon (1998) describe hierarchical levels of performance equating gorilla food-preparation in the wild and orangutan human-derived activities as cases of complex behavior that were learned through imitation. Some cases of human-derived orangutan activity (e.g., laundry stealing by boat, path-sweeping, and attempted fire-building) may either be interpreted as realistic goal-directed behavior or as pretend. These activities may be enacted in a manner similar to that in which young human children might enact hierarchical pretend play activities, for example pretending to stir and pour non-existent food, then serve it to dolls or play partners, or enacting a bathing the baby or going to bed routine. For the human children this is considered "play." While the imitative basis of such play is acknowledged, the significance of the activity is representational as the intention is not to achieve the real goal. In fact, if real food is introduced into a play meal, play may be disrupted. From the published descriptions of some nonhuman primate activities it is unclear whether the activities are "pretend" (and thus representational) or aimed at accomplishing a real goal. What is the relationship of these culturally based activities in the orangutans to culturally-based pretending in infants? Species differences require independent interpretations – it is not clear that "pretend" versus "real" is an appropriate distinction for the orangutans. The importance of this behavior may be that in order to enact such a complex set of behaviors they need to have in mind (that is mentally represent) the situation in as complex a way as human children do when pretending.

Summary: a comparison

Examination of the evidence for mental representation indicates that despite the need to overcome extensive technical difficulties, researchers have demonstrated representational abilities in apes, suggesting they may exceed the abilities of monkeys in this domain. There has been some attention to the question of whether representational activities observed in apes would qualify as "pretend," as "play," or both (e.g., Mitchell, 1987, 1990, 1994a).

Based on the evidence of representational capacity in apes the following comparisons with human children can be suggested. First, the occurrence and extent of activities that might qualify as pretend in captive apes varies with extent of interaction and depth of relationship with human caregivers. Second, occurrences of freely occurring hierarchical and representational behavior among rehabilitant orangutans and wild chimpanzees may be in the service of real goals, in contrast with children's clear distinction between real and pretend activities. Third, the limited observations of potential pretending in wild chimpanzees are closely tied with species-typical social activities (e.g., nest building, termite fishing), although if they were not we might not identify them!

Clearly cross-fostered apes show representational capacity, and some have learned human-devised linguistic systems. The view that language rests at least partially on representational capacity in other domains has been discussed previously (e.g., Sinclair, 1970; Lezine, 1973; Parker & Gibson, 1979; McCune-Nicolich, 1981; Mitchell, 1994a; McCune, 1995; Parker & McKinney, 1999). We would hypothesize that, like human children, apes' potential for language use is consonant with their representational capacity expressed in play or other domains. However, in apes, pretend play does not predict language learning and use. Viki who did not learn to produce language (except, possibly, for mama, papa, and cup, learned by operant conditioning) was a complex representational player. The Hayeses clearly spoke to her, and she may have comprehended some English in context, but this was never tested. Kanzi, a bonobo, has demonstrated varieties of representational play and comprehension of English sentences, as well as production of messages using a lexigram system. The relationship between play and language is hypothesized to rest on shared task constraints, specifically the capacity for mental representation, while each domain depends on additional abilities and circumstances. For language these include exposure to a language system and facility in some means of linguistically useful production system. Koko (gorilla), Chantek (orangutan), and Washoe (chimpanzee) all learned and utilize elements of American Sign Language, and all exhibit pretend play. The extent to which their sign productions are syntactic remains controversial, but the facts of their capacity for both pretend play and language is in keeping with the early phases of development in human children.

In considering ape pretending, a vast difference in frequency is apparent in comparison with human 2- to 5-year-olds' great preference for this type of activity. The function of play and of pretending in human children remains an area of controversy (see Smith, *PIAC9*). It may be that pretend play functions to sharpen the children's representational capacity using this external means during an age range when brain development remains fairly rapid, and internal representational ability (that is purely mental representation as opposed to representation supported by play or other actions) remains limited. Human children have a much longer childhood and enormous amounts of culturally specified information to acquire. In contrast, apes are motorically self-sufficient within months after birth and forage for themselves within a very few years. Their survival depends upon efficient food acquisition and processing at a much earlier age than that of human children. Thus there may be a premium on using their representational skills in the wild to serve these basic needs. Unlike their cross-fostered peers and human children, they may simply not have enough time on their hands to develop a large and frequently used repertoire of pretend play activities.

Part II

Pretense and imagination in children

EDY VENEZIANO

4

Language in pretense during the second year: what it can tell us about "pretending" in pretense and the "know-how" about the mind

Pretend play is a fascinatingly complex behavior from which psychologists have drawn information on a wide range of children's functionings and developments. Piaget considered it as an important window through which to glimpse the incipient representational capacities of the child. Indeed, in pretend play, the abridged and schematized enactment of activities (or events) outside of their habitual context consists in the *signifier*, the trace evoking the real activities, the *signified*. The playful attitude, the abridged actions, enacted sometimes in an exaggerated manner, the inanimate co-participants, the miniature objects, the repetition of actions lacking material results, the simulation of physical sensations in the absence of relevant physical stimuli, the displacement of the activity relative to its habitual setting, etc., are all indices that have been taken, separately or in combination, by authors trying to define and/or identify early manifestations of pretend in very young children (e.g., Piaget, 1945/1962; Inhelder *et al.*, 1972; Nicolich, 1977; McCune-Nicolich, 1981; Veneziano, 1981; Musatti, 1986; Lillard, 1993a).

More recently, pretend play has attracted researchers' attention for the potential developmental links it may have with components of "theory of mind." Indeed, the representational aspect of pretend play implies children's ability to consider one object as simultaneously having the properties it has in "real" life and those that it has by virtue of the meaning transformation it has undergone in pretend. The ability to hold double representations about a single entity is also necessary for a theory of mind. Moreover, the enactment of certain pretend activities may require the child to take someone else's perspective in playing characters, in making inanimate behave as animate (Inhelder *et al.*, 1972) with increasingly animate attributes (Wolf, Rygh & Altshuler, 1984), sometimes requiring alternation from one perspective to another. Such an ability is also a required component of theory of mind.

The developmental décalage between the time children engage in pretend play and that in which they start to provide first adequate responses to experimentally set theory-of-mind tasks (particularly those involving false belief) have placed pretend activity under close scrutiny. Thus, there is some debate as to the level of representation involved in pretending, whether implicit or, to the other extreme, self-declarative (e.g., Leslie, 1988), and questions have arisen as to the actual representational status of early pretense[1] (e.g. Perner, 1991; Lillard, 1993a). All authors agree, however, that pretend play implies some displacement from the reality plane and thus meaning transformations relative to the meaning that the actions and the objects involved would have were they considered "literally" (Fenson, 1984; Musatti, 1986; Pellegrini, 1990; Stambak & Sinclair, 1990; Veneziano, 1990; Verba, 1990; Musatti, Veneziano & Mayer, 1998). These basic characteristics may suffice to look for emerging components of a practical understanding of other people's mind if the representational aspect of pretend is considered within a communicative perspective. Indeed, given the subjective nature of pretend, the intended meanings of the child's play are not necessarily evident for a third party and sometimes only their verbalization may provide clarifying or even essential information to understand the child's pretend.

In this paper we propose to study, from a developmental point of view, which aspects of pretend play children choose to verbalize. Do children verbalize, from early on, aspects that, for their symbolic transformation, would be difficult for a third party to understand, like the new functions attributed to objects in play? Or, would the verbalization of such aspects start out to be as probable as other less specifically informative verbalizations?[2] If developmental changes towards the verbalization of the more hidden meanings of pretend were observed, they could be taken to manifest children's increasing feeling of a difference between their own state of knowledge about the pretend plot and that of a third party, and their grasp of language as a useful means to make these meanings available to an onlooker. Thus, the goal here is not to study the manifold functions that language may have in carrying out pretend play scenarios (for example, in creating the plot or in organizing its enactment) (e.g., Fein, 1981; Giffin, 1984; Garvey, 1990; Musatti *et al.*, 1998), nor in determining the relative importance of gestural versus verbal expressions (see, for example, Fenson, 1984; Musatti, 1986), but to capture the communicative function of language by analyzing how children's verbal behavior *relates to* interested onlookers' points of view.[3] This doesn't mean that language is viewed as a parallel representational plane superimposing itself onto the enactment of pretend scenarios. Pretend play is made up of the intertwining of nonverbal and verbal activities and language can thus be part and parcel of pretense (e.g., Harris & Kavanaugh, 1993; Musatti *et al.*, 1998). However, given that verbalizations inherently and simultaneously have a communicative counterpart, even when they are an integral part of the pretend play plot, children's verbalizations may at the same time be other-oriented.

Reciprocally, developmental changes in the verbalization of aspects of naturally occurring pretend play might also help our understanding of the pretend status of the play activities for the child. For example, in the case of a baby doll placed in a plastic container, if the child chooses to say "bath" instead of, for example, "baby" or "here," the child's verbalization makes explicit and brings into focus a particular aspect of her intended pretense activity, an aspect which, to be understood, requires that the other makes the same meaning transformation as the child has imagined. If, over the different pretend events, this kind of choice is made overall more often compared to that of the less informative verbalizations, then it may be inferred that the child considers the mentally constructed pretend meanings of play more salient and less clear in themselves than those aspects of play whose meaning is stable and shared with other people. In the context of make-believe (in the example, the absence of water and of any real outcome of the child's action), children's selective verbalizations would strengthen the interpretation that, even if the substitute object is only a functional substitute of the real one (Perner, 1991), rather than its *signifier*, it would not be a simple functional equivalent, whose meaning would be clear from the goal-oriented role provided by the result looked for (and probably obtained). Given that in pretense its use doesn't lead to any result, it would be a *make-believe* functional equivalent, that is, a functional equivalent for the imagined fiction and, as such, its meaning transformation deserves to be underscored: indeed, it is a crucial part of the total activity that helps identifying the episode as pretend and as a particular type of pretend event (in our example, as a pretend "bathing" event).[4]

If the developmental change discussed above was observed in pretend play, its interpretation in terms of an underlying informative thrust would be strengthened if it could be related to the emergence and early development of informative uses of language in other domains and communicative situations. To this effect, the developmental results relative to the occurrence of informative types of language in pretend play will be related to those concerning the appearance and early development of making reference to past events and providing simple explanations/justifications. Indeed, both of these uses are also of the "informative" type. References to the past are temporally displaced relative to the situation in which they are uttered and thus aim to inform or to direct the partners attention onto something which is not immediately perceptible; explanations/justifications are not given as such in the world, but are mental constructs implying a retroactive movement from what is to be explained, to the cause or reason that is temporally or logically behind it (Piaget, 1923; Grize, 1996; Schlesinger, Keren-Portnoy & Parush, 2001).

Method

Subjects and method of data collection

The data come from the longitudinal study of four mother–child dyads audio- and video-recorded in their home environment for approximately one hour, at regular intervals (usually every two weeks), while interacting naturally in free-play and book-reading situations. The families lived in Geneva, Switzerland, and were all French-speaking. Their socio-educational background can be considered middle-class. At the beginning of the study the ages of the children varied between 1 year 3 months and 1 year 5 months, and at the end, between 1 year 8 months and 2 years 5 months. A total of 48 hours of video recordings for the four dyads was obtained.[5] (See Table 4.1 specifying for each child the age range and the number of hours of video recording analyzed.)

During the observational sessions, mother and child used their own toys and books or those brought by the observers – these included toy babies and cradles, dolls, plastic boxes, a truck, a wooden spin top, a mechanical frog, wooden blocks, cotton balls, tissue, plastic chips, some toy dinnerware, Fisher-Price style family dolls and assorted beds, chairs and highchair, miniature food products, a cashier, toy plastic bottles in a container, a large-sized cartoon book depicting daily life events. Mothers were told that our interest was in their children's acquisition of language and that they should interact with them as normally as possible. The episodes of pretend play analyzed here, therefore, occurred naturally.

Table 4.1. *Dyads, age range of the study (in months), number of sessions and hours of recording per dyad*

Dyads	Age range (in months)	No. of sessions	Hours of video recordings
Dyad Cag	15–26	16	15 h 30 min
Dyad Cha	16–20	7	6
Dyad Ama	17–24	10	8 h 30 min
Dyad Gae	16–27	18	17

Method of data analysis

Each dyad's entire corpus was examined for the occurrence of pretend play episodes, all of which were then closely analyzed. They were transcribed from the time children started preparing the objects used in the pretend activity till the end of the pretend activity (by completion of the activity or by a change in the child's attention), noting in detail the actions and the objects used, the verbalizations and the temporal relation existing between the two, as well as the direction of gaze while engaging in these activities. For more information as to criteria and related examples, see Veneziano (1990) and Musatti *et al.* (1998).

Verbalization categories

In order to evaluate the "informative" potential of language used in pretend, we distinguished verbalizations according to their content, to the temporal relationship between the verbalizations and the occurrence of its referential meaning, and to the nature of the meanings expressed in language in relation to the activities and objects involved in pretend play. The categorization presented here elaborates an earlier proposal (Veneziano, 1990) and it has been partly used in a larger study of the contributions of language to pretend play (Musatti *et al.*, 1998).

Verbalizations are differentiated according to the extent to which they bring essential information on the specific nature of pretend. From this point of view four categories can be distinguished:

1. *Nonpretense* verbalizations – these are verbalizations that refer to the "literal aspects" of actions and objects involved in the pretense activity. For example, *là* "there," said while placing a baby doll into a toy cradle; *bébé* "baby," said while pretending to feed a baby doll; *gros là* "big there," while placing the bigger of two dolls on a toy pillow.

In the three remaining categories, verbalizations refer to the pretend aspects of the play and may:

2. *Duplicate* pretend meanings – these verbalizations, though referring to aspects of pretense, add little to the pretend meanings already conveyed by the child's actions, gestures and/or objects used. For example, *dodo* "night night," while placing a baby doll into a toy cradle; /bwa/[6] *boit/boire* "drink," while pretending to feed a baby doll with an empty toy bottle.

3. *Enrich/specify* pretend meanings – these are verbalizations that contribute decisively to the meaning of the pretend play, and to making it clearly understandable. This is particularly the case when the object's usual function needs to be radically transformed into a nonobvious one (for example a toy wagon becoming a bed), or when objects do not have a specific use and can be taken as any other thing (for example, a block, or a plastic chip). For example, *dodo* "night night," said while placing a baby doll into a toy wagon;[7] *biscuit* "biscuit," while bringing a bit of paper to her mouth; *chocolat #*[8] *ça* "chocolate # this," said while offering the interior part of a match box to the observer; *salade* "salad," said while pretending to feed a mechanical frog with a plastic chip; *shampoing* "shampoo," lightly shaking a toy coke-bottle above the head of a doll.

4. *Create* pretend meanings – these verbalizations contribute to the meaning of the pretend play even more decisively since they refer to pretend aspects that are created by the mere fact of stating them. This is, verbalizations referring to emotional or internal states of the play characters: for example, /plœ/ *pleure* "cries," referring to a baby doll lying in a box, followed by /dodo o'pa/ *dodo ne veut pas* "night night doesn't want," serving as a justification of the previously stated emotional state of the baby doll; /afwa/ *a froid* "is cold," said just before placing a toy quilt over the baby, where the statement of the internal state of the doll sets the stage for the child's action. This is also the case of verbalizations that create absent objects or entities (for example, *goutte*, for nose drops, while the child touches with her index the face of a baby doll just placed in a toy crib), attribute properties to absent objects (for example, *chaud ça* "hot this," referring to pretend water flowing from a toy bathtub faucet), or create states of objects (for example, *plus* "no more," meaning that there is no more to drink, looking at an empty toy bottle from which the child had pretended to drink previously). In other cases verbalizations are used to pretend exchanging greetings and salutations (*bonnes vacances* "good holidays," *à bientôt* "see you later") with fictitious interlocutors (a doll, a puppet or a stuffed animal).

These different kinds of verbalizations may refer to ongoing actions and activities, they may announce a pretend activity to come or yet refer to the pretend meaning of an activity already acted out.

We considered "informative" all those verbalizations that enrich/specify and create pretend meanings (categories 3 and 4 above), whether they accompany the play or recount it, as well as those verbalizations that announce pretend activities. Indeed, in all these cases their occurrence constitutes for the onlookers, or the potential participants, a determinant source of information about the pretend meaning(s) of the child's play activities.

Verbalization counts

All distinct verbalizations were counted separately unless they referred to the same pretend meaning. For example, *bébé dodo* "baby night night," said while placing a baby doll into a toy cradle, was counted as one verbalization and scored as category 2, the higher category of the two words. The child's subsequent verbalization /opa/ *veut pas* "doesn't want," said before removing the baby doll from the toy cradle, was counted separately as it referred to another piece of the larger pretend event.

Results

Figure 4.1 shows the proportion of informative verbalizations over all verbalizations used in pretend play episodes, for each of the four children.

For all the children we can observe the existence of two major periods:

1. A low-informative period, the length of which varies between two to six months, depending on the child, during which more than 50% of the children's verbalizations refer either to nonpretense aspects or to pretend meanings that have a clear counterpart in the actions and/or objects acted upon (categories 1 and 2).
2. A high-informative period, starting between 18 and 23 months, depending on the child, during which more than 50% of the children's verbalizations are used to specify, enrich, create or announce pretend meanings, contributing decisively to make them understood.

How does this change in the choice of verbalizations in pretend play relate to the emergence and early development of references to past events and of explanatory behaviors? Figure 4.2 plots the proportion of informative language in pretend play against the number of spontaneous

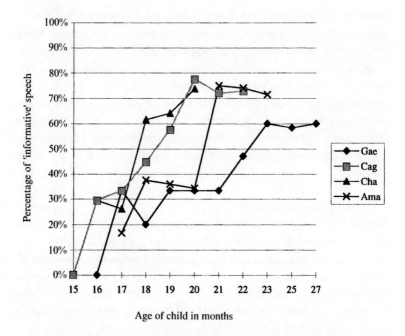

Figure 4.1. Percentage of informative language over total verbalizations in pretend play episodes, per child and per age.

references to the past and of justifications, summed over the observations available for each month, for each of the children (Veneziano & Sinclair, 1995). For all four children the increase in informative language in pretend play co-occurs or closely follows the appearance of references to the past and of explanations/justifications, and occurs when these behaviors become more frequent. There is thus a close temporal relationship in the developmental course of informative uses of language in these different domains.

Discussion

Our results show that while children use language in their pretend play from early on, the kinds of aspects of pretense that they verbalize change with development. Children tend systematically to choose verbalizations likely to be informative for a third party, a change that has to be considered as a specifically pragmatic acquisition, independent of children's advances in lexical or morphosyntactic language knowledge. Indeed, the same lexical item may be used more or less informatively, and lexical items used informatively later on may have been part of the children's vocabularies from very early on (see the examples of *dodo* "night night" in the data analysis section, and endnote 7). What changes is the *relationship* that verbalizations have with the situation in which they are uttered.

How can we explain the occurrence of this developmental change? We propose to understand it within a *"know-how about the mind"* model that envisages an intuitive understanding about representations and about the need to communicate them in order to share them with others.

By definition, pretend meanings are mental representations that differ from the meanings of the activities and objects carried out or existing in the real world, and they are created individually by the author of the pretense scenario[9] (e.g., Vygotsky, 1933/1978; Piaget, 1945/1962). Thus it can be supposed, at least at the developmental age related here, that the contents of both the reality and the imagined planes are well known to the child who conceives and gives existence to pretend play. The imagined, representational meanings, however, can be supposed to be unknown or difficult to find out for an onlooker. We think that the shift towards the systematic choice of informative type of language reflects children's increasing intuitive understanding that:

- representations about pretense differ from representations about reality;
- another person may not have access to the subjectively created pretense representations; and
- language is a good means to let others know about these pretense representations.

The communicative goal of children's pretend verbalizations may appear clearly when children address to their mother and/or to the observer verbalizations that they had already produced while intent in the acting out of pretense, or that elaborate a piece of pretend. For example, the child says *pleure* "cry(ies)" just after leaving a baby-doll in a plastic box; then she turns towards her mother and, looking at her, says again *pleure* "cry(ies)," and then confirms by nodding at her mother's request of confirmation *il pleure?* "he cries?" Later, in similar circumstances but using the box as a bed and referring to another baby doll, she says *pleure* "cry(ies)," then she goes to where her mother was sitting and says /o'pa/ *ne veut pas* "doesn't want" and continues immediately after by saying /odo o'pa/ *dodo ne veut pas* "night night doesn't want," providing the reason for the doll's pretend crying.

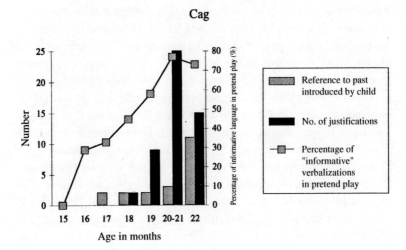

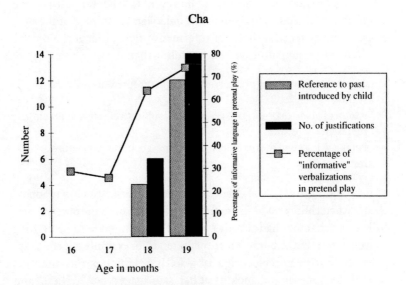

Figure 4.2. Developmental relations between references to past, explanations/justifications, and proportion of informative language in pretend play, by child and by age.

Language in pretense during the second year 69

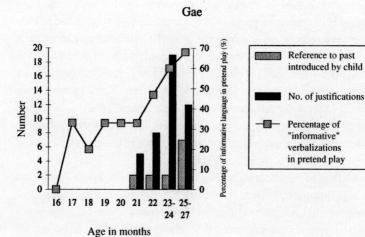

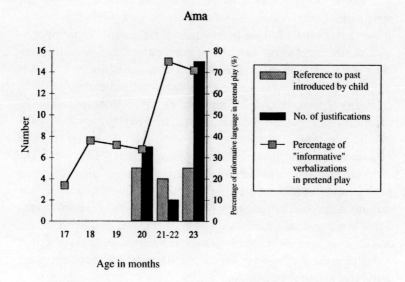

Figure 4.2 (*cont.*).

Children's intuitive understanding of the double representation/ reality (literal) aspects of pretend play are captured nicely by the following, naturally occurring, example showing that children may bring their attention to, and underscore verbally, the double status of objects used in pretend play. The child says *glace* "ice cream" just before bringing a block to her mouth and pretending to lick it. After confirming the mother's confirmation request *c'est une glace?* "is that an ice cream?," the child removes the block from her mouth and says *plot maman* "block mommy"; she then brings the block again to her mouth and pretends licking it as if it were an ice cream.

According to our interpretation, the change in the kinds of pretend meanings that are verbalized reflects both a greater understanding of the representational/real planes inherent in pretense and the beginnings of a practical, implicit, understanding that other people in their environment have mental states that may be different from their own. This interpretation finds support in the diversity of domains in which increased uses of informative language are observed. Indeed we have shown that the systematic change towards information-bearing verbalizations in pretend play closely follows developmental synchronies in the appearance of references to the past and of justifications, and co-occurs with their increased production. It is also consistent with findings of other research studies on children's early communicative behavior showing children's adjustment to their interlocutor's state of knowledge, both in naturalistic (e.g., Bretherton *et al.*, 1981; Bretherton, McNew & Beeghly-Smith, 1981; Dunn, Bretherton & Munn, 1987; Dunn, 1988, 1991; Golinkoff, 1993) and in experimental settings (e.g., O'Neill, 1996). The results obtained in the longitudinal studies of children's language in pretense point to capacities more complex than those implied by taking into account the partner's attention or his/her engagement in the situation at hand. In our pretend situations the adult partner was always present and had the same kind of objective information at his/her disposal as the child had; she/he did not, however, have access to the *mentally created* meanings attributed to it by the child.

We thus think that the developmental pattern observed strongly points towards the emergence of children's capacity to see their interactional partners as other egos having, like themselves, internal states – intentional, emotional and mental – that need to be taken into account when they want to share mental states (like the subjective interpretive properties of "objective" reality) or attain a particular goal.

As in the case of other fundamental notions of intelligence, at this level of development this attitude towards others should be viewed as a practical and intuitive understanding, a *know-how about the mind*, rather than a theory of mind (e.g., Bretherton *et al.*, 1981; Bretherton, McNew & Beeghly-Smith, 1981; Dunn, Bretherton & Munn, 1987; Dunn, 1988, 1991; Perner & Wilde-Astington, 1992; Golinkoff, 1993; Veneziano & Sinclair, 1995; Budwig, Stein & O'Brien, 2001; Veneziano, 2001), manifesting itself with the support of contextual reference points, and while children carry out their own goals and projects. It is only a first implicit acquisition that is probably not yet available to consciousness, nor necessarily accessible when children are placed in situations requiring different kinds of distancing from the natural communicative context (for example, when they are spectators or when they are asked reflective types of questions). Supposing this early differentiation between self and others on this practical level is consistent with the achievements of the sensorimotor period that ends with what Piaget calls a "Copernican revolution" placing the child in a universe where "persons become other 'egos' at the same time as the ego itself is being constituted and becoming a person" (Piaget, 1945/1962, p. 207), achievements which, according to Piaget, serve as a basis for reconstructions at higher representational levels.

When the representational aspects of pretense are considered jointly with its communicative implications, as we have done here, pretend play turns out to be an even more powerful domain of inquiry for children's *emerging know-how about the mind* and for an understanding of the first developmental steps towards the later-to-come "theory of mind," which itself should be considered as progressively evolving, in terms of the contents to which it can apply and the degree to which it can be explicitly accessed (Chandler, 2001).

Endnotes

1. Are the objects, gestures and actions "symbolic substitutions" representing other (real) objects and activities or are they "hypothetical substitutions" where the child is simply carrying out the activities using objects that can fulfill the function needed in the activity (see Perner, 1991)? Although this difference is theoretically important (see Lillard, 1993a, for an extended discussion), once the activity is considered pretending, it might be difficult to find the behavioral correlates of "I'm using this chip as salad" versus "I make the chip stand for salad." Lacking a real-life outcome, using the chip as salad and making the chip stand for salad are inherently linked in pretend. An intermediate and safe interpretation seems to me to consider the chip as a *make-believe* functional equivalent, given that its use does not lead to any result and

that it is part of the total activity that helps identifying the episode as pretend salad-eating (see also 4 below).
2. The term "informative" is used here in the general sense of making "the receiver aware of something of which he was not previously aware" (Lyons, 1977, p. 33).
3. In this study the child's mother and two observers were present in the situation and all were interested in the child's activities.
4. A *make-believe* functional substitute thus differs from a simple functional substitute because the latter is by definition result-oriented – not having a spoon I use a knife to stir, the best I can, real coffee or to eat, the best I can, my yogurt. However, it might not be a full representational signifier either because the child may not use the plastic box to signify the bathtub (to represent it) but simply as the best surrogate of a bathtub for the purpose of acting out as well as possible the intended or imagined activity or event (rather than evoking the real one).
5. The difference in age among the children reflects the fact that the study was originally designed to investigate the transition from single to multiword speech (Veneziano, Sinclair & Berthoud, 1990), and that the length of the study for each child was determined by language criteria attained at different ages by the children. At the beginning, all the children had only a few words in their lexical repertoire, and at the end, all the children produced several multiword combinations.
6. When children's verbalizations differ from standard French they are provided in phonetic form before their corresponding interpretation.
7. As can be seen from the two examples of *dodo* provided (categories 2 and 3), it is not the verbalization in itself that determines its information-bearing status but the *relationship* it holds with the pretend play scene.
8. The symbol # indicates that the following word is uttered after a pause of between 0.5 and 1 second.
9. In some cases it is the adult who proposes pretend meanings and the child may agree to carry them out. In this study, however, pretend meanings initiated by the adult's suggestion were not considered in the verbalization counts.

PIERRE-MARIE BAUDONNIÈRE, SYLVIE MARGULES, SOUMEYA BELKHENCHIR, GWÉNNAELLE CARN, FLORENCE PÈPE, AND VÉRONIQUE WARKENTIN

5

A longitudinal and cross-sectional study of the emergence of the symbolic function in children between 15 and 19 months of age: pretend play, object permanence understanding, and self-recognition

The emergence of the symbolic function at the end of the sensorimotor period plays an important role during the development of the human infant. It is the beginning of the attainment of signs and symbols – depicting an item (an object, a person, an event, etc.) by a differentiated signifier (language, mental image, symbolic gesture, etc.) specifically used for this particular representation.

Although the symbolic function cannot be tested directly, various behavioral manifestations, implying the use of differentiated signifiers, reflect it. Authors generally agree on the various behavioral manifestations appearing during the second semester of the second year of life:

- "generative" language (Piaget, 1936);
- deferred imitation (Piaget, 1945) or real imitation (Guillaume, 1925);
- the communicative function of immediate imitation (Baudonnière & Michel, 1988; Asendorpf & Baudonnière, 1993; Hart & Fegley, 1994);
- pretend play (Piaget, 1945; Kagan, 1981);
- stage 6 understanding of object permanence (Piaget, 1936);
- recognition of the specular self-image (Zazzo, 1975, 1977, 1985; Lewis & Brooks-Gunn, 1979).

These behavioral manifestations allow us to study the emergence of the symbolic function. The question is to determine whether these new competences emerge simultaneously, resulting from a common transformation, or successively. If the symbolic function appears as a singular modification influencing many behaviors, it should result in the simultaneous emergence of the various behavioral manifestations. On the contrary, if this emergence proceeds step by step, the different behaviors should emerge not synchronously but progressively, eventually in a hierarchical way.

The various behavioral manifestations have very different natures. "Generative" language is the ability to build new statements never heard before. Deferred imitation consists in reproducing a previously observed behavior, but in the absence of the model and in another context. The communicative function of immediate imitation is the predominant communication mode during the third year of life which allows peers to interact by doing the same thing at the same time and in a reciprocal way (Nadel & Baudonnière, 1982). Pretend play allows the child to refer to objects or situations by mimicking their own and others' actions in relation to these objects or situations. This evocation will progressively take place without the aid of any props. Stage 6 understanding of object permanence is the transition to a real understanding of permanence – even when the displacements of an object are not visible but must be deduced, the child remains capable of finding it.

Self-recognition, using the specular image, is more controversial and has given rise to numerous debates concerning the best way to detect it (Mitchell, 1993a, 1995; Heyes, 1994; Gallup *et al.*, 1995; de Veer & van den Bos, 1999). It had long been thought that for toddlers, being able to name their image using their first name or by saying "me" was a proof of self-recognition (Gesell & Thompson, 1934; Boulanger-Balleyguier, 1967). Verbal labeling is now considered insufficient because the child might have learned that the image bears the same name as him- or herself without having understood that both names correspond to the same person, and because this naming implies a certain level of language acquisition prior to test success.

It was then hypothesized that children knew the properties of reflecting surfaces of mirrors before "recognizing themselves" (Preyer, 1887; Wallon, 1942/1970; Zazzo, 1977, 1981; Gouin-Décarie, Pouliot & Poulin-Dubois, 1983). The test used was the turn test – the child is placed in front of a mirror, then an object appears behind him or her without the child's knowledge, until its reflection appears in the mirror. The child should turn to the object only if he or she understands that the mirror is reflecting his or her image. But it is only around 6–7 years of age that children fully understand the reflecting properties of mirrors (Jézéquel & Baudonnière, 1987). If the 18-month-old child recognizes his or her specular image, it is not because he or she understands the reflecting properties of mirrors. Moreover, a long habit of reflecting surfaces is not a compulsory condition for recognizing oneself. Bedouin children, who have never seen any reflecting surfaces, recognize themselves at the same age as other children, after only a few minutes in the presence of these surfaces (Priel & de Schonen, 1986).

The privileged option, nowadays, is the mark test set up by Gallup (1970) in order to test self-recognition in chimpanzees, and simultaneously invented by Amsterdam (1972) for use with children. It consists in applying a mark onto the subject's cheek, without his or her knowledge, and observing how the child attempts to wipe it off – by wiping the mirror or his or her own cheek. By this latter action, we discern that the child infers that the specular image is the image of his or her own face (Asendorpf & Baudonnière, 1993). This test was revised by Asendorpf, Warkentin & Baudonnière (1996) because of a double ambiguity – the mark applied to the cheek raises a problem of laterality because of the right/left inversion of the mirror, but also because the procedure – asking the child no question, only observing his or her reactions in front of the mirror – was often not interpretable when the child did not spontaneously wipe the mark off. The revised mark test consists in applying the mark onto the child's forehead and, if the child does not respond, the experimenter looks at the mirror and explicitly asks the child to wipe the stain off. This revised procedure allowed us to decrease considerably the number of ambiguous responses. (Ambiguous responses occurred when children "looked at least once at their mirror image without gross body movement for at least 5 s and . . . did not try to touch the mark or . . . touched [unmarked parts of their face]" – Asendorpf et al., 1996, p. 315).

Relationships among self-recognition, object permanence understanding, imitation, pretend play, and language are known to be somewhat variable. Stage 6 object permanence understanding develops before self-recognition and pretense (Bertenthal & Fischer, 1978; Chapman, 1987; Mitchell, 1993a; Warkentin, Baudonnière & Margules, 1995), and appears only weakly correlated with language acquisition (Corrigan, 1976). By contrast, the presence of pretend play around the second year of life seems essential to language acquisition (Sinclair, 1970; Lézine, 1973). Deferred imitation appears before self-recognition (Meltzoff, 1990), whereas extensive synchronic imitation seems concomitant with self-recognition (Asendorpf & Baudonnière, 1993; Hart & Fegley, 1994). A generalized ability to match between kinesthetic and visual modalities has been posited as essential for synchronic and deferred imitation, self-recognition, and pretend play (Mitchell, 1993a; 1997a), but whether self-recognition develops prior to pretend play (as stated by Chapman, 1987, based on longitudinal data) or the reverse (as stated by Warkentin *et al.*, 1995, based on cross-sectional data) is unclear. The purpose of the present study is to determine the relationships among three symbolic functions – object permanence understanding, self-recognition, and pretend play – using both cross-sectional and longitudinal data on the same children. We chose these three of all the behavioral manifestations of the symbolic function because we could reliably manipulate them during successive sessions.

With cross-sectional data each child is tested only once at a given age, such that it is impossible to know precisely when different behavioral capacities emerge if they are all present at that age. Thus, a cross-sectional methodology is insufficient to test the hypothesis of a hierarchy of emergence of these various capacities, or that of a simultaneous emergence. Moreover, a longitudinal study testing the same behavioral capacities is also problematic, because simple repetition of tests might decrease the acquisition age (if the capacities are influenced by learning, and are not purely maturational). We chose a longitudinal and cross-sectional experimental protocol, with three groups of children, G_{15}, G_{16}, and G_{17}, aged, respectively, 15, 16 and 17 months at the beginning of the experiment, who were tested every two weeks. It is thus possible to compare the children's performances at the same developmental age with different exposure rates (e.g., at the age of 17 months, G_{15} is at the fifth session, G_{16} is at the third session, and G_{17} is at the first session). This method has three advantages. It should allow: (1) precise dating of the various acquisitions; (2) control of test repetition effects; and (3) knowledge of whether the age of acquisition is independent of the age at which the tests begin. In addition, the limitation to a period of two months for the five sessions decreases the risk of children dropping out of the sample.

Hypotheses

If the acquisition of these different competences depends on maturational mechanisms only, then our results should agree with a maturational model (Figure 5.1A). On the contrary, if a periodic testing of the children about self-recognition, pretend play, and object permanence generates learning, then we can expect our results to agree with a learning

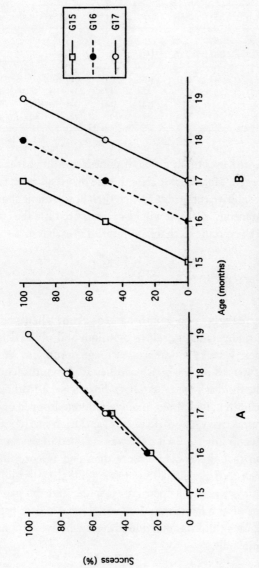

Figure 5.1. Theoretical results for maturational (A) and learning (B) models.

Table 5.1. *Mean ages and number of subjects (in brackets) at each test session*

Session	Age (months)								
	15	15.5	16	16.5	17	17.5	18	18.5	19
G15	15.1 (14)	15.5 (11)	16 (14)	16.4 (13)	16.9 (14)				
G16			16 (14)	16.5 (14)	17.0 (14)	17.4 (13)	17.9 (14)		
G17					17.0 (14)	17.4 (14)	17.9 (14)	18.3 (14)	18.8 (14)
N (total)	14	11	28	27	42	27	28	14	14

model (Figure 5.1B), with eventually different slopes according to the age at which the tests began. Moreover we hypothesize that if all three tests require a common cognitive competence, then their acquisitions should be practically simultaneous. If not, it will be necessary to raise the question of the nature of the specific mechanisms implied in each test.

Methods

Subjects

The age period selected was 15–19 months, an age about when stage 6 object permanence understanding, self-recognition, and pretend play usually emerge. There were 14 children in each age group: one group began at 15 months (G15: 8 boys, 6 girls); another at 16 months (G16: 7 boys, 7 girls); and the third at 17 months (G17: 6 boys, 8 girls). (Although we started with 52 children, two, three, and five children dropped out of G15, G16, and G17, respectively. The main cause of exclusion was repeated absence of a child due to illness.) Each child was included in the sample under the condition that there was no more than one absence; these absences were compensated by a blank session (see Table 5.1). All children attended the day-nursery of the Salpêtrière Hospital, and the parents gave formal consent. After a familiarization period (several days before the filming started), each child was individually tested in five sessions at 15-day (+/−2 days) intervals.

Procedure

Three female experimenters conducted the tests, and a given child was always tested by the same experimenter. Each session, preceded by a vocabulary agreement phase (see below), consisted in a succession of three tests:

- pretend play (PP);
- stages 4, 5 and 6 of object permanence (OP);
- specular image recognition (SR).

The order of the PP and OP tests was counterbalanced from one session to another. The SR test was always performed last, however, in order to avoid possible emotional aspects that could bias the other tests.

Vocabulary agreement phase

This very important phase allows, beyond a familiarization with the new situation, an agreement about a common vocabulary (the words "spot," "to show," "to wipe off") used in the SR test. This agreement is achieved using a doll, the forehead of which is stained with a green spot. This phase ends when the experimenter is sure that the child understands the vocabulary used.

Stages 4, 5 and 6 of object permanence understanding

We used a device comprising three soundproof removable hiding cylinders (15 cm high, 8 cm diameter) adjusted onto a rectangular stand at the child's hand-level. At half-height, a cross-sectional partition of each cylinder allows a small ball dropped by the experimenter to enter, but prevents the child from seeing inside.

The experimenter invites the child to play with the device and teaches him or her how to take the hiding cylinders off. The experimenter then shows a small ball to the child but does not let him or her take it. To test for stage 4 understanding, the experimenter places the small ball in her right hand and holds it out, closed, to the child. To test for stage 5 understanding, the experimenter holds the small ball between two fingers and drops it into either the left or right (in a counterbalanced order) hiding cylinder.

The following stages are similar to the preceding two except that the visible displacement (from the left hand to the interior of the right hand which then closes) is followed by a supplementary invisible displacement (from inside the closed right hand to inside the hiding cylinder). To test for stage 6 understanding, the experimenter then discreetly drops the small ball into either the right or left hiding cylinder, then holds out her closed hand to the child. Finally, to test for the generalization of stage 6 (called stage 6g), the experimenter discreetly eliminates the small ball from her right hand by putting it into her pocket, then successively proceeds over each of the three hiding cylinders, and holds out her closed hand to the child. We test the systematic search by the child. And the experimenter says "Oh, the ball is disappeared."

Pretend play test

This test consists of a series of three items showing the degree of pretend-play understanding. In each case, success is defined as the child's re-enacting the experimenter's pretense. If the child does not succeed on the first question, the experimenter asks the same question twice more.

- A new function is attributed to an object with a predefined function: the experimenter takes a small pan (25 cm × 25 cm) and puts it on her head pretending it is a hat ("Look, I put the hat on the head") and offers the pan to the child.
- A function is attributed to an object with no predefined function: the experimenter takes a wrapped rectangular closed box (20 cm × 10 cm × 10 cm) and pretends to play with a lorry ("Look, I am playing with the lorry") and offers the box to the child.
- Pretend play without any object: the experimenter takes an imaginary piece of cake from her hand and brings it to her mouth ("Look, I am taking a cake and I eat it; it is nice. Will you take a piece of cake and eat it?") and the experimenter offers her empty open hand to the child.

Recognition of the specular image

The experimenter calls the child away from the mirror in order to wipe his or her nose and discreetly marks the child's forehead with a green spot. She then goes back next to the mirror and invites the child to come and play again. The procedure is:

- the experimenter draws the child's attention to his or her reflection by manipulating a toy in front of the mirror;
- in the case of no reaction, the experimenter says: "Oh watch the spot on the face! Do you see the spot on the face?";
- if no answer, the experimenter asks: "Will you show me the spot on the face?";
- if no answer or if pointing to the mirror, the experimenter asks the child "Will you wipe the spot off?" and gives him or her a piece of cotton wool.

This test is successful when the child moves his or her hand to his or her forehead and visually observes the gesture in the mirror (for more details see Asendorpf & Baudonnière, 1993; Asendorpf et al., 1996).

The children were videotaped, without their knowledge, with two remote-controlled mobile video cameras, in a familiar room reorganized and standardized for the experiment. A mirror (1 m × 0.8 m) was placed vertically at 0.5 m away from the wall, facing one of the cameras, in such a way that the child and his or her specular image were both visible. The experimenter was supplied with a discreet audio receiver which allowed her to hear the instructions given by the operator located in another room.

Results

Development of performances on the various tests

We assumed that from the moment a child succeeds on a test, the required capacities have been acquired. We ignored failures in performance which sometimes occurred during sessions following initial success when these failures appeared to derive from interfering factors such as temporary tiredness or inattention, family disturbance, or not-yet-diagnosed illness. In fact, such failures represented less than 7% of all sessions. In the specular recognition test, children who were "ambiguous" even after the revised test (3% out of the 210 sessions) have been classified with those who do not recognize their specular image.

Object permanence test

We first wanted to check out whether or not the children had acquired object permanence stages 4 and 5, which respectively appear at 7–10 and 10–15 months of age. At all ages, the children had already acquired stages 4 and 5, if we ignore first sessions in which children were sometimes inhibited.

The comparisons among the performances of G_{15}, G_{16} and G_{17} at object permanence stage 6 (see Figure 5.2) shows that, in agreement with the results of previous studies, object permanence stage 6 understanding seems to be acquired early – by the second session, the success rates of each group are close to 80%. It is thus necessary to examine the results obtained on the test we added – stage 6g. Comparisons among the performances of G_{15}, G_{16} and G_{17} at object permanence stage 6g understanding gives a surprising result – the similarity of performances of the three groups at the first session (see Figure 5.3).

At 16 months, the G_{15} and G_{16} success rates are identical. It thus does not seem that the G_{15} children have progressed faster because of previous sessions. Performance similarity between these two groups is maintained

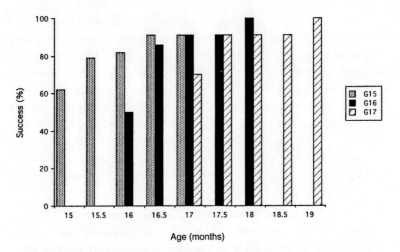

Figure 5.2. Percentage distribution of success rates at object permanence stage 6 for the three age groups.

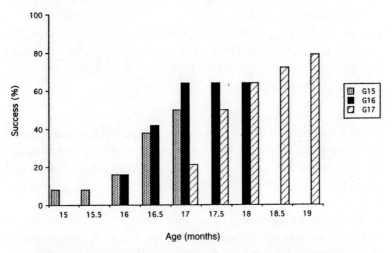

Figure 5.3. Percentage distribution of success rates at object permanence stage 6g for the three age groups.

at 16.5 and 17 months. At 17 months, the G16 success rate (64.3%) is significantly higher than that of G17 (21.4%). At 17.5 and 18 months, however, the G16 and G17 success rates are very much alike, showing that the difference at 17 months is probably a first-session effect for G17. These results are compatible with a maturational model, as the success rates of the three groups seem to be independent of the number of previous sessions, and seem to improve regularly with age.

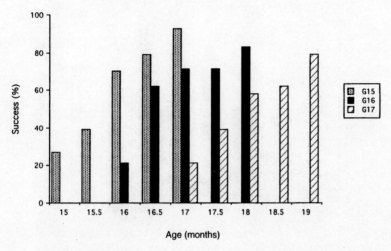

Figure 5.4. Percentage distribution of success rates at the pretend play test for the three age groups.

Pretend play test

We present here the results obtained on the "cake" item only which, in our opinion, is the only one that requires real symbolic processing. The two other items are initiators only, and ensure a proper comprehension of instructions.

At 16 months, the G15 success rate is significantly higher than that of G16. This superiority decreases as soon as 16.5 months, since we observe no more significant differences between these groups either at 16.5 or 17 months (see Figure 5.4). At 17 months, the G15 success rate (92.9%) is slightly higher than that of G16 (71.4%), but the difference is not significant. By contrast, G16 shows a significantly higher success rate than G17. This difference in favor of G16 compared to G17 is maintained at 17.5 and 18 months. Although these data thus favor a learning model, it should be pointed out that the acquisition rhythms of G15 and G16 seem faster than that of G17.

Specular image recognition test

As shown on Figure 5.5, the results are in agreement with a learning effect, since we observe similar performances during the first sessions (whatever the age) and quite similar improvements. Nevertheless, the improvement slope for G16 is steepest, and their success rate at 18 months is the highest.

At 17 months, 71.4% of G16 recognized their specular image, but only 35.7% of G17 did. At 18 months, the G16 success rate (92.5%) is again

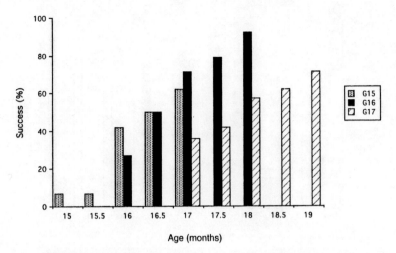

Figure 5.5. Percentage distribution of success rates at the self-recognition test for the three age groups.

significantly higher than that of G17 (57.1). At 19 months, the G17 success rate is noticeably higher than at 18 months, but the difference is not significant. Note that we obtained similar results with the specular image recognition test and the pretend play test.

Comparisons among the different acquisitions

Numerous authors consider the three tests we used as indicators of the symbolic function's emergence. We need to know whether or not these indicators appear simultaneously. The answer to such a question is not simple, and a global correlation between tests would probably make no sense. The answer depends on the performances of each child, and interfering factors are always possible (e.g., temporary tiredness or inattention). We thus decided to solve this problem indirectly.

For a given test, children can be classified into three subgroups: those who succeed as soon as the first session (G++); those who succeed between the second and the fifth sessions (G−+) and those who never succeed (G−−). Clearly the competences required for a given test are acquired by the G++ group, and not by the G−− group. The 42 subjects have thus been classified into these three subgroups, test by test, without taking their age into account (see Table 5.2). For the three tests on average, most of the children (56.3%) are in the G−+ sub-groups, 21.4% are in the G++ group and 22.2% are in the G−− group. This confirms the relevance

Table 5.2. *Distribution of children in each subgroup (G++, G−+, G−−) for each test*

Test	Subgroup		
	G++	G−+	G−−
Self-recognition	10	24	8
Pretend play	11	26	5
Object permanence	6	21	15

Table 5.3. *Number of relevant sessions for each subgroup for each test*

Test	Subgroup			
	G++	G+	G−	G−−
Self-recognition	50	57	63	40
Pretend play	55	73	57	25
Object permanence	30	61	44	75

of the chosen period of age, as the G−+ subgroup corresponds to the children acquiring the necessary competence during the longitudinal experiment. The performances of the G−+ children can be further separated into two subgroups: the sessions performed before the first success (G−) and the other sessions including the shift (G+). Table 5.3 shows the number of sessions present for each subgroup.

We can hypothesize that if the performances on two given tests require similar competence, the children's behavior will be predictable from one test to another taking into account the experimental error. There are three conditions that must be satisfied to accept that a similar competence is necessary.

The first condition requires making sure that the number of successful sessions on one test (e.g., SR) is higher for those children who achieved G++ on a second test (e.g., PP) than for those who achieved G− on the same second test (i.e., PP). This means that most subjects succeeding on one test generally succeed on another test, and conversely most of those failing that one test also generally fail the other test. (This checking is essential – if it were not satisfied, the following analyses would have a limited meaning.) The second condition concerns acquisition synchrony during the study, and requires making sure that the number of successful sessions on one test (e.g., SR) should be higher for G+ sessions on the

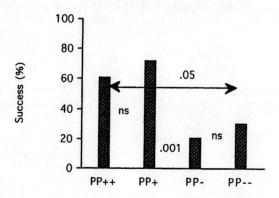

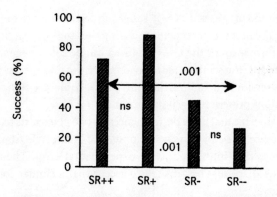

Figure 5.6. Covariation between the self-recognition test and the pretend play test (P-values based on ANOVA).

second test (e.g., PP) than for G− sessions on the same second test (i.e., PP). Finally, in case of a positive answer to the two preceding conditions, the third condition requires making sure that the G++ and G+ subgroups on the one hand, and the G−− and G− subgroups on the other, show similar success rates. Such similarities would confirm the practically synchronous aspect of the acquisitions in both tests.

Figure 5.6 shows the relationships between the specular image recognition test and the pretend play test. The three analyses are satisfied in

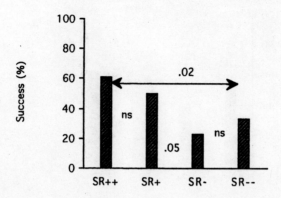

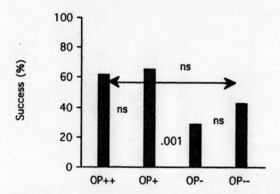

Figure 5.7. Covariation between the self-recognition test and the object permanence test (*P*-values based on ANOVA).

both directions. It thus seems that the competences necessary to succeed on these two tests are acquired practically simultaneously, and that both tests probably require common cognitive tools.

By contrast, the results about the relationships between the specular image recognition test and the object permanence test are far less clear (see Figure 5.7). Whereas the self-recognition test seems to predict the successful performances on the object permanence test (all three conditions are satisfied), the reciprocal is only partially true, as the performance

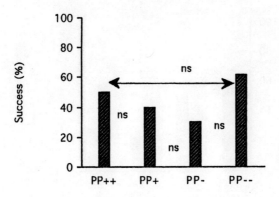

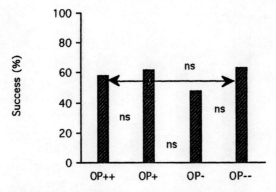

Figure 5.8. Covariation between the object permanence test and the pretend play test (P-values based on ANOVA).

on self-recognition tasks for groups OP++ and OP-- are not significantly different. It thus seems that some link between self-recognition and object permanence can be observed, but it is noticeably less than the one observed between self-recognition and pretend play.

Relationships between pretend play and object permanence are nonexistent (see Figure 5.8). The two main conditions (G++ > G-- and G+ > G-) are satisfied neither for the object permanence test in relation to the pretend play test nor for the reverse relationship. It is clear that these two tests require different cognitive tools.

Discussion

A first conclusion can be drawn – all our results are more in agreement with a learning model than with a maturational one, whatever the test considered. Nevertheless, it should be noted that the procedure we used – not a real teaching procedure, but a simple questioning repeated every two weeks – is still capable of improving the performances, sometimes considerably, in less than two months. This result is not without consequences for our representation of the true competences of such young children, and implies that single session studies give only partial information about these competences, more especially when children are young.

The second important result concerns the learning velocities, particularly in self-recognition and pretend play. Whereas the 16-month-olds learn faster than the 17-month-olds, the 15-month-olds seem to improve their performances in the same way as the 16-month-old children. This seems to indicate that beginning the tests at 15 months would not improve the performances more than starting at 16 months. These results are compatible with the hypothesis of a sensitive period around 16 months of age, during which confronting the children with symbolic tests is particularly efficient. Such a result, on such a small sample, is important, but it needs to be confirmed. The question must also be raised about short-term and eventually medium-term general effects concerning all the behaviors related to the symbolic function, and particularly language. The problem is to find the appropriate experimental paradigms.

The third result concerns the simultaneity of acquisition of the three tests. Our longitudinal and cross-sectional methodology allows a precise dating of the acquisitions. Self-recognition and pretend play emerge practically simultaneously, contrary to object permanence. This means that the former tests would require common competences, whereas understanding stage 6 object permanence occurs earlier. These results agree with claims by previous authors that object permanence understanding develops before self-recognition and pretense (Bertenthal & Fischer, 1978; Chapman, 1987; Mitchell, 1993a; Warkentin *et al.*, 1995). It is probable that understanding object permanence does not require processing as elaborate as that required by self-recognition and pretend play. In order to succeed on the object permanence test, children could learn such covariation rules as "the hand over the hiding cylinder indicates the place of the object." Such rules would be relatively simple for children to process and progressively generalize. Consequently, the representational "tools" required for understanding object permanence do not seem to be as complex as those necessary in the two other tests.

In conclusion, this longitudinal and cross-sectional study shows that the age of 16 months seems to constitute a sensitive period during which repeating the various test-situations (self-recognition, pretend play, object permanence) is particularly efficient – the different cognitive capacities can be more precociously acquired by simply repeating the tests, without any need for explicit teaching. The emergence of these different competences thus does not seem to result from nervous system maturation only.

Moreover, it is clear that self-recognition and pretend play emerge practically simultaneously. Cognitive capacities of the same nature thus seem to support these two behaviors. The most probable hypothesis is the one according to which the child succeeds on these two tests only if he or she accesses a secondary representation – he or she would be capable of representing himself (herself) in situations detached from the direct perceptive reality (primary representation). Indeed the child must associate his or her specular image (primary representation) and his or her own face that he or she cannot see directly (secondary representation), in order to succeed at self-recognition. Similarly, the child must associate a symbolic gesture (primary representation) with the intrinsic properties of the evoked object (secondary representation) (Asendorpf *et al.*, 1996), in order to succeed at pretend play. These two behavioral manifestations imply the building of an elaborate mental representation and can be considered truly symbolic.

6

Caregiver–child social pretend play: what transpires?

Much of the early research on children's pretend play focused on the young child's ability to generate pretend actions. This research tradition owes its origins to Piaget's (1945/1962) seminal observations of his own children at play. Rich in description, and accompanied by a strong conceptual framework, Piaget's work inspired a series of subsequent studies of children's ability to produce pretend actions during the course of solitary play (for reviews, see Fein, 1981; Bretherton, 1984). Frequently, these studies drew on Piaget's notion of the emerging semiotic function, in particular the young child's ability to distinguish between signifier (e.g., wooden block) and signified (e.g., bar of soap), to explain the accomplishments witnessed in solitary pretend play (Nicolich, 1977; McCune-Nicolich, 1981).

While not repudiating Piaget's approach directly, a different tradition emerged in the latter part of the twentieth century. The focus of researchers in this tradition is pretend play in a social context, particularly pretending in mother–child and child–child dyads. The accompanying conceptual framework, though less explicit than Piaget's, emphasizes the social–cognitive gains that derive from collaborative interactions (see Rogoff, 1990; Cole, 1996). To be sure, an emphasis on social context in symbolic (pretend) play is in itself not new, owing a debt to Vygotsky (1930–1966/1978) and Werner & Kaplan (1963) among others. However, more recent empirical work on pretending in context has added important information about the particular factors that influence social pretend play. Researchers have noted that the age of the child (Kavanaugh, Whittington, & Cerbone, 1983; Haight & Miller, 1992; Haight, Parke, & Black, 1997), who the child plays with (e.g., sibling/peer or parent: Dunn & Dale, 1984; Farver, 1993), and the culture in which the child lives (Farver, 1993; Gaskins, 1996; Haight *et al.*, 1997; Göncü *et al.*, 1999; Haight *et al.*, 1999) all impact the frequency and quality of pretending with others.

In this chapter, I focus on social pretend play in mother–child dyads. Social pretend play can certainly be conceived more broadly. There is a fairly extensive literature on shared pretend play among siblings and peers (see Howes, 1992) as well as recent work comparing mothers and fathers as play partners discussed toward the end of this chapter (Haight *et al.*, 1997). Nonetheless, in European–American populations mothers are frequent participants in young children's pretend play (Haight *et al.*, 1994). Furthermore, we now know a good deal about the role that mothers play both in initiating and in maintaining social pretend play with young children. I review and synthesize that evidence here. However, because much of the available data comes from studies of middle class, European–American mother–child dyads (Farver & Howes, 1993), I conclude the chapter by considering the larger implications of cultural differences on the development of social pretend play.

The interactive context of early pretending

Although in many Western cultures mothers are frequent participants in social pretend play, clearly children pretend while playing alone. To evaluate the effect of shared pretending on children's pretense abilities, several investigators have compared solo pretend play to pretending during joint play with mothers. These studies have shown consistently that maternal involvement increases the duration (Dunn & Wooding, 1977; Slade, 1987a; Fiese, 1990; Haight & Miller, 1992), complexity (Slade, 1987a), and diversity (O'Connell & Bretherton, 1984) of children's pretend play. Furthermore, there is little doubt that mothers assume the leading edge in these joint play episodes by employing a number of "strategies" that support higher levels of child pretense. During the course of dyadic play, mothers provide prompts and explicit direction (Miller & Garvey, 1984; Haight & Miller, 1993), offer descriptions of the child's pretending (Kavanaugh *et al.*, 1983), demonstrate and describe their own pretend gestures and actions (O'Connell & Bretherton, 1984; Kavanaugh & Harris, 1991), insert supportive commentary and "how to" suggestions (Dunn & Dale, 1984; O'Connell & Bretherton, 1984), request pretend actions from the child (e.g., "What do you want to feed the baby?") and repeat/expand on child pretend utterances (Kavanaugh *et al.*, 1983; Haight & Miller, 1993).

In addition to these observational studies, several experimental investigations have shown that an adult's modeling of pretend actions leads to an increase over baseline in the frequency and complexity of children's pretend actions (Watson & Fischer, 1977; Fenson & Ramsay, 1981; Bretherton *et al.*, 1984; Fenson, 1984). Importantly, some of these modeling studies offer rough analogues to mother–child play. For example, Fenson & Ramsay (1981) found that older infants increased their frequency of two-sequence doll play (e.g., placing a doll in the supine position and then covering it with a cloth) after watching an adult model such actions.

Taken together, the observational and modeling studies suggest that a supportive social context can provide children with the opportunity to increase the frequency and complexity of their pretend play. Nevertheless, we still know relatively little about the parameters that are most effective in raising the level of the young children's pretend play. It is not clear, for example, whether some maternal behaviors (e.g., demonstrating a pretend action versus describing the child's pretend actions) are more effective than others, whether particular pretend actions matter less to the child than the sheer number of opportunities to engage in shared pretense, or whether the intensity and quality of the mothers' involvement is the critical factor in advancing the child's pretend play.

A closer look at shared pretend play

Slade (1987a,b) conducted two studies that offer insights into how mothers' participation in dyadic play enhances pretending among young children. In one study (Slade, 1987a), she observed mothers and their children in free play at bi-monthly intervals from the child's age of 20 to 28 months, and noted the effects on children's pretense of two different levels of mothers' involvement. One was a somewhat passive interaction involving verbal commentary only, while the other constituted more active involvement (e.g., explicit suggestions for pretend activities). Compared to the child playing alone, both levels of mothers' involvement increased the length and complexity of the children's pretend play. However, complexity of child pretense was highest when the mother both initiated pretend themes and remained actively involved in the play. A related study (Slade, 1987b) not only supported the claim that maternal involvement enhanced children's pretending during dyadic play but investigated the impact of attachment on symbolic (pretend) play. At the outset of this study, children were assessed in the Strange Situation, classified as secure or anxious, and then observed in free play with their mothers at regular intervals from 20 to 28 months of age. The principal finding was that secure children had the longest episodes and spent more time in the abstract and planful pretend play than anxious children. As Slade (1987b) noted, in Vygotskian terms secure children seemed to function at the higher end of the zone of proximal development. It is impossible, of course, to tease apart the nature of the interaction between mother and child during dyadic play. In this particular case, longer pretend episodes among secure children could be due to the greater persistence and willingness of secure children to explore their environment (Main, 1983) which, in turn, could lead to an increase in their mother's interest to engage with them in pretend play.

At bottom, then, what we can conclude is that dyads comprised of secure children "worked better" (Slade, 1987b) in the realm of pretense and symbolism. Still, it is noteworthy that mothers of secure children were more likely to be actively engaged in pretend play while mothers of anxious children were more likely to be involved more passively, e.g., primarily through commentary on their child's play (Slade, 1987b). It appears, then, that mothers' active involvement constitutes an important element in promoting higher levels of pretend play among young children.

Interestingly, mothers' activity during pretend play with their children has been contrasted with the pretending that occurs during joint play with siblings (Dunn & Dale, 1984; Farver, 1993). In a frequently cited study, Dunn & Dale (1984) found that during the course of joint pretend play siblings often assumed the role of "complementary actors" (e.g., planning, exchanging roles) whereas mothers were more likely to participate as spectator–commentator. By contrast, other investigators (Kavanaugh *et al.*, 1983; Slade, 1987a,b; Haight & Miller, 1993) have observed mothers who participated actively in shared pretend play with their children. This discrepancy may be resolved in part by Dunn's (1991) observation that siblings are particularly likely to engage in rich and complementary pretending when there is a strong affectionate bond between them. This observation would seem to parallel Slade's (1987b) finding that during shared pretend play mother–child dyads with secure children established a comfortable relationship in which children interacted with their mothers as though they were playmates. Thus, one could argue that Slade (1987b) and Dunn (1991) converge on the quality of the relationship between the play partners, more than age differences per se, as the principal factor that influences the nature of shared pretending. To the extent that play partners have established an intimate and affectionate bond they are more likely to engage each other in rich and complementary pretend play.

It seems that active involvement in pretend play has a positive effect on both the frequency and quality of young children's shared pretend play. But can we unpack this notion and say more specifically what active involvement entails? Broadly conceived, I believe there are two crucial dimensions to active involvement: initiating and maintaining pretend scenarios.

Initiating pretend episodes

Slade's (1987a,b) work reveals that one element of active involvement is direct engagement in pretend sequences with the child (e.g., suggesting pretend actions, adopting complementary roles) as opposed to verbal commentary on the child's pretend play. This suggests that the mothers in Slade's studies assumed a role that Haight & Miller (1993) have referred to as "initiator." As the term implies, the initiator is the first person to direct pretend actions toward the play partner. Typically, when mothers initiate they direct the child away from solo play or restructure joint non-pretend play so that it involves pretending. In considering this role, it is important to note that initiations can be either verbal (e.g., "You be the baby. I'll be the Mommy.") or non-verbal (e.g., pretending to feed baby) (Haight & Miller, 1993). For instance, in the example above the mother's remarks are not a passive commentary on the child's pretend play. Rather, they amount to an invitation to begin a pretend sequence. Thus, in the course of adult–child dyadic play, verbal statements are not necessarily associated with lower levels of child pretense. What seems critical is that either through verbal or non-verbal interactions the caregiver enters into rather than simply comments on the child's play.

There is, however, one important caveat about mothers' direct engagement in pretend play: it must be non-intrusive. In her study of mother–toddler play, Fiese (1990) noted the negative impact of "maternal intrusions" on children's pretending. These were mistimed initiations of new pretend activities, e.g., just as the child began to engage the mother in a pretend scenario the mother initiated a different pretend action. Such actions and comments have the effect of overriding the child's own initiatives and of interfering with the turn-taking and reciprocity that is crucial to mutually rewarding social pretend play.

If timely direct engagement is one important element in initiating shared pretend play, a second is maintaining a supportive framework for pretense. With young children (under 2 years of age) particularly, shared pretend play frequently amounts to brief episodes woven into the larger fabric of sensorimotor exploration of objects. In the midst of this literal play, mothers often take the lead in establishing a pretend scenario (Kavanaugh *et al.*, 1983; Haight & Miller, 1993). They stipulate the make-believe content of objects, e.g., "This [block of wood] can be our soap," and suggest a scenario in which the prop can be used in pretend fashion, e.g., "Let's wash our hands for dinner."). Among European–American dyads, there is often a striking persistence in the mother's attempt to induce pretend play. For example, consider the mother's behavior in the following excerpt from data I have collected of free play between mothers and children:

Mother	*Child* (24 months)
Opens empty box of crackers and says, "Want a cracker?"	No response
Makes eating noises, offers box to C., saying, "Mommy's going to have a cracker. Mmm. Delicious."	No response
"You want a cracker? There's none in there? Well, pretend there's a cracker. [Offers box] Take a bite."	Takes box from M

This example illustrates a common occurrence in my data, particularly with children under 2 years of age, where the mother attempts to forge a pretend scenario with the child. Often the child does not make a complementary pretend response, as in the example above, but this rarely dissuades the mother. Quite frequently mothers simply shift to a different pretend theme. What the child takes away from these interactions is less clear, of course, but over time one could expect that even unrequited maternal initiations have the effect of priming the child for pretend play. At a minimum they signal the mother's willingness to explore the realm of make-believe.

Maintaining pretend episodes

Beyond age 2 years of age children become increasingly likely to initiate pretend episodes during the course of joint mother–child play (Kavanaugh *et al.*, 1983; Haight & Miller, 1993). The strongest support for this claim comes from Haight & Miller's (1993) intense longitudinal study of 1- to 4-year-old children interacting at home with their mothers. This study revealed a steady increase over age in the percentage of child-initiated pretend play episodes, from a low of 1% at 1 year to a high of 49% at 4 years of age. Additional support comes from Kavanaugh *et al.*'s (1983) free play, cross-sectional (laboratory) study of mothers and their 12- to 15-month, 18- to 21-month, and 24- to 27-month-olds where children in the oldest group undertook considerably more pretend initiatives than children in either of the two younger age groups.

Child-initiated pretense, then, provides an opportunity for mothers to respond in a manner that facilitates further pretending. The question is how this might be accomplished. Haight & Miller (1993) provide a clear illustration in detailing what they refer to as "maternal elaborations" – statements that added new content to the child's pretend initiation. They offer the example of a 2-year-old who begins a pretend sequence by turning to her mother and saying, "Whoosh, I turned you into a frog." The mother then replied, "That's quite a magic straw." Note that in contrast to the maternal intrusions observed by Fiese (1990) that disrupted child pretense, here the mother's statement accepts and carries forward the child's pretend theme. Kavanaugh *et al.* (1983) noted a similar tendency for mothers to enhance the child's make-believe initiatives by offering descriptions of the child's pretend actions. For example, as the child held an empty cup and tipped it upward toward the mouth of a doll, the mother said, "You're giving the girl a drink," or as a child pushed a block across the floor saying, "Broom, broom," the mother said "You're making the car go fast!"

It appears that by accepting and elaborating on the child's pretend initiatives mothers create a context that is conducive to pretend play and, more particularly, to an increase in child-initiated pretense. Investigators using longitudinal designs report that child-initiated pretense is likely to increase over time in samples where mothers elaborate on young children's pretend scenarios (Haight & Miller, 1993), and that both the frequency and complexity of pretending increase during the second and third years in dyads where mothers take an active (non-intrusive) role in children's pretend play (Slade, 1987a,b).

The cultural context

As noted at the outset, the great majority of studies reviewed in this chapter have involved social pretend play among European–American mother–child dyads. I have argued that these studies show: (1) that mothers often assume a crucial role in the interaction by initiating and structuring pretend play episodes, particularly among children 2 years of age and younger; and (2) that as children become more competent pretenders, mothers play an equally important role in maintaining and enhancing the child's pretend initiatives. These two maternal "strategies" are well-suited to mothers who believe that pretend play is beneficial to their child's social and intellectual development.

Although caregivers in middle class European–American families often share this view (Haight *et al.*, 1997), it is far from a universal perspective. There are substantial differences between and within cultures about the value of pretend play in children's lives (Göncü *et al.*, 1999). For example, Feitelson (1977) reported that in the former Soviet Union, where the view prevailed that pretend play developed only in interaction with adults, preschool teachers were instructed to socialize actively with children by introducing elements of imaginative play. By contrast, Feitelson noted that Middle Eastern mothers not only avoided modeling pretend actions with children but actively interfered with any imaginative play they happened to observe. Similarly, in the Mayan culture, where pretending is not considered central to the child's development, adult participation in children's pretend play is not viewed as particularly appropriate (Gaskins, 1996). Within the United States, the Mennonites hold strong negative opinions about pretend play viewing it largely as wasteful of children's time. Not surprisingly, adults in this community offer little encouragement to children to engage in pretend play (Taylor & Carlson, 2000). In other settings, notably in two studies of Mexican children, pretend play was supported but siblings and older children were more likely than parents to assume the role of play partner (Zukow, 1989; Farver, 1993). In a particularly revealing direct comparison of mother–child dyads drawn from economically similar communities of northern California and central Mexico, Farver & Howes (1993) found that the American mothers guided and directed pretend play with toddlers far more often than the Mexican mothers who engaged in more functional play with their children during the course of daily work activity.

Haight *et al.*'s (1999) longitudinal study of Irish–American and Chinese children pretending at home highlights a cultural difference of particular interest to the central thesis of this chapter. Haight and her colleagues found that although both American and Chinese caregivers initiated pretend play episodes, American parents were likely to do so in order to involve children in fantasy whereas Chinese parents were likely to use pretending to teach children about proper conduct. The implications of these stylistic differences may be quite important. If pretend play serves a relatively specific didactic purpose, such as inculcating particular social values, then we might expect caregivers who hold this view to behave differently than those who believe that pretend play facilitates intellectual development more broadly conceived. Specifically, we might predict that in communities where adults believe that pretend play primarily serves a socializing role, caregivers would be less likely to elaborate on the child's pretend initiatives, especially initiatives that do not involve themes about societal values. Accordingly, we should expect more focused, caregiver-directed pretend play in dyads where the caregivers assign high priority to the moral and instructional value of pretense. Indirect support for this prediction comes from the finding that Chinese caregiver–child interactions reflect less mutuality than those of European–American caregiver–child interactions (Miller, Fung, & Mintz, 1996). Moreover, even within communities where adults subscribe to the view that pretending promotes intelligence and creativity, the strength of the caregiver's belief is related to the frequency of pretend play with children. Thus, in their study of the beliefs and actions of middle class European–American caregivers, Haight *et al.* (1997) found a significant correlation between mothers' ratings of the importance of pretend play to their child's development and time spent in pretend play with the child.

In short, it appears that cultures vary substantially in how they conceptualize pretend play and its role in children's development. These differences influence the overall nature of caregiver–child social pretend play, e.g., whether it is oriented primarily toward fantasy exploration or toward moral instruction (Haight *et al.*, 1999). It is possible that cultural differences may also influence the broader dimension of caregiver-directed versus child-focused social pretend play.

Summary

The purpose of this chapter is to review caregiver–child social pretend play with a particular emphasis on the interactions between European–American mothers and their young children. I have argued that during dyadic interactions this population of mothers may perform two important functions that have the potential to enhance children's pretend play. First, they can initiate pretend episodes by suggesting make-believe themes and by demonstrating specific pretend actions. Second, once child-initiated pretending takes hold, they may enhance and elaborate on the child's pretense which, initially, can be quite fleeting and sporadic. It is possible that some of these maternal behaviors may generalize to European–American fathers who have been found to enjoy participating in social pretend play (Haight *et al.*, 1997).

Still, it is important to note that given the observational/correlational nature of the data on social pretend play we can draw no firm conclusions about the causal impact of mothers' (caregivers') pretend play skills on the emerging pretense abilities of young children. For example, mothers' (caregivers') play style could be a response to, rather than a cause of, children's ability to pretend. However, the question of causality might be resolved by comparing the pretense abilities of two groups of children whose mothers (caregivers) differ markedly on how skilled they are at social pretend play. Note that Slade's (1987b) study used this type of design, but ideally one would like the two groups to be as similar as possible, e.g., not differ on quality of attachment. If maternal (caregiver) play style is instrumental in fostering the development of pretend skills in young children, we would expect the offspring of mothers (caregivers) who are highly skilled at social pretend play to demonstrate stronger pretense abilities. This might take the form of greater skill in understanding someone else's pretense overtures. In a series of studies Paul Harris and I have shown that before age 2 years 6 months children rarely understand another person's pretense (Harris & Kavanaugh, 1993). Would children who interact regularly with skilled pretend play partners understand the non-literal intentions of others at an earlier age?

Finally, recent studies on social pretend play reveal important differences between and within cultures in how communities organize and support children's exploration of fantasy. Haight and colleagues have offered an important way to conceptualize these differences by distinguishing between variable and universal dimensions of social pretend play (Haight *et al.*, 1999). As their analysis suggests, there is still much to be learned about the full impact of culture on the imaginary world of children. One question yet to be investigated is how the context in which children learn to pretend affects their later development of imagination. For example, do children who grow up in an environment that encourages open-ended, child-directed fantasy come to appreciate and explore associated non-literal domains (e.g., fiction, arts) in a dramatically different way than do children who learn to pretend in a more purposeful, didactic context? We might wonder also whether there are implications for other areas of children's lives – such as creativity and success in school – that relate to the early expression of pretend play in context. Future research on the interrelationship of cultural values and social pretend play should provide answers to these and other important questions about the development of fantasy and imagination.

7

Just through the looking glass: children's understanding of pretense

> "When *I* use a word," Humpty Dumpty said in a rather scornful tone, "it means just what I choose it to mean—neither more nor less."
>
> "The question is," said Alice, "whether you *can* make words mean so many different things."
>
> "The question is," said Humpty Dumpty, "which is to be master—that's all."
>
> (CARROLL, 1871/1946, p. 238)

Alice's world through the looking glass, though actually a world of dream, is in some respects like the world of pretend. Just as Humpty-Dumpty can be master of words, making them mean what he wants them to mean, in pretend play children designate what various objects and activities denote. Pretending is a special frame that organizes the activities within it. What goes on in the frame has its own reality, stipulated as the players choose, such that a behavior acted while pretending may have alternative significance while not pretending (Bateson, 1955/1972). This impressive ability to deal in framed worlds is fundamental to pretense (Fein, 1981; Bretherton, 1984). Framing also appears fundamental to other human activities, such as reflecting on dreams and understanding others' minds. Like pretense, another's mental world is framed by that person's perspective. When children learn to pretend, they learn to deal with worlds that are framed differently from the real world. In the looking glass world, nursery rhyme characters come to life, words mean anything you want, and memory can work both backwards and forwards.

The pretend world is not entirely unconstrained. In some ways it reflects the world children are actually in. For example, in both the looking glass and the real world, there is a temporal relationship between a pin prick and a scream – it simply happens backwards in the world through the looking glass. Likewise, in the world through the looking glass and in real-world chess alike, there are queens and kings, and queens move every which way directly and quickly while kings do not do much of anything. But some things are changed from the real world to the looking glass one. For example, looking glass chess pieces can talk, and one has jam every other day (which means never, since each day is itself and not "an other" day). As Kavanaugh (PIAC6) and his colleagues have shown, children can manage such stipulated worlds quite well, at least by 2 years 6 months of age. Even before preschool, children seem to appreciate that pretense is different from reality, and has its own frame.

Yet, at this early age children seem not to understand the framed world of the mind, at least as regards belief. Instead, children under about 5 years of age seem to think beliefs and reality must coincide, as though beliefs were not framed (Flavell & Miller, 1998). One pertinent question is whether children understand the framed world of the mind as regards pretense. To clarify, pretense is itself a framed world, and children appear to understand this when they pretend. But the framed world of pretense emanates from minds, making pretense framed in a second sense. Do young children understand that pretenders think their pretend thoughts, or do they simply know that they act out pretense activities? Some have credited children with appreciating the second sense in which pretense is framed, or that pretense is generated by the pretender's thoughts (e.g., Woolley, PIAC8).

If a child pretends a chess queen is a real queen, the child does more than see herself in some alternative universe in which the chess piece's identity has become a "real queen." She also sees herself as mentally representing the chess piece as a real queen, as having "real queen" in her mind, and applying it to the chess piece. Certainly this projection of a real queen representation is what children are doing when they engage in such a pretense. But a body of experimental evidence suggests that most preschool children are not aware that pretense worlds are mental. I begin by discussing what pretending entails, then discuss research on how children conceptualize pretense.

Defining pretense

Pretend play has several defining and one characteristic component (Lillard, 1993a; 1998a). First, there must be a pretender, a person or at least an animate being of active mind. Second, there is a reality that is pretended about. Since reality is omnipresent, this feature is not hard to come by, but what is pretended is generally different from reality. I cannot pretend to be my very self at this very moment in time and space (Austin, 1958/1979; Leslie, 1987; Lillard, 1993a). Third, pretense is guided by a mental representation, an ideation of the alternative state of affairs. A fourth defining characteristic is that the mental representation must be projected onto reality. If one is simply imagining a real queen, without projecting it onto the chess piece, then one is not pretending the chess piece is a real queen. Pretense is an act of projective imagining. The fifth defining feature of pretense is awareness (Anscombe, 1981; Leslie, 1987). A pretender must be aware of the actual situation and the nonactual, represented one, or else one is mistaken, not pretending. Perhaps one sees a small white object in the corner of the room, and one believes it is a chess piece, when in fact it is a stone. In this case one is projecting a chess piece representation onto a stone. This is a case of false belief, not pretense, because one is not aware of the reality. Sixth, one must project the representation intentionally; to do so without such intention is not pretense (Austin, 1958/1979; Searle, 1975). In contrast, when a psychotherapy patient (Charles) projects his mother (Elizabeth) on to the therapist (Eliza), he may be aware he is doing so yet not be doing so intentionally. We would not call that act pretense (Lillard, 1993a).

Finally, there is an important characteristic yet optional feature of pretense that children may often prioritize over its defining mental components. This is the feature of external manifestation, like action or costume. One might well make a doll wave at throngs of admirers if one was pretending it were a queen. This action is characteristic of queens. However, one need not move the doll at all, and still be pretending it is a queen. The doll might simply sit on an imaginary throne. As long as to the pretender's mind the doll is a queen, then, regardless of action, for that moment in time, that person is pretending the doll is a queen. In pretending there is potential for and sometimes even expectation of action, although it is not necessary (but see Nichols & Stich, 2000).

In sum, pretense involves a pretender, a reality, and a mental representation; awareness of the divergence between what is represented and reality; and intentionally projecting the representation onto reality. External manifestations like action or costume are frequent accompaniments to pretense, but they are not necessary to it. Over the past 10 years my colleagues and I have conducted a systematic program of research to examine when children understand some of these features of pretense.

Pretense as the domain of animates

One set of experiments has concerned when children understand that only animates pretend (an aspect of the first defining feature). Children know at least by age 3 years that animates (people and possibly animals) think and that inanimates (like chairs and rocks) do not (Gelman, Spelke & Meck, 1983; Dolgin & Behrend, 1984). To examine when children understand this constraint for pretense, we showed 3- and 4-year-old children pictures of various inanimate entities and people, the prototypical animate (Carey, 1985; Lillard *et al.*, 2000). We focused on people and not other animals for the animate category, since whether animals pretend is controversial. For each person or inanimate entity we asked whether it could pretend, think, breathe, move, or get wet. The findings were clear: although children were somewhat confused with vehicles, for the most part even 3-year-olds understood that inanimates do not pretend, and people do, when the stimuli were static pictures of those entities.

To check on how solid this understanding was, in further experiments we presented actual objects, rather than pictures, and had the objects act or be made to look like other objects, thereby creating "suggestive instances." For example a bottle was presented with a horse costume on, and a top was shown spinning like a ballerina. Children's understanding was robust even to this challenge. However, when this challenge was coupled with a second one, even 4-year-olds' performance plummeted. When the experimenter noted that the bottle looked like a horse, or that the top looked like a ballerina, even 4-year-olds often claimed that the object was itself pretending. So although ordinary objects that manifest other ones are not readily construed as pretending, once given verbal description of the alternative identities that the objects manifested, children readily conferred pretense on inanimate objects. Importantly, we also asked if the entities could think, and children performed well in all conditions. Thus children conferred pretending to entities to which they denied thinking, suggesting they at least believe that thought is not essential to pretending. To conclude, even 3-year-olds demonstrate a fairly good understanding of the fact that inanimates do not pretend, although their knowledge is easily disrupted when both the inanimate and an adult suggest that the object is like something else.

Pretense as involving minds

Beyond knowing that inanimates do not pretend is the knowledge that when an animate does pretend, his or her mind and brain are involved.

Pretending involves more than an animate body; it also involves an animate mind. To examine this, in one experiment I simply asked children whether one needs a brain to pretend (Lillard, 1996). This drew on Johnson & Wellman's (1982) finding that preschoolers know brains are necessary for cognitive acts like thinking, but believe brains are unnecessary for bodily actions like hopping or brushing teeth. I asked children whether one needs a brain to pretend, and only about 40% of children under 6 years of age said yes, although the vast majority of children realized that thinking, remembering, and imagining all require a brain. Later experiments in this line asked children to put various activities in one of two boxes. One box was for activities they could do just with their bodies, without using their minds at all, and the second one was for activities they could do just with their minds, without using their bodies at all. About 60% of 4-year-olds opted to put most or all pretense items (such as "pretend you are a kangaroo") in the body box. Importantly, thinking ("think about your birthday party") almost always went in the mind box, and physical events ("get wet in the rain") almost always went in the body box. In some experiments a third box for activities that needed a mind *and* a body was added, but this made no difference to children's choices for pretend, which still was usually destined for the body only box (Lillard, 1996; Sobel & Lillard, 2001b). So for younger children, pretending was more often categorized with physical activities in requiring only a body, than with cognitive states, that children of these ages know require a mind and brain. Despite its being fairly well restricted to animates, pretending was not seen as requiring a mind by many children until 6 (65%) or 8 (85%) years of age.

Understanding pretense intentions

Another aspect of pretending that children might understand early is the sixth defining feature – that pretenders intend to convey their pretense acts. Children appear to understand some aspects of intentions and desires, like their independence from reality, earlier than they understand those same aspects of belief (Bretherton & Beeghly, 1982; Wellman & Woolley, 1990; Lillard & Flavell, 1992; Bartsch & Wellman, 1995). To examine when children understand that pretenders must be trying to convey the pretense entity they enact, we presented a doll hopping, and we told the child that the doll was hopping like a rabbit, but that she was not trying to hop like a rabbit. Children confirmed this information in their answers to two control questions ("Is she hopping like a rabbit? Is she trying to hop like a rabbit?") and then answered the test question ("Is she pretending to be a rabbit?"). Under these circumstances, again only about 40% of 4-year-olds revealed understanding that lack of intention to be like something precludes pretending to be it. The other 60% claimed she was pretending, usually on all four of their four trials (Lillard, 1998b).

Other experiments examined whether children would do better on this task when given a forced choice, rather than a yes–no question. A picture board was used, portraying a cartoon of a troll doll, connected by dotted lines to two bubbles. Four-year-olds were taught that one bubble represented mental states, showing what the troll was trying to be like, and that a second bubble represented action, showing what the troll was actually being like. In test trials, children were told, for example, that the troll (who was actually flying about) was trying to be like a bat but was actually being like a bird, while bat and bird pictures were placed in the appropriate bubbles. Even under such facilitative conditions, most 4-year-olds usually claimed that the doll was pretending to be what she was actually being like, not what she was trying to be like (Lillard, 1998b). Although for pretense, intention is more important than what one is actually looking like, under these test conditions most 4-year-olds privilege appearance.

Some results challenge these findings. Joseph (1998) suggests that when actualization is not at odds with intention, children may appreciate that pretense involves trying. In his work, children were presented with a doll who was really sneezing, and another doll who was just pretending to sneeze, and were asked which one was trying to sneeze. Despite 4-year-olds' poor understanding of the fact that sneezing is unintentional (Smith, 1978; Lillard & Joffre, 1999), they tended to choose correctly the pretend-sneezer and not the real-sneezer as the one who was trying to sneeze. Suspecting this might be due to children linking positive events ("trying and succeeding" with "pretending"), we replicated Joseph's procedure with positive unintentional behaviors, like laughing at a clown and slurping up a milkshake. This made no difference: children still asserted that the pretender was the one who was trying (Lillard & Joffre, 1999). This finding might suggest some inkling of understanding that pretense involves trying, when one's behaviors are consistent with one's intentions (unlike my troll's). However, children's replies to follow-up questions cast some doubt on this conclusion. When asked why children had responded that the pretend-sneezer was trying to sneeze, children gave responses like, "Because he was not really sneezing," rather than answers confirming that pretending involves trying. In everyday life, young children often use "trying" to suggest lack of true accomplishment, as in, "I'm trying to zip my coat!" The pretend sneezer was not accomplishing actual sneezes; perhaps this is why children so consistently chose the pretend sneezer as trying. Future work should clarify children's understanding of trying, then reconsider children's understanding of how trying is related to pretense.

The understanding of "want" by 3- and 4-year-olds seems to conform more fully to adult usage (Bartsch & Wellman, 1995) than does "trying." A recent study by Ganea (P. A. Ganea, A. S. Lillard & E. Turkheimer, 2001, unpub. data) suggests some early understanding of desire's relationship to pretense. When shown a person who declared that she wanted to be like an elephant, but then proceeded to act more like a different animal, 4-year-olds still often claimed that she was pretending to be an elephant (which was what she said she wanted to be). In this case children treated personal and positive desire statements as being more relevant to pretense than action. In contrast, when the experimenter described someone's intention ("He's trying to be like a bird.") in Lillard (1998b), 4-year-olds privileged action in determining what the person was pretending to be.

Pretense as relying on thought

The third and fourth defining features of pretense concern mental representation – one mentally represents the pretense situation, and one projects that representation onto reality. Pretending a banana is a telephone involves (intentionally and knowingly) thinking of the banana as a telephone. To examine when children understand the mental representational features of pretense, we showed children dolls who could not mentally represent a certain situation, but who were behaving as one would if one were pretending it. For example, Lillard (1993b, Exp. 3) presented children with Moe, a troll from the faraway Land of the Trolls, who had never heard of a kangaroo, and did not even know that they hopped, but who was nonetheless hopping like one. Since Moe knew nothing about kangaroos, he could not mentally represent himself as one. Over four trials as well as several variations on this experiment, about 65% of 4- and 5-year-olds have consistently claimed that the character was in fact pretending to be a kangaroo (see Lillard, 2001a for more details).

To act like something without trying to be like it, or without thinking about it, may seem odd. Aronson & Golomb (1999) maintained that children probably can imagine no reason why Moe would resemble a kangaroo other than because he was pretending to be one. They further claimed that children must revise the premises in such situations, so that they believe Moe is trying to be like and is thinking about the entity. Against this, most children correctly answer control questions affirming that he is not trying to be or thinking about being the entity. Second, this very circumstance actually does happen in natural circumstances. Children are sometimes told, for example, that they (or their siblings) are acting like babies or pigs when they were not intentionally acting like or even thinking about them, hence are not pretending to be them. Nonetheless, we conducted an experiment in which we gave children an alternative, reasonable explanation of Moe's action, to see if this would help them to deny pretense (Rickert & Lillard, 2001). For example, we explained that the pavement was very hot, and that Moe was hopping so as not to burn his feet. We then mentioned that his hopping resembled that of a rabbit, and that he had never even heard of a rabbit and did not even know that rabbits hop. Although most 4-year-olds remembered in follow-up questions why Moe had been hopping, they still claimed that he was pretending to be a rabbit. These data suggest that a failure to conceive of other reasons why Moe is hopping is not responsible for their errors.

Another possibility is that it is difficult for children to reconcile the conflicting action with the mental state information in the Moe task. Perhaps a noncontradictory paradigm like that used by Joseph would enable better performance. To this end, in another experiment (Lillard & Joffre, 1999), 4-year-olds were shown two dolls and were told, "One of these girls is pretending to be a horse and one isn't." One doll was made to trot while the experimenter said, "She has a toy horse." The second one was made to trot and the experimenter said, "She's thinking about a horse." Answers to control questions indicated that children were clear about which doll was which. Children were then asked which doll was pretending to be a horse, with a reminder that only one was doing so. Granted either of the dolls could be pretending to be a horse in this situation, but if one is sensitive to the mental representational component of pretense, the best answer is the one who is thinking about a horse while trotting. Having a horse doesn't have much to do with whether one is pretending to be one in any given instance, but thinking about one certainly does. Four-year-olds choose the correct doll on just 45% of these trials, suggesting that even in Joseph-like circumstances in which no contradictory action was present, children were not sensitive to the cognitive (as opposed to the intentional, in Joseph's work) underpinnings of pretense.

An alternative possibility is that the direction of reasoning requested of children in the Moe experiments and the one just mentioned is hard for children. These experiments told children about mental contents, then asked children to reason about pretense. Joseph (1998), and some others who have obtained better levels of performance on tasks designed to assess this understanding (Custer, 1996; Hickling, Wellman & Gottfried, 1997; Gerow, Taylor & Moses, 1999; Davis, Woolley & Bruell, 2001), told children that someone was pretending, and asked children to reason about the person's mental content. To test whether this latter direction of reasoning is easier, in another condition in the experiment just described, children were shown two dolls and were told, "One of these girls knows what a frog is and one doesn't." One doll was made to hop, and the experimenter said, "She's pretending she's a frog." Then the second doll was made to hop, and the experimenter said, "She looks like a frog." Children were reminded that only one of the two dolls knew what a frog was, and were asked which one knew. The direction of reasoning is thus from pretense to knowledge state. In this case, the pretender must have been the one who knew, since only one knew, and pretending requires knowing. But 4-year-olds chose the pretending doll as knowing on just 40% of trials, suggesting that a direction of reasoning explanation does not resolve why some other tasks have proven easier for children.

Self before other

At this point, it appears that young children may appreciate that a pretender is more likely to be trying than is someone engaged in the same behavior for other, unintended reasons, and that by 4 years of age they have some appreciation of the intention component of pretense relative to reflex-type acts. Supporting this, Mitchell (2000) had children engage in various actions, like reaching across a table to pick something up, and commented that they had looked like something else, say a cat, when they did it. For other children, a confederate experimenter was engaged in the same behavior, with the same descriptive comment following. Interestingly, when asked if these were cases of pretending to be a cat, about 60% of 4-year-olds claimed it was for the confederate (consistent with Lillard, 1993b), but only about 40% did so for the self. Hence, unlike the case for false belief (Wellman, Cross & Watson, 2001), understanding pretense may be advanced for the self. One possible reason for this could be that one's own intentions are experienced. Although children do not recognize that Moe's not trying is relevant to his not pretending, perhaps this recognition is achieved somewhat earlier for the self.

Is it just the word?
One might question whether children's problem in some experiments is only with the word "pretend" (Woolley, 1995a; P. Mitchell, 1996). Perhaps children simply mismapped the word pretend to the characteristic component of the activity, while neglecting the defining one (Lillard, 1993a), but they are well aware, when watching people pretend, that minds and even mental representations are involved. To test this, David Sobel and I presented 4-year-olds with videos of people engaged in various actions, and sometimes described those actions with the word "pretend" and sometimes did not. If the word were throwing children's judgments off, by leading them to focus on the action component of the behavior, we reasoned that children would correctly infer mental involvement when the word pretend was not used. Children were asked, of each video, if it should go in a mind or a body box as described earlier. The word made no difference; 4-year-olds usually chose the body box regardless of condition (Sobel & Lillard, 2001b). Perhaps as soon as children realize that pretending involves the mind, they enlarge their definition of the word pretend to suit that new meaning. Interestingly, 4-year-olds consistently perform significantly better in both the Moe and the mind/body box paradigms when the pretended-about entity is a fantasy character (like the Lion King) (Lillard & Sobel, 1999; Sobel & Lillard, 2001a). This also suggests that the word pretend is not the main source of the problem. Furthermore, attempts to train children to pass the pretense understanding tasks (by discussing the fact that pretenders in videotapes are thinking about their pretense) have failed, also suggesting the issue is not merely definitional (Lillard & Joffre, 1999).

Contradictory studies
Other studies have been taken to suggest that children appreciate pretense not just as involving intention (as in Joseph, 1998) but also as involving thoughts (e.g., Woolley, PIAC8). My own reading of these studies, discussed at length elsewhere (Lillard, 2001a), is that several tasks could be passed by resort to lower level knowledge, like simply knowing that pretense is not real (Flavell, Flavell & Green, 1987; Perner, Baker & Hutton, 1994; Custer, 1996). As discussed in my introduction, children appear to appreciate early in their pretense careers that pretend is framed,

separate from the real world. Other studies do appear to reveal early appreciation of the fact that pretense is subjective, and possibly also that pretense content is importantly linked to thought content (Hickling *et al.*, 1997; Bruell & Woolley, 1998; Woolley, PIAC8). However, the methods used to elicit this understanding from young children – thought bubbles and imaginary content – do not characterize the everyday pretending of 3-year-olds. (Consistent imaginary object play emerges somewhat later; Overton & Jackson, 1973; Ungerer *et al.*, 1981; but see Baudonnière *et al.*, PIAC5.) Therefore this evidence does not challenge the assertion that very young children do not appreciate that pretense is fundamentally mental. Although they appear to appreciate that pretense situations are framed, most young children do not generally appreciate that those frames emanate from minds.

Links to understanding nonpretending minds

The evidence just reviewed suggests that preschoolers usually realize that certain types of entities (inanimates) cannot pretend, but this understanding is easily shaken by adult suggestion. Most young children do not realize that pretending entities are using their minds, or that pretending requires a mental representation of the pretense scenario. Regarding intention, 4-year-olds seem to understand that pretending involves trying when the pretense is consistent with an action. But when an alternative behavior, which the character's behavior is said to resemble, is presented, and intention information is stated by the experimenter not the actor, then intention information is ignored. Two other defining features – difference from reality and awareness – have yet to be systematically studied. Some research suggests that children are very likely to understand the first of these (Harris & Kavanaugh, 1993), and less likely to understand the latter (Peskin & Olson, 1997). Children's understanding of pretense's defining features hence appears to be very limited before elementary school.

Despite this, pretending does appear to be linked with children's social cognition. Children who pretend more, or earlier, pass social cognition assessments earlier (Astington & Jenkins, 1995; Youngblade & Dunn, 1995; Taylor & Carlson, 1997; Schwebel, Rosen & Singer, 1999; Lillard, 2001b). Such correlations might be explained in three ways: understanding minds facilitates pretending; pretending facilitates understanding minds; or some third variable like interest in people facilitates both

(Lillard, 1998a). Further work is needed to indicate which is the best model (see Smith, *PIAC9*).

Assume for the moment that the middle path is the right one – that pretending facilitates social understanding. Many have expected that this is the case, and that it is so because in pretense children gain practice reflecting on their own mental representations. Children then learn to consider mental representations outside pretense as well (Flavell, 1988; Forguson & Gopnik, 1988; Taylor & Carlson, 1997). The data presented here suggest that this is not an avenue by which pretending might facilitate social understanding. Instead, facilitation may occur because of the kinds of events that children enact in their looking glass worlds. The correlational studies just mentioned generally do not link all types of pretending to understanding minds; indeed several studies found no correlation with solitary pretense measures. The most reliably correlated measures are those that tap social pretense: role play; assigning identities to objects verbally; and having an imaginary companion (thereby adopting the companions' mental stances towards the world). I speculate that this practice at considering others' mental worlds, when pretending to be or pretending with other characters, leads to understanding minds. It would thus be the process of simulating other mental beings (Harris *et al.*, 1991) that would lead to understanding minds. Of course, it might instead be the case that only those who understand minds well can engage in the intense negotiation of social pretense (de Lorimier *et al.*, 1995; Howe, Petrakos & Rinaldi, 1998), or that a third factor underlies both skills.

Conclusion

In pretense children create an alternative world that in some respects reflects their real world. Pretense is like the world of Alice in the looking glass – much remains the same, but some crucial parameters are changed. Perhaps mirror self-recognition and pretense co-emerge (Baudonnière *et al.*, *PIAC5*) because in both cases, children must understand that there can be a world apart that is related to but is not the same as the world they are in (Mitchell, 1994a). Mirrors reflect a world that is much like the one that one is in, but handedness, visibility, and dimensionality change. Children understand pretending as a world through the looking glass, different but related to the world they are in. Given that children engage in so much pretense with such a limited understanding of its mental components, perhaps the possibility that animals may pretend is not far-fetched.

Elsewhere I have suggested that the pretend world for children is similar to Twin Earth for philosophers (Lillard, 2001a) – a place that is in many respects just like the real world, but in which some parameters are changed. Despite not understanding its mental origins, within this pretense world, children reason at more sophisticated levels than they reason outside of pretense[1] (Kuczaj, 1981; Dias & Harris, 1990; Markovits *et al.*, 1996), and may thereby come to better understandings of the real world, just as philosophers do (Pessin & Goldberg, 1996). These improved understandings may involve social cognition only to the extent that children explore social cognitive issues in their pretend play.

Endnote

1. This sort of effect is apparently not specific to pretense, but happens under a variety of circumstances when children are asked to consider alternative situations (Harris, 2000; Smith *PIAC9*).

Acknowledgments

The studies described here were made possible in part by NSF grant #DGE-9550152 and NIH grant #R15–HD30418.

8

Young children's understanding of pretense and other fictional mental states

> the gift of fantasy has meant more to me than my talent for absorbing positive knowledge.
>
> ALBERT EINSTEIN (cited in Colburn, 1985)

Awareness of non-realities

Our capacity for fantasy has often been touted as one of the unique attributes that defines us as human. As early as 18 months many human children have entered the realm of fantasy through their engagement in pretend play. Although we are not certain of this, even younger infants may enter this realm via their dreams. And certainly way into old age we enjoy reading novels, going to plays, and engaging in elaborate daydreams. The question addressed in this chapter is the nature of our conscious understanding of this fantastical realm, in particular the understanding that young preschool-age children display. The understanding of this age group is the focus, because many believe that at this age children begin to consider the fantastical and to make a conscious division of their world into real and not-real, a distinction that is basic to and deeply entrenched in our adult thinking.

There are many levels of reality and many varieties of non-realities. As Austin (1962, p. 67) put it, "That may not be real duck because it is a decoy, or a toy duck, or a species of goose closely resembling a duck, or because I am having a hallucination." In previous work (Woolley & Wellman, 1990) I have investigated the extent to which children talk about the many levels of reality in their everyday conversation. To obtain access to such conversations, we searched the CHILDES database (MacWhinney & Snow, 1985) for use of the words *real* and *really*. The CHILDES database contains transcriptions of recordings of naturally occurring conversations between young children and their parents, collected and compiled by a number of investigators in the field. These early spontaneous conversations can prove a rich source of information about when and how children begin to divide the world into real and not-real, and can also reveal the variety of non-realities children are aware of. Our investigation of seven children from this database revealed that these children began contrasting reality with various alternatives shortly before their third birthday. Between the ages of 3 and 5 years they frequently contrasted reality with toys (e.g., "it's not a real bowling ball; it has little red things on it"), pictures (e.g., "that ain't a real skunk, that's only in the book"), and pretense (e.g., "you're not really dead; we're just playing"). In a controlled experiment based upon these naturalistic data, and involving a larger sample of children, we found that children of this same age made clear distinctions between reality and these other alternatives.

Given that young children clearly are dividing the world into real things and not-real things, the next step is to understand exactly how children are conceiving of these non-realities. I will begin by discussing my findings regarding children's understanding of imagination and dreams. Then I will review some of the conceptual similarities among various fictional mental states, and end with a report of some recent empirical work on young children's understanding of pretense.

Understanding imagination

Earlier work on children's "theories of mind" (Wellman & Estes, 1986; Estes, Wellman, & Woolley, 1989) has revealed that children as young as 3 years of age are aware of many of the properties that distinguish mental entities from real physical objects. For example, young children understand that dreamed-of and imagined entities do not afford behavioral-sensory evidence in the way that real things do. That is, they understand that whereas one can see and touch a real cookie, one cannot do the same with a mental one. They also understand that certain mental entities afford particular properties that real things lack. For example, one can imagine a pair of scissors and then make them open and shut "just by thinking about it." This same process will not make real scissors open and shut, and children as young as 3 years are aware of this.

Children also demonstrate a sophisticated understanding of other important aspects of imagination. In particular, I have probed children's understanding of three aspects: (1) the correspondence between imagination and reality; (2) the representational nature of imagined entities; and (3) the origins of imagination. Regarding the first, my initial studies on this question (Woolley & Wellman, 1993) revealed that young children understand that imagination reflects reality less accurately than does knowledge. However, many young children in these studies appeared to believe that simply imagining something could potentially produce real-world consequences. They claimed that, after looking into an empty box and imagining an object inside, the object would really be there. Thus, although they seemed to understand that imagination is less veridical than knowledge, they still weren't drawing a clear line between the two.

However, in follow-up work (Woolley & Phelps, 1994) in which children were offered the opportunity to make a behavioral response to indicate their beliefs, very few children appeared to be confused about imagination-reality relations. Children were presented with a series of boxes and asked to imagine a particular object (e.g., a pencil) inside one. After imagining the object in the box, children were asked by an unfamiliar adult who had just entered the room, whether there were any of those objects in the room. Children were instructed to give to this adult any box that they thought would be helpful. The results were that children rarely gave any boxes to this adult, indicating a clear understanding that their imagination indeed had not created a physical replica. It is possible that, in the earlier studies, children's verbal responses reflected their wishes or desires rather than their actual beliefs about the content of the boxes. Alternatively, as Subbotsky (1993) has suggested, different levels of understanding may be reflected in verbal and behavioral responses.

Regarding the representational nature of imagined entities, much research has shown that children younger than 4 years of age have difficulty understanding that an epistemic mental representation (e.g., a belief) can differ from reality (e.g., Gopnik & Astington, 1988; Perner, Leekam, & Wimmer, 1987), and researchers have proposed a watershed at this age in understanding mental representations. One important difference between beliefs and imagination is that beliefs purport to represent reality whereas imagination does not. Thus, I reasoned that children might have an earlier understanding of situations in which imagined contents contrast with reality than of situations in which a belief contrasts with reality. To address this I (Woolley, 1995b) conducted a series of studies in which children experienced parallel false belief and "false" imagination scenarios. For example, an adult either expressed a belief that an object was in a box or explained to the child participant that she was imagining that something was in a box. Then the adult was shown what was really inside the box. In both cases the object that was in the box was different from the expressed content of the adult's mental state. Children performed significantly better when asked to state what the experimenter was imagining than when asked to state what she believed. These findings indicate that children have a precocious understanding of the representational nature of imagination, and thus possibly other fictional mental states as well.

Third, I have addressed whether children are able to distinguish mental representations resulting from imagination from those resulting from other sources. In these studies (Woolley & Bruell, 1996) children either looked into a box, imagined something in a box, were told what was inside, or made a guess about the contents of a box. Then, after a short delay, they were asked to recall the origin of their mental representation. Results indicated that children as young as 3 years are able to differentiate mental representations based on fiction (e.g., imagination) from those based on fact (e.g., seeing, being told). Thus very young children are aware that imagined things and experienced things are different in important ways.

Overall, these studies on children's understanding of imagination indicate that young children share with adults some very basic notions about imagination. They understand that it is independent from reality and that it does not create it, that it involves mental representation that can sometimes differ from reality, and that it arises differently from other mental states. These findings together suggest that a similar pattern might be evident regarding children's understanding of dreams and pretense.

Understanding dreams

Paralleling these findings, children also demonstrate a mature understanding of certain central aspects of dreams by age 3 years. In one study (Woolley & Wellman, 1992) we presented 3- and 4-year-old children with stories about children who were either dreaming about an object, playing with an object, or looking at a photograph of an object. Most adults in Western society consider dreams to be non-physical, perceptually private, and individuated. Based on this we asked children three key questions: (1) whether the focal story character could see the object; (2) if someone else came into the room, whether he or she could see the object; and (3) whether the focal character could act upon the object. Children were also asked whether another person who was sleeping in the same room as the focal character would dream about the same object the character was dreaming about. We also probed children's understanding of the potentially fictional nature of dreams, by presenting them with both real entities (e.g., an ant crawling on the ground) and fictional entities (e.g., an ant riding a bicycle), and asking the following questions: (1) whether the child had seen one; (2) whether it existed; (3) whether the child could think about one; and (4) whether the child could dream about one.

Our results were that children judged dream entities, photographs, and physical objects to be appropriately different in terms of physical versus nonphysical properties and in terms of perceptually public versus private status. Children understood that dreams are nonphysical, and that they are perceptually private (i.e., others cannot see them). They also understood the fictional nature of dreams. Thus, many important aspects of the adult understanding are in place by age 3 years.

Importantly though, these findings should not be taken to indicate that children's understanding of imagination and dreams is necessarily entirely adult-like. Two pieces of evidence point to some confusion regarding the nature of dreams. First, recall that in this study we also were interested in whether children believe dreams are shared or individuated. Results of the questions on this aspect indicated that whereas the majority of 4-year-olds conceived of dreams as private fantasies, the majority of 3-year-olds either conceived of them as shared fantasies or were transitional between the two levels of understanding.

Second, work by Woolley & Boerger (2002) indicates that young children have very different beliefs from adults about the level of conscious control people have over their dream content. Children aged 3 to 11 years were presented with various scenarios in which characters were said to be attempting to control their dreams. Characters were either trying to dream about a particular thing, trying to prevent a certain type of dream, trying to continue a good dream, or trying to stop a bad dream. Until the age of 9 children exhibited strong beliefs that one can exert control over one's dreams.

Understanding pretense

Certain very basic aspects of both imagination and dreams are understood as young as age 3 years. Other arguably more complex aspects of dreams are understood later. What might we expect regarding children's understanding of pretense given what we know about their understanding of imagination and dreams? From an empirical standpoint, we might expect children to know that pretense is fictional – that it differs from reality in important ways. They might know that one can pretend to do things that one cannot really do, that pretending something does not make it really happen, that one can pretend something that is different from what really exists. But do they understand that pretending involves the mind? Or can they understand all this without being aware of the role of the mind?

From a conceptual standpoint there are a number of compelling points of overlap among imagining, dreaming, and pretending. First, all are fictional mental states, in that they do not purport to represent reality accurately. Although these mental states may indeed accurately represent reality, as in the case of someone faithfully imagining the beach where he or she recently vacationed, it is not a defining characteristic of fictional mental states as it is of epistemic ones (e.g., beliefs). Imagining and pretending both involve maintaining a conscious awareness of the real world while engaging in a mental event. They also both involve deliberate, planful, constructed, and controlled processing. Dreams and pretending, however, do not overlap in these ways. Dreams are unconscious, and they are not planned. Imagining and dreaming, however, share a slightly different set of features, many of which distinguish them from pretense. Both, unlike pretense, have the option of being nonpropositional; whereas one can imagine a lollipop, or dream of a boat, one always must pretend that one thing is something else. Both, unlike pretense, lack an action component. Pretense also has a much greater potential for sharing with others than does imagination or dreaming. So, overall, there are important similarities and important differences that make predictions from a sophisticated mentalistic understanding of dreams and imagination to a similar understanding regarding pretense questionable. On the one hand, the many shared features between imagination and pretense in particular suggest continuities in understanding. On the other hand, the differences outlined above certainly leave the door open for somewhat different levels of understanding.

Perhaps one of the most important differences concerns the role of action. As outlined above and as Lillard describes (1993a,b), pretense differs from dreams and imagination in having a strong action component. This difference may be salient enough to throw a wrench in any similarities we might hope to find between these mental states. In fact, this is what Lillard finds in her research (1993b; 1996; 1998b; PIAC7). Both sets of studies presented in this chapter were developed in response to the results of Lillard's (1993b) Moe the Troll studies. In Lillard's studies, children were presented with a troll doll that was, for example, jumping up and down "like a rabbit." Children were told, however, that the troll did not know about rabbits. Children had to determine whether or not Moe was pretending to be a rabbit. Children younger than age 6 years were found not to understand that one must know about an animal if one can correctly be said to be pretending to be that animal. Thus Lillard concludes that young children lack a mentalistic understanding of pretense.

One concern we had with this paradigm was that it may have been difficult for children to maintain what they were told about the troll's mental state (e.g., that he didn't know about rabbits) in the face of a salient action (in this case, jumping) that apparently contradicted it. In other words, the action component of the characters' pretense may have been much more accessible to children than was the mental state component. Although this is sensible from an ecological standpoint, it seemed unwise to conclude from this sort of situation that children entirely lack an understanding of the mental state component. We attempted, in our first set of studies, to equate the salience of action and mental state by using drawings of characters instead of dolls and by depicting characters' thoughts in thought bubbles. The second set extends this methodology in using live models along with thought bubbles. Both studies point toward the same conclusion: children as young as 3 years are aware of the mental aspects of pretending.

Understanding the roles of knowledge and thinking in pretense

Our focus in the first set of studies (Davis, Woolley & Bruell, 2002) was children's understanding of the involvement of knowledge and thinking in pretense. We included two types of knowledge tasks, one, like Lillard's, involving one animal, and another involving two animals. The purpose of using these two different types of tasks was to ascertain whether or not the use of just one animal might cause children to infer that the character must be pretending to be that animal. In other words, in the first type of task (one-animal tasks), and in Lillard's studies, children were told that the characters were acting like a certain animal that the character did not know about. They were then asked if the character was pretending to be that animal. As children were not given an alternate explanation as to why the character was acting in such a way, it seems conceivable that they might say that the character was pretending for want of another explanation. The use of two animals in the second type of task (two-animal tasks) was designed to address this concern by providing children with an opposing motive for the story character's action.

One-animal tasks

Children were shown a picture of a character that was said to be acting like an animal the character did not know about. For example, Gleep (a character said to be from another planet) was described as hopping like a bunny rabbit hops, but as not knowing what a bunny rabbit was. A bunny rabbit was depicted next to Gleep. The critical test question asked whether the character was just performing an action or pretending (e.g., "So, what's Gleep doing, is he pretending to be a bunny rabbit, or is he just hopping?"). This task was meant to be similar to Lillard's (1993b) tasks, and was used as a comparison to both her tasks and the two-animal tasks in the present study.

Two-animal tasks

Here children were presented with a character that was said to be performing an action that corresponded to two different animals, only one of which was familiar to the character. The character Gleep, for example, was said to be from another planet where creatures called "mins" reside. Children were told that Gleep was wiggling his nose like mins and bunny rabbits wiggle their noses. Gleep knew what a min was, but did not know what a bunny rabbit was. Both the character and the two animals were depicted. The test question asked which animal the character was pretending to be (e.g., "What would you say Gleep is pretending to be, a min or a bunny rabbit?"). Importantly this task format allowed children to point to their answers, an option not possible in the one-animal tasks.

Overall, children performed surprisingly well. The results did reveal a significant increase in performance with age, with 3-year-olds performing at 67%, 4-year-olds at 83%, and 5-year-olds at 88% correct. (These numbers are averaged across the one- and two-animal tasks. For the one-animal tasks, 4- and 5-year-olds performed significantly above chance, for the two-animal tasks all age groups' performance was above chance.) Importantly we also found that children performed significantly better on the two-animal tasks (91% correct) than on the one-animal tasks (69% correct), suggesting that the lack of alternative in Lillard's tasks may indeed have accounted at least in part for children's poor performance.

Our second study introduced the use of thought bubbles, shown by Wellman, Hollander & Schult (1996) to be effective in assessing children's understanding of various mental states. We also embedded the pictures in stories, to make the questions seem less test-like and more like the sort of exchange that might happen during parent–child storybook reading. The focus of this study was children's understanding of the role of thinking in pretense. Children received two types of tasks: (1) pretense tasks, in which they were given information about characters' thoughts and asked to infer the content of their pretense; and (2) think tasks, in which children were given pretense information and required to infer thoughts.

Pretense tasks

Children were told stories in which a character was performing an action common to two different creatures, but were told that the character was only thinking about one of them. For example, in one story, Sara was depicted next to a butterfly. The children were told that the butterfly flew away, and Sara was then shown next to a bird. She was described as waving her arms "like birds and butterflies flap their wings." Next children were shown a picture with Sara next to a bird, and with a thought bubble indicating that she was thinking about a butterfly. Importantly, the second creature that was outside the thought bubble (in this case the bird) was always pictured in some sort of enclosure (e.g., a cage) in an attempt to make the pictures of the two creatures as similar as possible. For the test question, children were asked which creature the character was pretending to be (e.g., "Is Sara pretending to be a bird or is she pretending to be a butterfly?").

Think tasks

Children were again told that the main character was performing an action common to two particular animals, both of which the character was familiar with. In this story type, however, the children were told which animal the character was pretending to be instead of what the character was thinking about. For example, in one story, Stacy was pictured next to a lion. Children were told that she walked away from the lion and was then shown standing next to a horse. They were then shown a third picture in which both animals were pictured. Stacy was shown standing closest to the animal that she was not pretending to be (the horse). She was described as shaking her head "like lions and horses shake their manes." Children were told that "Stacy is next to the horse and is pretending to be a lion." Children were then asked which animal Stacy was thinking about.

Our reasoning was that if children were not aware that pretense has a mental component, there should be no reason for them to chose the animal Stacey was pretending to be over the one she was next to, thus predicting random performance.

As with the first study, performance improved with age. Three-year-olds were 72%, 4-year-olds were 88%, and 5-year-olds were 93% correct. Children performed similarly on the two types of tasks. All age groups performed significantly above chance on the pretense task. For the think task, 4- and 5-year-olds' performance was above chance and there was a trend for the 3-year-olds to be above chance. Thus overall, this set of studies revealed a high level of understanding of the mental prerequisites for pretense in young children.

Understanding representational diversity in pretense

Continuing this line of investigation, in a second set of studies (Bruell & Woolley, 1998) we probed preschoolers' understanding of representational diversity in the domain of pretense. Our thesis was that different objects inspire different ideas about what to pretend and that children know that. For example, upon seeing a large box in the classroom, for example, one child might jump at the chance to play "spaceship" while another might hope to play "house." Importantly, even though two characters may engage in the same action around an object, it is the way that they are thinking about the object that determines what we can say they are pretending.

We conducted four studies to address children's understanding of this fact (Bruell & Woolley, 1998). As with the earlier set of studies, we utilized thought bubbles in an attempt to balance the salience of mental state and action information. However, a potential concern of using thought bubbles and still pictures of people is that we may have tipped the scale in the opposite direction, making mental state information overly salient at the expense of action. In an attempt to remedy this problem, we created a novel paradigm, which retained both the movement of the characters central to Lillard's design, and the thought bubble approach to presenting mental state information critical to the Davis et al. (2002) design.

We did this by presenting 3- and 4-year-olds with video skits in which two characters pretended different things with the same object. For example, two children played with a rope, swinging it back and forth. Thought bubbles indicating what the characters were thinking about (e.g., a baseball bat or a broom) were superimposed over the actors' heads (see Bruell & Woolley, 1998, for a full description of the procedures). Children were randomly assigned to conditions in which both characters were acting in exactly the same way and in which the characters performed different actions. The test question required children to identify the content of each character's pretense. Our reasoning was that if children understand that thought bubbles show what someone is thinking, and also that thinking is involved in pretense, then they should be able to use the information provided in the thought bubbles to ascertain the pretense intended by the actors. Also, we predicted that, if action is most important or of sole importance, children will be more likely to respond that the characters are pretending different things when their actions are different than when they are the same.

Results indicated that both 3- and 4-year-olds appreciate the potential for representational diversity in pretense, and understand pretense to be a mental activity. Although 4-year-olds (89% correct) performed significantly better than 3-year-olds (70%) in answering questions about the content of the characters' pretense, both groups' performance was significantly better than chance. There was no difference in performance between the two action conditions, suggesting that children were not relying solely or even primarily on action to make decisions about what characters were pretending.

We conducted three additional studies to attempt to rule out alternative explanations of 3-year-olds' good performance. In our second study, we addressed the possibility that children might have simply found the thought bubbles attractive and chosen their content for their answers for this reason. This concern arose because performance on a question about the real identity of the object was lower than expected. To address this we made two changes, one in how we introduced the task, and one in how we asked the "reality" question (see Bruell & Woolley, 1998, for more details). With these minor changes, performance on the reality question improved from 56% to 74% while performance on pretense questions remained at the above chance level of study 1 (70%). Thus the results of study 2 replicated study 1, and argued against this alternative explanation for participants' good performance in that study.

In each of these studies, children both saw thought bubble contents and heard dialogue by the characters suggesting the content of their pretense. For example, a character playing with a rope but with a thought bubble of a baseball bat would say "wisht, wisht, I'm swinging real hard." We felt it was important to isolate each of these potential cues to the contents of each characters' pretense. Only this way could we be sure that children were relying on the thought bubble information. So in studies 3 and 4 we compared the unique contributions made by the dialogues and the thought bubbles by systematically removing one at a time. Our reasoning in removing the thought bubbles was that if children were drawing inferences about the actors' pretense from information in the dialogues, and not using information about mental states conveyed by the thought bubbles, then they should still perform well when no thought bubbles were present. Children performed nearly identically on the reality questions as in study 2, but performed more poorly on the pretend questions – only 33% correct, compared to 70% when the thought bubbles were visible (studies 1 and 2). This was not significantly above chance, and indicates that children in this study were not able to understand representational diversity when provided with information in the dialogues only. More to the point, information contained in the dialogue, by itself, was insufficient to allow children to ascertain the pretend identities attributed to the objects by the actors.

In study 4 we reintroduced the thought bubbles, but eliminated the dialogue. With only thought bubble information serving as a cue to the content of each character's pretense, children performed significantly better than chance on both the reality question (92%) and the pretend question (75%). Thus we concluded that it is predominantly the information about mental state as conveyed by the thought bubbles and not information provided by the dialogues that guided children's understanding of representational diversity in our studies.

In summary, in this set of studies we provided children with a number of potential cues to the nature of another's pretense: action information, dialogue information, and mental state information. Results of the first two studies indicated that action was not playing an important role in children's understanding of the characters' pretense stipulations; participants' understanding that the two actors were pretending diverse things was not affected by the actions in which the characters engaged. Rather, these results indicated that children considered information about mental states to be most informative. The results of studies 3 and 4 indicate that children relied more on mental state information than on dialogue information when trying to ascertain the content of another's pretense. Here, children were far more accurate in their descriptions of pretense when they were provided with meaningful thought bubbles alone than when they were provided with meaningful dialogues alone.

Thus, of the three types of information children had available to them, mental state information played the most important role in children's attributions of pretense stipulations. From this we concluded that children as young as 3 years of age have an understanding of the mental nature of pretense.

There are a number of reasons why our studies on children's pretense understanding may have uncovered a more sophisticated understanding than has been demonstrated previously (Lillard, 1993a,b; *PIAC7*; but see Mitchell, *PIAC1*, for an alternative perspective). I'll mention two here. First, our tasks relied much less on children's linguistic ability than did previous research. Children were given the option of pointing instead of solely relying on verbal responses, mental information was conveyed by thought bubbles rather than verbally, and stories were shorter. Second, we attempted to provide a more balanced presentation of action and mental state information. Equating the salience of action and mental state information may have helped children convey their understanding of the greater importance of the latter.

Having noted these factors, it is also important to consider the more global issue of assessing what children may typically do versus what they can do under certain facilitating conditions. In their everyday lives children probably often do use others' actions to make judgments about what they are pretending. Certainly mental states are not as easily accessible to outside observers as we made them in our studies. However, children's tendency to make judgments about pretense based on action should not be taken to imply that they are wholly unable to think about the mental aspects of pretense. Because young children's understanding about the mind may be less stable than that of older children and adults, sensitive methods such as those used in our studies are necessary if we are to capture young children's nascent understanding about the mind. Although it is important to know what children may typically do when observing pretend activities, it is no less important to know what they might actually understand about the role of the mind in pretense when presented with novel pretend situations.

Conclusion

The process of getting young children to reflect on the contents of their own or another's mental state is a difficult and challenging one. Although children begin pretending often as early as 18 months, they do not begin to talk about the mental realm until they are almost 3 years old (Shatz, Wellman, & Silbur, 1983; Bartsch & Wellman, 1995). It is very possible that children do first understand pretense simply as acting in a way that is different from the norm. Yet as children's understanding of the mind blossoms late in their third year of life, they come to talk and understand a lot about imagination, dreaming, thinking, and knowing. Might they then come to understand pretense differently, as reflecting mental contents? As researchers working with young children know, it can be very difficult to design tasks that tap this understanding. One way to resolve some of the confusion that may result from the apparently contradictory findings in the literature may be to appeal to Siegler's (1996) focus on variability in children's conceptions. Perhaps the different levels of performance displayed in the studies reported in this chapter and those reported by Lillard (PIAC7) reflect different levels of understanding. That is, no one study is right or wrong; they are simply tapping into different aspects of the phenomenon. When children first start to realize the mind's involvement in pretense, their understanding may be most apparent in their spontaneous talk. Somewhat more mature concepts might be revealed through methods such as ours, whereas only fully robust levels of understanding are revealed by tasks like Lillard's troll task.

Let me conclude with a personal observation (congruent with work by Mitchell, 2000). When my son was 4 years old, I took him to our lab to participate in a study of young children's understanding of the roles of knowledge and thinking in pretense. Much to my dismay, my bright sophisticated son completely flunked both of Lillard's (1993b) troll tasks and a task using thought bubbles. Sadly he displayed no knowledge of the roles of mental states in pretense, and seemed to base his responses on what he, rather than the story character, knew and liked. Not entirely willing to give up so easily, on the way home I casually posed him the following question, "So, do you know what a shnoogle is? "No," he replied. "Oh, well, could you pretend to be a shnoogle?" I asked. "No." "Why not?" "'Cause I don't even know what a shnoogle is!"

9

Pretend play, metarepresentation and theory of mind

> In symbolic play, young children advance upon their cognitions about people, objects and actions and in this way construct increasingly sophisticated representations of the world.
>
> (BORNSTEIN et al., 1996, p. 2923)

> [O]ur intuition is that extensive fantasy experiences help children develop an understanding of mind.
>
> (TAYLOR & CARLSON, 1997, p. 452)

Pretend play as a human characteristic

Fully developed pretend play, including role play and sociodramatic play, seems universal in all human societies and uniquely human, with only rudimentary pretence shown by some apes (Schwartzmann, 1978; Slaughter & Dombrowski, 1989; Smith, 1996). Cross-cultural variations exist in amount and type of pretence (Smilansky, 1968; Roopnarine, Johnson & Hooper, 1994), which appears dependent upon children's physical and cultural context. For example, among hunter–gatherers, children in mixed-age peer groups used sticks and pebbles to represent village huts and herding cows (Konner, 1976; Eibl-Eibesfeldt, 1989), and in the Marquesas Islands of Polynesia, children made mud bananas (Martini, 1994). Such symbolic object-use comprised half of the "fantasy play" episodes of the Marquesan children; the other half involved more complex scripts and roles, such as "'ship' play, 'fishing,' 'hunting,' and 'preparing feasts'" (Martini, 1994, p. 84). These episodes "follow the same fantasy scripts from one performance to the next," and often involve some imitation of adult actions.

The ubiquitousness of pretend play in human children suggests biological processes at work (Slaughter & Dombrowski, 1989; Harris, 1994), and raises the question of its biological function. The most persistent hypothesis is that "children's social and pretend play [is] an evolutionary contribution to human psychological growth and development" (Slaughter & Dombrowski, 1989, p. 290). In this chapter, I examine the role of pretend play in human development, particularly in relation to metarepresentational abilities and theory of mind.

Models of pretend play's developmental role

The anthropological evidence suggests a role for pretend play in development. Figure 9.1 (a, b, c) suggests three models for such a role.

1. Pretend play is a by-product of other aspect(s) of development, with no important developmental consequence(s) of its own (Figure 9.1a).
2. Pretend play is a facilitator of developmental consequence(s); it can help bring about important developmental consequence(s) but it is not essential for this if other expected developmental pathways are present (Figure 9.1b).
3. Pretend play is necessary for important developmental consequence(s); in the absence of pretend play, these developmental consequences will not happen or will at least be significantly held back (Figure 9.1c).

A few caveats about the nature of these three models should be noted:

- these are models of individual development, not of possible functional roles of pretend play in a societal context;
- through the course of human bio-cultural evolution there might have been shifts from one model to another (for example, from models 3 to 2 if cultural change provided more pathways to development, or from models 1 to 2 if an existing ability can be put to new uses ("exaptation");
- in models 2 and 3 there remain issues about threshold effects – whether certain amounts, frequencies or types of pretend play are necessary or sufficient to have developmental consequences.

What evidence do we have regarding these models? I describe two phases in evidence gathering: the 1970–1980s and the 1990s.

Research in the 1970–1980s

During the 1970–1980s several developmental functions of pretend and sociodramatic play were suggested, particularly cognitive and social-emotional.

(a) pretend play as by-product

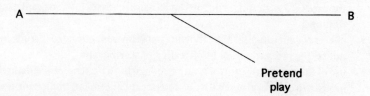

(b) pretend play as one facilitator

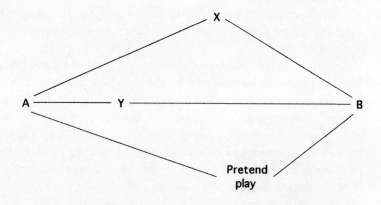

(c) pretend play as essential

Figure 9.1. Three models of the role of pretend play in development; A, represents an early point in development of some important acquisition (such as theory of mind), and B, a later point.

For cognitive functions, evidence suggested that pretence assists conservation learning (Golomb and co-workers, 1977, 1981, 1982) and associative fluency (Dansky, 1980), and (in the form of sociodramatic play) is vital for language development, cognitive development, creativity, and role-taking (Smilansky, 1968; Smilansky & Shefatya, 1990). These ideas were tested by experimental studies, including "play tutoring" studies. Consistent with the period's "play ethos" (Smith, 1988) – a strong expectation of positive developmental effects of (pretend) play – results appeared to support all these functions of play. Unfortunately, this first generation of studies exhibited several methodological drawbacks:

- selective interpretation of results: doing multiple tests, but only highlighting significant findings; or highlighting trends ($P<0.1$) that go the expected way; and giving methodological excuses for nonsignificant findings;
- effects of experimental bias: allowing experimenter effects to intrude by not taking precautions for blind testing or scoring; and
- use of inappropriate control groups: comparing a pretend play enhanced group to a control group with no pretence enhancement, but having less verbal stimulation and adult involvement.

When these methodological flaws were corrected, researchers failed to find clear benefits of pretend play (Smith & Syddall, 1978; Smith, Dalgleish & Herzmark, 1981; Christie & Johnsen, 1985; Smith, 1988; Hutt et al., 1989).

Social–emotional functions derive from attachment theory. Infants form an attachment to caregivers by 7–9 months of age, and are more secure exploring the world when in proximity to an attachment figure. A "secure" attachment functions well, such that the proximity of a caregiver provides reassurance and comfort to the infant; an "insecure" attachment functions less effectively. Securely attached children show more elaborate, socially flexible play, with more benign resolution of pretend conflicts, and insecure–avoidant children have more aggressive and fewer nurturant themes, and may become obsessive in play. If pretence is a way to explore and master emotional difficulties (e.g., fear of the dark, family conflicts), securely attached children could benefit from pretence more than could insecurely attached children. Ironically, if true, the proposed function of pretend play is *"least open to those most in need of it"* (Bretherton, 1989, p. 399; italics in original). Indeed, children who have experienced emotional trauma show more nonresolution of negative affective experience through pretend activity; noncoordination and disorganisation of play objects and activities; perseveration of activity and repetition of single schemes; and global inhibition of pretend play (Gordon, 1993). These findings suggest that pretend play may be diagnostic of a child's emotional condition, but does not promote emotional mastery for children who need it.

The 1990s – pretend play and theory of mind

In the 1990s, a new generation of studies on the cognitive benefits of pretence have linked it to theory of mind (ToM). ToM involves understanding (representing) the knowledge and beliefs of others. Since knowledge and beliefs are "representations" of reality, ToM requires a representation of a representation – a second-order representation or "metarepresentation." Interpreting whether or not children have metarepresentational knowledge is difficult. For example, if you ask a child if another person X knows or believes something is true ["does Robert think the sun is hot?"] and they answer correctly ["yes"], the answer might reflect ToM, or it might just reflect the child's own knowledge or belief. So possession of ToM is usually measured by the understanding that another person may hold a false belief, for example by unexpected transfer tasks and unexpected object tasks (P. Mitchell, 1997).

There are good reasons to suppose that pretend play might be important for ToM acquisition:

- both appear absent in nonhuman species, with possible exceptions of simple forms in the great apes;
- both appear present in all normal humans and societies;
- both may be absent or severely held-back in cases of autism and some other clinical syndromes;
- both have a characteristic developmental trend, with main features developing in the period from 2–3 years to 5–6 years;
- the defining features of mature pretence (Lillard, 1994; PIAC7) – particularly a mental representation that is projected onto reality, with awareness and intention on the pretender's part – imply metarepresentational abilities similar to those in ToM.

In fact, it has been argued that even simple pretend play involves metarepresentation (Leslie, 1987, 1988).

Pretend play and metarepresentation

In Leslie's (1987, 1988) view, simple pretend play (such as object substitution) is an early indicator of metarepresentational abilities (starting at around 18 months of age). This suggested a leading role for pretend play, which emerges at around 18 months, compared to 3–4 years of age for abilities in ToM.

Leslie's argument proceeds as follows. Suppose a child pretends a banana is a telephone. The representation of the real object – this is a banana – is a primary representation. When engaging in object substitution, there is a second representation – this banana is a telephone – which is "decoupled" from reality. Leslie argues that the decoupled expression is

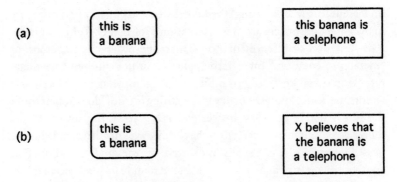

Figure 9.2. Leslie's (1987, 1988) comparison of metarepresentation in object substitution pretence (a), and theory of mind (b).

a representation of the primary representation – a metarepresentation (Figure 9.2a). He draws a parallel between the metarepresentation present in pretence and ToM (Figure 9.2b).

Critics of this view argue that individual pretence is only metarepresentational if children have some awareness of their pretending, rather than just imitating actions seen earlier or responding to the stimulus properties of the object (Lillard, 1993a; Jarrold et al., 1994). Even the capacity of very young children to engage in mutual pretence can be explained without recourse to metarepresentations, if the child relies on scripts or quite generalized knowledge about behavior sequences and scenarios, in order to join in a mutual pretend play episode. The supposition that it is unnecessary to impute metarepresentation to young children engaging in pretence is born out by several experimental studies. Children have some recognition that pretence is mentalistic and subjective only by 3 years of age (Hickling, Wellman & Gottfried, 1997), and are able to infer that mental states are necessary for pretence no earlier than 4 years (Lillard, 1994; PIAC7; Joseph, 1998). When children (aged 3, 4 and 5 years) who observed video clips of realistic or pretend sequences were asked if the scenes were real or pretend and (if they got this right) what the characters were thinking about and what they were "really, really" doing in the situation, 3-year-olds recognized pretence, but only by 5 years did they consistently infer the pretender's thoughts; and only this latter was related to ToM abilities (Rosen, Schwebel & Singer, 1997).

These experiments suggest that children's ability to recognize and take part in pretence is independent of (and precedes) their capacity to represent mental states in pretence. All a child need do is imagine what the pretend world could be like, and reason from there (Harris, 1991). This does not require metarepresentations, but the ability to simulate counterfactual, hypothetical reality by setting aside what one knows about the world. Experiments by Harris & Kavanaugh (1993) led them to conclude that "children might be able to engage in pretense, including joint pretense, without diagnosing the mental state of their play partner or of themselves" (p. 76).

Naturalistic studies of developmental changes in pretence also support this conclusion (Howes, 1992), but suggest connections between aspects of pretence and ToM. At 25–30 months, "Child offers storyline or script. Mother requests creation of new elements and prompts child to a more realistic or detailed enactment" (p. 15). In peer–peer social pretend, "Each partner's pretend reflects the same script but their actions show no within pair integration. Partners inform each other of the script by comments on their own pretend and telling the other how to act" (pp. 15–16). In neither of these characterizations does metarepresentation seem to be necessarily involved. At 31–36 months, "Mother praises and encourages independence. May engage in joint pretend" (p. 68). Peer–peer play is elaborated to "Joint pretend with enactment of complementary roles. Children *discriminate between speech used for enactment . . . and speech about enactment*. Children assign roles and *negotiate pretend themes and plans*" (p. 68; my italics), which seems likely to be metarepresentational. By 37–48 months there is little ambiguity about the metarepresentational nature of pretence: "Children adopt relational roles, are willing to accept identity transformations and generate or accept instruction for appropriate role performance. Children *negotiate scripts and dominant roles and use metacommunication to establish the play script and clarify role enactment*" (p. 68; my italics).

These characterizations suggest that metarepresentation in social pretend is emerging from 31–36 months and clearly present by 37–48 months. These age ranges are summary values. More exact values are given in Howes & Matheson (1992) for "complex social pretend play" which involves metacommunication in the form of naming roles, explicitly assigning roles, leaving a role to modify the play script, proposing a play script, and prompting the other child. These data suggest a first appearance of complex social pretend at 30–35 months (<1% of natural play time), increasing at 36–41 months (about 3–4%) and 42–47 months (about 5–16%), by which time at least half the children engaged in this form of play. In summary, naturalistic studies provide fairly unambiguous evidence for metarepresentation in pretence, i.e., verbal metacommunication of roles and pretence structure, by 42–47 months, and some children show signs of it from 36 months and possibly even earlier, at 30 months.

There are, then, good reasons for rejecting Leslie's (1987, 1988) view that even simple pretence implies metarepresentation. Rather, it is around 3 or 4 years that metarepresentation starts to become evident, in more complex pretend play. This is also the period at which ToM abilities emerge. In concert with Leslie's idea that pretend play which involves metarepresentation is a springboard for the metarepresentation in ToM, might complex pretend play have a role in facilitating ToM at this age? Possible evidence comes from social context effects, from correlational studies, and from training studies.

Can pretend play explain facilitative effects of social context on ToM acquisition?

Two separate sets of studies open the possibility that pretend play explains the facilitative effects of attachment security, and of older siblings/playmates, on ToM development.

Attachment security

Securely attached children scored better than insecurely attached children (all 3 to 6 years old) on a false belief task, even with chronological age, verbal mental age, and a measure of social maturity controlled for (Fonagy, Redfern & Charman, 1997). Secure attachment may be effected by the sensitivity and consistency of the caregiver's responses to the infant's signals. For example, sensitive caregivers show "mind–mindedness"; that is, they "treat their children as 'mental agents', taking into account their comments, actions and perspective" (Meins, 1997, p. 108), and these caregivers had children who succeeded better on ToM at 4 years of age. Why might there be a link between security of attachment and ToM abilities? The reasons advanced in both these studies are, broadly, the quality of pretend play; quality of conversational exchange and elaboration; and competence in peer group interaction. Secure attachment might encourage these, and they in turn might facilitate pretend play.

Older siblings/playmates

There is a nearly linear effect of increasing number of older siblings (from 0 to 3) on ToM, with each older sibling having an equivalent benefit of about 6 months of age/experience (Ruffman et al., 1998). The effect is only found for younger siblings from the ages of 3 to 6 years old. Benefits from older children for a younger child's ToM understanding are not limited to siblings. In an extended family environment in Cyprus, the "sibling" effect on ToM in 3- to 5-year-olds extended to include the number of adults and older children with whom the child was in regular/frequent contact (Lewis et al., 1996). Such data support the idea that ToM develops through "apprenticeship" with older people (Lewis et al., 1996), but even similar age children (best friends) may help (Brown, Donelan-McCall & Dunn, 1996).

Conversational exchanges, especially about feelings and mental states, are also implicated in ToM development. For example, 4-year-old children hear a lot of mental state terms when interacting with their mother, an older sibling, and a best friend, and they use these terms more with the last two interactants (Brown et al., 1996). Pretend play provides repeated opportunities for such conversational exchanges. In fact, children's use of mental state terms (a lot of which occurred during pretend play) correlated with scores on a false belief task (Brown et al., 1996). In addition, dyads (comprising siblings 5–6 years old) who engaged in more frequent pretence were more likely to use internal state terms, especially in high-level negotiations about play (Howe, Petrakos & Rinaldi, 1998), and children were more intensely involved in negotiations during pretence than during nonpretence (de Lorimier, Doyle & Tessier, 1995).

Pretend play is only one means through which attachment security and older siblings/playmates might influence ToM acquisition, but it has been argued to be "perhaps our best candidate for a cooperative activity which furthers the eventual understanding of false belief" (Perner, Ruffman & Leekam, 1994, p. 1236). A plausible case for some link between pretence and ToM exists, but the evidence does not distinguish among models 1, 2, and 3. As Astington & Jenkins (1999) persuasively argue from longitudinal data, more developed linguistic skill might be an underlying factor which explains sophisticated pretence and ToM skills.

Correlations between pretence and ToM

Children at particular ages vary in how much pretence they engage in. Some studies have reported positive correlations between measures of pretend play and ToM. In most studies, one or a few aspects of pretence, rather than frequency or occurrence of pretence itself, correlated with success on ToM tasks. For example, the overall amount of fantasy play in 4- to 5-year-olds' conversations did not correlate with success on ToM tasks, but the frequency of joint proposals ("let's make cookies") and explicit role assignment ("You be the mommy") during pretence did (Astington & Jenkins, 1995). Success of 3- to 5-year-olds in differentiating reality and pretence did not correlate with success on ToM, but success in inferring the mental state of the pretender did (Rosen et al., 1997). And success of 3- to 5-year-olds on ToM tasks was not significantly correlated with success on four different pretend tasks, but was significantly correlated ($r = 0.39$) with success on the Moe task concerning children's understanding of pretence (Lillard, 2001b). (In the Moe task, a troll figure is apparently hopping like a kangaroo, but is not thinking it is doing this; the child is asked whether Moe is pretending to be a kangaroo.) Further binomial analysis showed that more children had a low Moe but high ToM score, than had a high Moe but low ToM score, suggesting that pretence ability did not help ToM – model 1. However, most commentators consider the Moe task difficult, so conclusions from a low Moe score might not reflect conclusions from easier pretend measures.

Composite measures of pretend/fantasy and ToM were not correlated for children at age 3 years, but were correlated for children at age 4 years (Taylor & Carlson, 1997). The ToM composite was based on tasks evaluating understanding of appearance-reality, false belief for self and other, and interpretive diversity; two pretend/fantasy composites, a high/low fantasy rating and a principal fantasy component rating (both showed similar results), were based on information from child and parent interviews, and data about imaginary companions, impersonation, and level of pretend play. The correlation for the whole sample (controlling for age and verbal intelligence) was modest: $r = 0.16$ for the correlation of principal fantasy component with ToM – significant, but accounting for only 2.6% of the variance. (For 4-year-olds, the corresponding $r = 0.27$ accounted for only 7.3% of the variance.) Taylor & Carlson (1997, p. 451) suggest that their methods for assessing variability in fantasy might have been more appropriate for older than for younger children. The assertion that their results "provide strong evidence that there is a relation between theory of mind development and pretend play in 4-year-old children" appears too positive in the light of their actual findings.

Sometimes only some aspects of pretence are related to future ToM success: frequency of role enactment in the pretence of children observed at 33 months during play correlated significantly with their ToM performance tested at 40 months; but participation in pretence, diversity of themes in pretence, and frequency of (verbal) role play did not (Youngblade & Dunn, 1995).

Overall, the correlational findings appear rather mixed in their outcomes. If pretend play has a strong causal role in ToM (model 3) one might expect a stronger and more consistent pattern.

Training studies

Two studies by Dias & Harris (1988, 1990) looked at effects of make-believe play on deductive reasoning. In the first (1988), 5- to 6-year-old children were given syllogisms (e.g. "All fishes live in trees. Tot is a fish. Does Tot live in water?"). In a verbal group condition, children were told: "I am going to read you some little stories about things that will sound funny. But let's pretend that everything in the stories is true." In a play group condition, they were told: "Let's pretend that I am in another planet. Everything in that planet is different. I'll tell you what's going on there." The children in the play group condition scored better on the syllogisms at both ages, on correct responses and justifications. The second study (1990) had a similar design, but contrasted three paired sets of conditions: with/without imagery ("make a picture in your head"); with/without a planet context ("in another planet"); and with/without make-believe intonation. The clearest result is the poor performance of the without-imagery/without-planet/without make-believe intonation group, with imagery appearing the strongest cue.

Dias and Harris argued that setting current reality aside and imagining a fictive alternative may be important in understanding false beliefs, and ToM development. However, there are several problems with this interpretation. The studies suffered from some of the limitations described for earlier studies. In the 1988 study, there was no blind testing, though scoring was done blind. In the 1990 study, there was neither blind testing nor, apparently, blind scoring, and there were only six children in each group. In both studies, all conditions employed the "let's pretend" instruction, so there was no control group in which imagination was not promoted. Subsequent studies within this paradigm (Leevers & Harris, 1999) require reinterpretation of the earlier work. Apparently, it is not the fantasy or pretend component, but simply any instruction which prompts an analytic, logical approach to the premises, which helps at these syllogistic tasks.

The only direct training study on pretend play and ToM to date is by Dockett (1998; and pers. comm.). Children (mean age 50 months) who attended morning and afternoon sessions at the same preschool were pre- and post-tested on measures of shared pretence (from observations) and ToM (several standard tasks). The morning group of children received play training for three weeks; this consisted of play focused around a pizza shop. The afternoon (control) group experienced the normal curriculum without the sociodramatic play associated around the pizza shop materials. From an equal baseline, the play training group, relative to the control group, significantly increased in frequency and complexity of group pretence, and improved significantly more on the ToM tests both at post-test and follow-up three weeks later.

This study provides the best evidence yet for a causal link from pretend play to ToM (models 2 or 3), but a number of shortcomings are present (also S. Dockett, pers. comm.). The groups were small ($n = 15$ and 18 respectively) and were not specifically matched for relevant characteristics; the training period was short; and blind testing was not used. The latter may be the most serious reservation, given previous findings about experimenter effects in studies of play and problem solving (Smith & Whitney, 1987; Smith, 1988).

What the evidence tells us so far

There is a case that pretend play has been selected for its developmental significance – models 2 or 3; but it could be a by-product of other aspects of development – model 1. Studies during the 1970s and 1980s failed to pinpoint a clear developmental function for pretend play, and suffered from significant methodological flaws which did not allow a proper test of these models.

The last decade has seen a particular focus of attention on whether pretend play may facilitate ToM acquisition. Leslie's hypothesis that early pretend play (at about 18 months of age) is metarepresentational, and therefore a precursor of ToM, would support model 3, but it is almost certainly incorrect. There are correlations between pretend play and ToM, particularly for more advanced aspects of pretend, and for understanding mental states in pretence. But these studies do not discriminate among models 1, 2 and 3. The nature of pretend play is such that it may give more practice in mental state term use than some other kinds of behavior; at best, this might support models 2 and 3 over model 1. Few training studies have been done, and Harris now provides an alternative explanation of the Dias and Harris results. Dockett's study suggests a causal link, but was small scale and lacked blind testing.

In my view, the evidence to date favors model 2, so far as ToM acquisition is concerned. The reason to reject model 1 would be the findings that pretend play gives particularly favourable opportunities for mental state term use. The reason to reject model 3 is the patchy findings from the correlational studies, which should be much more consistent if pretence is a major or sole pathway. As noted above, even if model 2 were to hold now, this does not imply that it necessarily did so through much of our evolutionary history. It might be a relatively recent circumstance that pretence facilitated ToM in this way. In modern societies (from which most of the evidence has been gathered) pretend play appears to be encouraged and fostered by adults, and in nursery schools, to an extent probably greatly exceeding what children experienced in traditional societies (Haight & Miller, 1993). Only more experimental and anthropological research that avoids the mistakes of the past can answer the many questions raised in this chapter.

GRETA G. FEIN, LYNN D. DARLING, & LOIS A. GROTH

10

Replica toys, stories, and a functional theory of mind

From the Barbies and Batmobiles of the primary school to the cuddly pups and bulldozers of the preschool, an impressive amount of pretend play happens with objects designed to be used in pretense. On one day, the rage is for Cabbage Patch dolls and Ninja Turtles and on another Beanie Babies and Nano fighters. Some modern toys are promotional and come with an adult-authored story line. Others represent generic characters (babies or pets) or real creatures from field, zoo, or farm.

Representational toys have been found in archeological digs, pyramids, and ancient graves (King, 1979). They have found their way into museums recording the development of playthings throughout the ages. Dolls were once made of corn husks or banyan leaves; toy canoes were scooped from fallen tree limbs; horses and tigers were fashioned from mud or carved from stone. Human culture has endowed these representational objects with immense significance as though these artifacts of childhood hold clues to the beginnings of human thinking, expression, and feeling. As early as the second year of life children spontaneously talk to and for these inanimate representations of real or imagined creatures; children make these objects act upon things and with one another. From an early age, human children gleefully attribute life, feeling, and thinking to inanimate things.

These attributions tend to be in keeping with the object's represented identity: babies are helpless and hungry; tigers dangerous and predatory; monsters evil and cruel; and mothers nurturing and protective, instances of the mandated imaginings proposed by Walton (1990) in his discussion of props as representations. Young children not only give these inanimate objects human mental and emotional attributes, they also give them personalities. What is remarkable is that after a mere 2 or 3 years of life, children have organized so eloquently a wealth of social observations into coherent patterns that can then be projected as properties of the internal mental and emotional lives of represented others.

Replica figures are often featured in studies of early mental development. The most intriguing of these investigate the age at which young children can be said to have a theory of mind. In one study, preschoolers are asked to consider the plight of a cookie monster who yearns for a cookie (Bartsch, 1996); the study hinges on the assumption that young children can understand that an inanimate object has desire and behaves in accord with that desire. In other studies, puppets Maxi and Mari search for chocolates that were put in one place but now are elsewhere (Carpendale & Chandler, 1996); young children presumably believe that these inanimate objects remember what they have seen and derive beliefs from this information. A somewhat different problem is illustrated by Luna and Moe, trolls from a far off place who have never seen and do not know about kangaroos or rabbits (Lillard, 1993b); here it is assumed that young children understand that alien minds might not know about things that truly exist. To complicate the assumption, the children must understand that when the investigator describes Moe as hopping like a kangaroo it is the investigator's description (not Moe's).

In these and other tasks, researchers take as given that the participating child meaningfully attributes complicated internal states to a variety of inanimate toy creatures. Matters of false belief, deception, or pretense seem secondary to the importance of the primary assumption that young children are capable of projecting mind into things. Young children may not have a contemplative theory of mind, but it is likely that they have a functional theory of mind. If children naturally and spontaneously treat their toys in this manner, their animating behaviors demonstrate the ability to impute internal states to others, especially others who otherwise would have tabula rasa minds. The stories children tell about replica play figures can clarify how these figures are used and their contribution to storytelling sophistication.

When judged by formal criteria, young children's stories often lack coherence. The characters behave and experience events, yet story sequences seem fragmented. Longitudinal studies of children's original storytelling find that young children are unable to construct narratives that satisfy the requirements of a good story (Botvin & Sutton-Smith, 1977; Applebee, 1978; Mandler, 1984). These requirements include four components (Morrow, 1986): setting (title, characters, time, place); theme (initiating event, goal, or problem); plot episodes (events leading to a goal); and resolution (solution, ending statement, long-range consequences). The first component children represent is setting. The presence of a main character is identified at 2 years of age with related characters appearing at 3 years. Initiating events and consequences come next (Sutton-Smith, Botvin & Mahoney, 1976). Action sequences are the most typical narrative form used by 4-year-olds (Trabasso, Stein & Johnson, 1981) who begin to include information about problems and affective themes, many of which are negative. Children of this age seem to appreciate what the characters know, think, and feel (Ames, 1966; Botvin & Sutton-Smith, 1977; Applebee, 1978; Feldman et al., 1990). Although children's stories improve after 4 years of age, their stories may still lack information about the setting, initiating actions, and goals, elements that are present by the age of 6 years (Botvin & Sutton-Smith, 1977; Applebee, 1978).

Children's stories are based upon two types of schemas: event schemas and story schemas. Event schemas are mental structures that describe generalized knowledge of frequently experienced routines, such as morning in nursery school. Event schemas yield chronicles – either personal narratives or ordinary happenings. In contrast, story schemas are mental structures that organize expectations about how a fictional sequence might proceed. The child's developing story schema forms the basis for the telling of tales, inventions decontextualized from ordinary experience. Children's early stories are more like chronicles than tales. The missing link in narrative development between 4 and 5 years of age may reside in children's ability to decontextualize problematic, troubling events. This link allows children to represent discordant events, a competency that is emerging but not consolidated in 4-year-olds.

In the theory of narrative thought recently offered by Bruner (1990), automated, expected social routines are called canonical cultural patterns. Canonical patterns set the stage for deviations. Sometimes, events are not what they should be in a particular place at a particular time. On these occasions, the ordinary has been replaced by the exceptional; the canonical has been breached (Lucariello, 1990). And so observers and participants ask: how come?; what do I do now? The human mind is uniquely endowed with the ability to ask and answer such questions (Bruner, 1990). In fact, the human mind can contemplate breaches that have not occurred and may never occur. This ability is founded on unique mental competencies derived from narrative thought.

Narrative thought permits the child to interpret relationships among problematic, troublesome events. At an extremely early age, children try to make sense of these troublesome events (breaches). In fact, children seem to be uniquely endowed with the ability to wonder about the causes and consequences of departures from the ordinary. Children use this ability when they comprehend or tell a story. Stories are vehicles for representing and communicating intriguing and provocative variations in the canonical-breach relationship (Fein, 1995). Narrative thought thereby gives children access to a powerful tool of human culture.

In one study, preschoolers were asked to tell stories about two kinds of picture sequences, one canonical, the other, breach (Shapiro & Hudson, 1991). For example, mother and child are making cookies. In the breach sequence they burn. In the canonical sequence, they bake with no mishap. On measures of coherence such as the number of narrative components used in the telling of the story, the breach version elicited more coherent stories. In Bruner's theory, narrative development begins with accounts of observable story events (the "landscape of action") to which are gradually added accounts of internal, psychological events (the "landscape of consciousness"). The breach version also encouraged more internal state reactions, a finding in keeping with Bruner's distinction (Lucariello, 1990).

The years from 4 to 5 seem to be a period of transition during which story performance is highly sensitive to training or context. Children of this age told better stories (more coherent, cohesive, and elaborate) under a variety of conditions – after extensive training in thematic–fantasy and sociodramatic play, when provided with environmental supports such as picture sequences, or after observing stories' enactments or readings (Saltz, Dixon & Johnson, 1977; Pellegrini, 1985; Pratt & McKenzie-Keating, 1985; Shapiro & Hudson, 1991).

From this perspective, replica play can be viewed as a naturally evolving play form that supports storytelling structures before they can function independently (Fein & Glaubman, 1992). When given replica toys as props, 4-year-olds told more tales than chronicles and their stories had a more mature structure than when no props were provided (Wright, 1992). In another study, children were able to ascribe speech, action, sensations, perceptions, psychological states, emotions, and obligations to replica figures while playing out imaginary sequences, elements typically missing from children's dictated stories (Wolf, Rygh & Altshuler, 1984). Conceivably, wisely chosen replica toys will create the conditions under which children reveal their highest levels of narrative competence.

Canonical versus breach

In order to examine the usefulness of the canonical–breach distinction, we created two sets of replica toys. The canonical set was designed to elicit stories about family life and ordinary, routine events. This set contained six toy figures: an adult male; adult female; boy; girl; baby; and dog. Another set was designed to elicit more sophisticated problem stories by presenting at least one character incompatible with the others. The breach set contained the first four figures, but an alligator and a 9-inch (*c.* 22.5 cm) piece of wool thread were substituted for the baby and the dog.

There are at least three reasons why these sets might make a difference. First, the figures might prime the storytelling process by supplying story characters; if so, replica figures function primarily as triggers to get the children started. If this explanation is correct, the composition of the prop set should have little effect on story structure or theme. Sets with different figures will yield the same results.

A second possibility is that the figures serve to remind children of a story happening. If so, the figures function more like topic markers than triggers. Topic markers identify what the story will be about – a birthday party, a trip, or some other familiar event. If this explanation is correct, sets with thematically different figures should have little impact on story structure, but a strong effect on story theme. Stories are more likely to be chronicles than tales (Shapiro & Hudson, 1991).

A third possibility is that the figures evoke Proppian problem types as analyzed by Botvin & Sutton-Smith (1977). Replica figures representing ordinary family characters, should yield stories about familiar events that are more like chronicles than tales. These script-like narratives will lack the nuclear dyads through which tales provide motivations and resolutions. However, when figures associated with problem states are included, children's stories should show structural advances through the addition of nuclear dyads and themes appropriate to the figures. Two types of plot dyads predominate in fictional narratives such as folk tales and children's stories (Botvin & Sutton-Smith, 1977). In one type, the characters encounter a need that is then eliminated (lack–lack liquidated); in another, the characters encounter a threat that is resolved (villainy–villainy nullified). Children's knowledge of ordinary events may be important in their daily conduct; but for storytelling or other narrative processes, it is disruptions or breaches of the ordinary that make events memorable and tellable (Bruner, 1990; Lucariello, 1990). Even though children's appreciation of ordinary, canonical events may be necessary for story comprehension and production, it is clearly not enough.

Study 1

Participants were 27 middle-class children (13 boys) 4 years of age, who attended child care centers. Two original stories were elicited on separate days, three or four days apart. On one day, the canonical (family) set was provided; on the other day, the breach (alligator) set was provided. Elicitation conditions were counterbalanced. In each condition the child was asked to tell a story about anything he/she wished using any or all of the figures.

We assessed story length, story genre, story grammar, plot dyad, and internal state attributions. Length was measured by the number of propositions (i.e. independent clauses including a subject and predicate). For genre, we distinguished between chronicles and tales. To assess children's knowledge of formal story structure, we used the story grammar categories of setting, plot episodes, and resolution (Morrow, 1986). For plot dyad, we expanded the Proppian model devised by Botvin & Sutton-Smith (1977) to accommodate the diverse stories we collected (see Fein, 1995 for a description). Levels 0 to 3 of the seven-tier scale contain no dyadic plot units whereas levels 4–7 contain at least one. Finally, we evaluated dyadic plot units according to their content: in lack–lack liquidated stories, the characters suffer a loss which is restored; and in villainy–villainy nullified stories the tension is between a threat and its elimination.

Three kinds of internal state attributions were evaluated: (1) the attribution of affect to a story character (e.g., a character is crying, sad, or scary); (2) the attribution of mental or motivational states (e.g., the character decides, thinks or wants); and (3) the attribution of states of consciousness (e.g., sleeping, waking, dead, or living).

A 2 (gender) × 2 (prop set) repeated measures MANOVA performed on story grammar scores for four components revealed a main effect for story grammar, $F_{(4, 49)} = 2.75$, $P < 0.05$. Table 10.1 shows how canonical and breach props affected specific components. Of the four components in the Morrow (1986) model, the only one to reveal a significant difference was theme. Breach props produced more theme components than did canonical props. A similar effect appears on the revised plot dyad measure.

Table 10.1. *Canonical and breach props: Measures of coherence*

	Canonical	Breach	F
Story grammar			
setting	1.4 (0.8)	1.7 (1.0)	2.61
theme	0.8 (0.9)	1.3 (0.8)	7.70**
plot	0.8 (0.7)	0.7 (0.8)	0.05
ending	0.8 (0.9)	1.0 (0.8)	1.00
Plot Dyad	2.1 (1.6)	3.1 (1.8)	11.79**

** = $P < 0.01$

Breach props encouraged more stories containing a complete dyad (66% for breach and 33% for canonical). Canonical props produced more themes of lack (78%) whereas breach props produce more themes of villainy (76%). For those children who received scores of 4 or better in both conditions, seven produced lack stories with canonical toys and of these, 4 switched to villainy with the breach toys. Of the 2 children who told villainy stories with the canonical set, none switched to lack with the breach set ($Z = 2.0$, $P < 0.05$). Finally, of the nine children who scored less than 4 with the canonical set but scored 4 or better with the breach set, seven told villainy stories and two told lack stories. Although the number of participants is small, the pattern indicates that props affected the problems developed in the stories.

Analyses of chronicles and tales showed that breach toys favored tale-telling. With canonical props, only 33% of the children told tales, but with breach props, 60% did so. When tallied for individual children, results are striking: of the 18 children who told chronicles with canonical props, eight switched to tales with breach props. In contrast, of the nine children who told tales with canonical props, only one switched to chronicles with breach props ($Z = 2.33$, $P < 0.05$). There are also striking differences between boys and girls. Regardless of prop type, girls tell more tales than do boys (16 vs. 8). Yet tale telling improves in both gender groups with the breach set: for girls, the change is from 6 to 10, and for boys, the change is from 2 to 6.

Measures of the children's attributions of affect, mind, or conscious state to story characters failed to reveal differences between the toy sets. However, our measures may not have captured the way children of this age express feelings and emotions in replica play. Children might identify the external behaviors and verbal expressions that signify emotion rather than the internal states that accompany them. In order to explore this possibility, we recoded the stories so as to separate observable actions that signify emotional states (e.g. cry, laugh, chase, smack, bite, pet, all alone, run away) from direct attributions of internal state (e.g. sad, love, mad, hate, lonesome, scare). For the canonical set, 13 children used one or more different action/affect terms; for the breach set 21 children did so. The data are even more striking when we consider the number of children who use more than one of these terms in a story: five children with the canonical set; and 14 children with the breach set. However, when internal state attributions are considered, the groups do not differ: nine children with the canonical set; and eight with the breach set.

In order to determine story length, stories were scored for the number of propositions. Scores ranged from 2 to over 100 yielding large variances and nonsignificant differences (Table 10.1). A better measure of canonical–breach effects is obtained when nonparametric analyses are used and only the sign (increase/decrease) of the change from canonical to breach is tallied. These differences are striking: 20 children told *longer* stories with the canonical set, five did so with the breach set and two showed no change ($Z = 3.0, P < 0.01$).

So far: toys and stories

Earlier, we presented three ways of thinking about the effect of replica props on 4-year-olds' stories. If young children simply have trouble getting started, props will serve merely as triggers with little effect on theme or level. The resulting stories might be longer, or require less prompting, but otherwise their quality and content are unaffected. Clearly our findings are incompatible with this view. The breach set produced higher scores on measures of story coherence and canonical stories were longer than breach stories.

A second possibility was that props are thematic markers. If so, props affect story content but not the structural level. With one exception, our findings do not conform to this view. Our comparisons yielded strong effects for both structure and theme. Breach stories were structurally more complex, and the presence of a toy alligator provoked more villainy stories. However, children had more to say about familiar, ordinary characters, a finding in keeping with theories that stress the reality base of narrative thought. However, the extent of the knowledge base did not enhance narrative quality, suggesting that narrative competence may be independent of ordinary experience (Shapiro & Hudson, 1991).

Our results are more compatible with the third view that narrative activity is organized around vivid, unexpected, or disturbing events that depart from the commonplace and ordinary (Fein, 1989). As Sutton-Smith (1986) notes, it is the dark, chaotic, and passionate side of human consciousness that fires imaginative thought. Children describe violent, painful, avoidant, unhappy, or affectionate behavior more readily than the internal states that accompany these behaviors. Children describe the condition of being "all alone" more readily than the internal state of "loneliness." Thus prop effects appear in the far corner of the "landscape of action," but not yet anchored in the "landscape of consciousness." Perhaps, it is necessary to posit a third landscape, that of "internal state behavior" in order to provide for a developmental transition from action to consciousness.

What is breachy about breach?

So far, the data suggest that stories are enhanced by replica toys that represent a character unlikely to be encountered in ordinary experience. An alligator with human figures presented a more extraordinary situation than did a dog with human figures. The question then becomes, "How extraordinary does the breach have to be?" Suppose we replaced the alligator with a less scary, more benign animal. For this we chose a realistic looking, plastic panda bear. We also added a third condition to find out whether the mere presence of props would make a difference.

Study 2

The three comparison groups were: breach (man, girl, panda, dog); canonical (man, woman, boy, girl); and noprop. Thirty-nine middle-class 4-year-olds (17 boys) were randomly assigned to breach, canonical, and noprop conditions. Children were tested twice about six weeks apart. We assessed story grammar, story theme, and story length using measures described in the first study.

Our luck and Bruner's contrast between canonical and breach gave out. No measure revealed differences between any of the toy sets at any time. Children's performance was moderately stable during the six-week period: $r = 0.48$ for story grammar; $r = 0.50$ for story length, and $r = 0.71$ for story theme. In retrospect, we realize that a panda, sweet and cuddly though it might be, is not an exciting story character for 4-year-olds. Some characters such as the baby or the alligator were more likely than a dog or a panda to evoke Proppian problems and the conflicts associated with them (Sutton-Smith *et al.*, 1976; Botvin & Sutton-Smith, 1977). The baby is helpless, the alligator is hostile. In contrast, the panda is a rather nice fellow, and so his contribution is benign and uninspiring. Here is one of the most sophisticated panda stories:

> Um, they went out to the woods. And a . . . one bad man came up to her. And she didn't know him. So she went with him. Then, ummm, then and then her pet Raja (the panda) came and he killed the stranger. And so, the lady went back to her house.

However, most children ignored the panda, whereas few ignored the alligator. So, although the panda seems to qualify as an unusual character, and therefore a potential breach, the children had difficulty constructing stories of lack or villainy in which the panda was a central figure. Perhaps the breach hypothesis works only when the character has dramatic, salient characteristics that declare themselves to the storyteller as unambiguously dangerous or deprived. More is at stake here than the distinction between canonical and breach (Walton, 1990). Toys that promote complex stories must have assignable dispositional attributes that easily translate into problem situations such as babies who are hungry or alligators who attack.

Replica toys and mental states

Our quest for the ideal set of replica toys is unfinished. However, we now have a corpus of original stories that tell us something about children's animations. One frequent theme was that of togetherness. In the following simple togetherness story the characters were represented as engaging in a shared, socially coordinated effort:

> We're collecting sea shells. (moves girl figure)
> Yeah. (moves adult female toward girl)
> Let's go collect sea shells together. (holds figures together)
> I can see a pink one. (moves girl)
> Oh, let's pick that one up. (moves woman)
> I'm all done.

Here the characters inform one another about the scene and arrive at a consensus about what they will do together. In another more eventful togetherness story, the characters experience love, dislike, forgetting, and remembering:

> Once upon a time, a woman and a man got married. (holds male and female)
> And they danced together. (figures dance)
> And he forgot to put up his hand. (moves male's hand)
> So he put up his hand. (moves hand up)
> And they danced some more. (figures dance together)
> And they loved dancing.
> And they danced all night.
> And danced all day.
> And danced until the whole world cracked up. (figures look at one another)
> And they lined up to defect. (dog behind female; panda behind male)
> That puppy belonged to the girl.
> And the koala bear belonged to the boy.
> Ummm... the girl and the boy loved the dog.
> But the girl did not love the koala bear.
> And she told the boy that she did not love his pet.
> And the boy walked away. (male moves away from female)
> And never married her again.
> But she remembered she loved him.
> And he still remembered they broke up.
> And he still wasn't polite to the girl...

The story goes on for another 43 lines, ending with:

> And they were dancing together. (figures dancing)
> But they still married.
> But now they know that the world,
> That the world only drops to the sea.
> And that's the end of the story.

What a lovely complicated tale! First, the world "cracked up." Later, the storyteller reflects on that event, attributing to the characters revised and corrected knowledge of the relation between land and sea at night fall. In this heady romance multiple perspectives are represented by the likings of the characters for one another's pets. Here we have the same event, a dog. Both he and she like it. But we also have the koala bear. He likes it and she doesn't. Same object, different views (Carpendale & Chandler, 1996). The views are irreconcilable and lead to the dissolution of the marriage. In the reconciliation, he decides to accept the fact that she doesn't like his pet. Keep in mind that these complex internal states are being attributed to inanimate objects.

Lack stories are often about lost people or animals and the consequent effort to find them. In the following excerpt, the storyteller describes the third round of hide and seek in which the dog is the seeker:

> The panda, he'll hide in the tissue box. (puts panda in box)
> And he's going to hide over there. (puts male behind box)
> And she goes in the tissues. (puts female in the tissue box)
> And then, "woof, woof!" (moves dog)
> "Is he in this box?" (dog looks in different box)
> "No." (says dog)
> "Is he in here?" (dog looks)
> "Now I am looking in the tissues!" (dog looks in tissue box)
> "I found you!" (said with glee)

How simple, yet how remarkable! The storyteller knows where the replica figures are hidden. Yet, she attributes ignorance to the dog who searches incorrect locations as if the correct location of the figures were unknown. Our hunch that original stories might provide a corpus of children's narrative devices marking a functional theory of mind looks promising. In addition, the stories revealed situations such as replica hide and seek that require children to demonstrate such a functional theory. Although story data are more complex than more traditional tasks and will require sophisticated analytical tools to complement current paradigms, the yield promises to be rich and informative.

At 4 years of age, children's narrative abilities are sensitive to context. Not only do children produce better stories with some replica toys than with others, it is clearly the *meaning* of the figures that affects story content and structure. If so, the issue for children's original stories is who, rather than where or what. Children may like nice animals. But nice animals do not promote elaborate or sophisticated stories. The study of replica toy effects and the resulting adventures of story characters promises to offer fresh data for theories of young children's narrative development and allied insights about how children come to enliven inanimate representations of animate beings.

11

Young children's animal-role pretend play

Animals are compelling presences in the world of young children, and yet the opportunities they offer for understanding topics in child development have often gone unexplored. The variations different animals present as social interactants can provide insight into a range of topics, such as children's understandings of intentionality, intersubjectivity, and sense of self-and-other. Pretense is another such topic – indeed one which draws on all of those listed above as well.

This chapter will describe the animal-role pretend play and imitation[1] of the children in a year-long ethnographic study of a class of 24 preschoolers, ages 3 years 6 months to 6 years. At these ages, children are already adept at pretend play and we have a good understanding of their capacities for understanding others' minds. In this chapter, the focus is on how these abilities are deployed, and to what consequences, in animal-role pretend play. The chapter asks: how much and when do children pretend to be animals?; what do children do when they pretend to be animals?; and what does animal-role pretense contribute developmentally? The episodes to be described enable some generalizations about children's concepts of animal pretense; about how an animal differs from a human as a role in pretend, and about children's conceptions of animals' similarities and differences from humans. Children's animal roles entailed less use of language than did other roles, including in negotiating the pretend frame itself. But this observance of animals' speechlessness also allows certain freedoms. Freedoms are also a theme in how an animal-role relates to socially regulated time, space, and behavior. On another level, any act of animal pretend requires translation of the body to "resemble" the animals', both objectively and subjectively. This translation reveals the importance of animals' agency, affect and coherence for children's conception of them and sense of relatedness to them. The fact that pretend play requires the self-reflective but nonverbal representation of these properties has important implications for the sense of self.

Frequency and occurrence of animal-role pretend play

Pretend play was observed naturalistically in several ways. Children's animal-related play was selectively sampled; at other times live animals were brought into the room and children were videotaped with them (during this they were also asked questions including what it would be like to be the animal, and this elicited much pretend). Parents were surveyed and 11 made home records, and a subset of children was observed in the spring using focal child sampling, representative of all times of the school day except napping. This section will characterize the frequency and patterns in which animal-role pretend play occurred, as background to the description and analysis that follows.

Pretend play (defined as involving some marked nonliteral use of objects or persons) accounted for 17% of the focal children's time. Of this, 13% (or 2% of total observation time) involved animal pretend play. (This may be an underestimate, because such play seemed lower in the spring, but this cannot be verified from the data.) These episodes involved 8 of the 10 focal children, and they ranged in duration from 30 seconds to 6.6 minutes. The average was 2.6 minutes, compared to 5.1 for non-animal pretend play. These are not high figures, but over the year a large amount of animal role play did occur.

Children varied in how much animal role pretend play they did, but this depended also on the context. Among children showing more extreme patterns, three were reported by parents to do extensive animal pretend at home but were observed doing only modest amounts at school (Chen, Billy, Rosa); two showed the opposite pattern (Dawn, Benson); three pretended frequently at school and at home (Drew, Yasmin, Joe), and six did little in either context (Chris, Adrienne, Dimitri, Irwin, Toby, Reuben). What might account for the variation in extent of pretend animal play? As one indicator, the number of animal identities listed by parents was examined in relation to age, gender, presence of siblings, and pet ownership. No relation was found, except for pet ownership – children with no pets were significantly more likely to have at least two pretend animal identities than those with pets in the home (Fisher exact $= 0.0476$).

Across all the children, the menagerie of animals spontaneously represented in pretend during school time or reported at home included alligators, an ape, bats, beavers, beetles, cats, crabs, dinosaurs, dogs, doves, ducks, elephants, ferrets, fish, flies, frogs, lions, mice, monkeys, rabbits, sharks, sheep, a shrew, slugs, snakes, squirrels, spiders, a swordfish, toads, turtles, and a wolf. In addition, cartoon, movie or literary animal personalities were used as models and sometimes blurred with real animals, but these are not included in this analysis. Similarly, unlike Corsaro (1985), who analyzed only animal pretend episodes that involved roles relevant to socialization, the focus here is on animals' roles, whatever the content.

Animals as pretenders

In pretense situations there are logically several positions: the pretender and his or her actions; the model or thing pretended; and possibly co-pretenders, props, and an audience (intended or not). While an actual animal potentially can occupy (or be cast in) any of these, most of our attention in later sections will be on animals as the model. But some comment can be made about children's placement of animals in the other positions too. Observationally distinguishing these different positions is difficult because children do not always communicate their intentions. Although the children were not asked if animals pretend, animals were given make-believe roles in some pretend episodes. For example, the guinea pig was held in an oatmeal container that was used as a vehicle of sorts, or it was placed in a building-block maze; or the turtle was the target of air-boxing "Ninja" boy-turtles. But in such episodes the animal is better regarded as a prop, not a participant, since children gave no indication of regarding the animal as capable of pretend, and the animal usually violated its possibly assigned role (i.e., by crawling away or sitting inert and unaffected). This does not, however, rule out the possibility that more compliant animals may "conform" to nonliteral mental acts attributed by imaginative children, affording a sense of quasi-joint pretense (Kavanaugh, PIAC6) with a quality different than play with an inanimate prop.

Actual animals were never explicitly cast in the positions of pretender, co-pretender or audience, which suggests that children did not expect animals to understand fictional mental states. But children do believe that animals have simpler mental states, and even understand conventional gestures (Myers, 1998). This latter understanding was attributed most liberally by some younger children, such as Billy. In the fall (age 3 years 8 months), he was observed to yell at bugs, "So they can hear." Later in the year he tried to get a dog to shake hands by telling it, "Give me your paw," rather than using the trained command word. Both examples imply that Billy believed the animals would interpret his intended meanings. Older children, such as Joe (5 years 4 months), showed a more sophisticated grasp of animals' limitations, denying humans and animals can entirely understand each other through language. But Joe did believe animals have their own languages, and he conceived of these as codes that allow sharing of mental states, importing a sort of naive theory of human intentional communication.

Given this age difference, we might expect younger children to be less discerning in whether they feel animals partake in pretend. In one case Billy (4 years 1 month) suggested that the monkey that visited class must play with its toys all day at home. When pressed on what these toys were, however, he did not explain them as being props for pretense, but gave examples of food items: "Sometimes like ah if he set a banana on, he would go, ah, and pick it up and peel it and maybe eat it" (Billy motioned peeling and then put hand to mouth, like holding a banana). Billy may have no explicit concept of pretend. One older child, Ivy (5 years 9 months) attempted to engage the monkey in give-and-take by offering it not just any toy, but a toy monkey. This was a game and not pretend, but did suggest she assumed that the animal would recognize its own resemblance.

Adults in the class also attributed mental states and linguistic understanding to animals. For example, one teacher assistant spoke to the monkey about not crossing a line on the floor as if it could understand the line's symbolic meaning as a boundary. And a dog's handler told several children who were pretending to be dogs, "Pogo's going to think you're a dog," implying that the dog might not see them literally. These examples suggest that the adults may believe that – or at least speak to children as if – animals have capacities for pretense. If we were to examine scientifically animals' capacities for pretense, we would need a theory of the development of a concept of pretend and its extension to nonhumans (see Mitchell, 1987, 1990; chapters in this volume). What is clear, however, is that animals' mental worlds are construed by all in the classroom in ways that suggest animals may partake in pretense.

Animals as models for pretend roles

Our major attention is to what children do when occupying an animal role in pretend play. As Harris (1998) and others have pointed out, rather than confusing fantasy and reality, pretend play demonstrates the transfer of knowledge from reality to fiction, not simply the ignoring of real world constraints. In the case of pretending to be animals, the relevant knowledge includes the range of intentions of other agents, the different intentions and other characteristics of different kinds of animals, and the nature of interpersonal and indeed interspecies coordination. What are children's conceptions of these things, as revealed in pretend? Although pretend need not adhere to the child's literal understanding, recurrent patterns across different play episodes give some confidence that underlying conceptions are being tapped. This section looks at the places of language and social conventions in these children's animal-role pretend play.

Language

Animals do not speak human language. How does children's play as animals approach this fact? Language usually has two main functions in pretend play: negotiation of the play frame itself; and enacting the make-believe. In the first function, children playing animal roles usually stepped out of their animal role to negotiate the frame, as in this example, where Yasmin was being a cat, a frequent role for her:

> Rosa asked, "Do you want to go in a big cage?" When Yasmin didn't reply Rosa said, "Little kitties can't talk." Yasmin: "How can kitties talk?" Rosa: "Can big kitties talk?" Yasmin explained: "They can go 'meow, meow'."

In human-role play, in-role words reiterate the frame continuously. In animal-role play, meowing does, but action can also do so. Thus, at times a pretend frame could be established simply by acting the animal part, such as when a group of children chased each other as "sharks," arm-jaws high and wide on the ocean-lawn. However, just as in human-role play, if the action undertaken is "improper" at the literal place and time, the play frame must be marked. Abeo illustrated this need once when she ate part of her lunch right off the table, of necessity declaring, "I'm a beaver," to the teacher.

The second role of language, as a medium to enact make-believe, also demonstrates children's distinctions among humans and animals. This is one way that cartoon or other fictive animal personalities differ from real animals as models – in the former, children spoke more freely than in the latter. Fully pretending to be an animal meant not speaking, which could become problematic:

Yasmin is (again) a cat, crawling behind Dawn who holds a yarn leash tied around Yasmin's waist. Dawn commands, "Go! Stop!! Go!" The play peters out as Yasmin stops to play with dolls. Then Dawn is on "skates" (a board equipped with small wheels), holding the leash and telling Yasmin, "Go, girl!" But Yasmin crawls just like when she was a cat. The process is complicated and they don't get far. Before long, Yasmin comes back crawling and meowing up at me, this time tied to the skate and following Dawn.

Lack of language offers compensatory freedoms to children in the pretend-animal role. An "animal" need not explain him/herself or respond to language and so can carry on in an idiosyncratic way, so long as other players tolerate the animal's cryptic and unaccountable behavior. Being an animal can also allow escape to a private realm. But usually children do not depart from their membership in the shared world of their classmates during animal-role pretend. They usually partake in various play scripts that involve familiar human settings, such as the home, and human roles and identities. Examples include especially the role of pet – whether obedient or unruly.

Socialized behavior
Nonetheless, there is a risk of portraying children's animal personages as just meowing people-with-fur. Other patterns of animal-role pretend demonstrate children's differentiation of animals' place in the human world. Specifically, children-as-animals enact divergent relations to time and space, social roles, and rules of proper conduct.

The space and time of the nursery school classroom was carefully and pervasively structured to facilitate both proximate ends of getting through the day, and ultimate ends of child development. Part of the latter is the need for open areas and times where children's own activity creates the structure. Animal pretend play is only one such activity, but it reliably prompted children to find uses of tables, blocks, cabinets, props and so on that diverged from the socially conventional meanings. The children re-calibrated their surroundings to their animal shapes and tempos. As a turtle, Katra (4 years 3 months) crawled up into a blanket shelf never before observed to have been occupied by a child. Slithering snakes examined previously ignored undersides of wheeled cabinets. Yasmin (4 years 7 months) once had an extended interaction with the class's caged doves. She gestured to them, imitated and anticipated their moves with her hands, and danced before them, then flew off – when she finally noticed all the other children had begun eating lunch.

Both children's pretend play and their statements about animals revealed that they saw the animal role as offering freedom from ascribed social roles and statuses. Animal forms allowed children to get dirty, move around differently, and push social conventions. They would use the animal role as justification for their action, as Abeo did as a beaver, or as other children did as they squirmed across the sand as "slugs," or carried on in disruptive play. While young children do not lack excuses for such behaviors, animal-role pretend offered a water-tight rationale. Several children noted that being one or another animal would allow them to avoid going to school or pressures from "mommies," or indeed the need to conform to anyone else's wishes. For example, Sam (4 years 9 months) explained that, as a monkey, "I would stay in trees and never come down... because I don't want to come down."

Because the animal model offers pretenders a relaxation of the demand to value proper over improper behavior, it provides opportunities to engage in improper behavior. Children may explore the forbidden (as two boys did in examining animal feces), or act aggressively (as did many a dinosaur or fighting cat and dog). Animal roles are not unique in this affordance, which seems common to other kinds of pretense, but because animals stand outside the realm of the proper already, there is additional safety. One can pretend to be a "bad person," or – with less risk of disapproval – be good at pretending to be a "wild animal," and try out similar behaviors.

All these forms suggest that pretending to be a noncultural animal affects the content of the pretend. Although children pretending to be animals allow their characters to follow along in human scripts, they also observe, and exploit, the fact that animals often stand outside the shared meanings that bind society. In this, children reveal that they clearly and with some detail do distinguish animals from the cultural human realm.

Translation of the body

Pervasive in all their pretend of animals was the children's physical and imaginative translation of the animal models' bodies into their own. This was observable even as children pretended to be the snake – clearly a challenge for a biped to embody. But they did so by squirming along on the floor, minimizing use of hands and feet; or by illustrating how it "makes a 's' – Ssssssss" with an undulating arm in the air. That particular snake was a Ball python, and two children independently curled themselves up in imitation to show others what it was like. This kind of attempted bodily fidelity was ubiquitous, and most examples discussed in this chapter include such active translation. To touch on just one other case, the turtle was challenging as a model for pretend. To "be" a turtle, children got on all fours, hunched over, and pulled their arms and legs in.

Not only are such acts physical, they are imaginative and subjectively charged. Despite the awkwardness of imitating a turtle, it was popular because of the feeling of safety that "having a shell" engendered. Children seemed equally focused on the feel of pretended-animal action as on the bodily appearance. The two work together, as in this episode:

> Yasmin, "meowing," is a kitty. Solly is "a dog, a space dog" on all fours and making "arfing" noises. Both are crawling about in the doll/playhouse area, and near the block play area. I help Solly with his suspenders which are coming loose. His role as a dog is not greatly interrupted by asking me for help: he spoke, but continued his posture and noises. He pulls away, goes to the play house door, and looks in. His pose evokes a dog's alert "sitting" posture.

The fidelity of Solly's posture contributes to the maintenance of the play frame and convincingness of the play. More basically, it may reflect an inner kinesthetic empathy or "meaningful actual or virtual ... enactment of bodily moves" (Shapiro, 1997, p. 278) of the dog's body. Such examples require "a capacity for matching between the kinesthetic ... sensations of one's own body's position and ... feeling, and visual images of ... others' bodies" (Mitchell, 1997a, p. 43). Thus, not only is Solly's posture a simulation or performance for his playmates; it is also a translation of body and being into dog reality – as Solly apprehends it.

The depictions discussed here had fidelity to the animal models' initiative and qualities of motion, their shapes and positions, and their affects. For example, Cassia (5 years 4 months) showed what the rabbit did: "This is what he was doing" (she puts her tongue out and makes a lapping noise, then lifts one foot up, standing on the other) "he was licking his shoe – yuk." In this pretense and practically all others, children highlighted how animals were the authors of their own often unique and striking motions. The bodily coherence (or unity, completeness, and distinctive shape) of the animal was also important in pretend, as we saw with the snake. Often this quality posed enticing challenges of representation. As mentioned, many children depicted the turtle by curling forward and holding their limbs in close, but tucking the head proved especially difficult. Some boys found another means:

> Solly (5:0) places both his hands on the back of Joe's (5:5) head and pushes it down in an exaggerated nodding motion three times. Each

> time, Joe brings his head back up by himself, to look toward Solly.
> Then, Solly grabs Joe's shirt, prompting Joe to try pulling it up over his
> head. Solly helps, but Joe's shirt only comes to his nose, and he pauses
> to look at Solly. Benson (5:4) sees this activity and immediately begins
> pulling his shirt up also. Solly ... crawls away behind Joe ... and comes
> around behind on all fours, putting hands and knees down
> deliberately and emphatically. This stroll lasts 27 seconds.

This example also shows the third quality mentioned above: affect. Note Solly's quintessential turtle-like plodding gate, the impacts of which were audible on the videotape over the noise of the class. Solly was a very careful observer of animals, and watched the turtles up close for an extended period. Interestingly, different species seem to offer characteristic vitality affects or patterns of arousal – from the slowness of the turtle or snake, to the jumpiness of the tarantulas, to the cuddly liveliness of the dog or the wiry liveliness of the monkey – all visitors to the classroom, and all models for pretend.

The qualities highlighted in the examples above – motions, body shapes, and affects – are significant because they are also invariant cues which humans, from very early in infancy, use to distinguish self and social other (Stern, 1985; Bråten, 1998; cf. James, 1890). In other words, self and other are distinguished and at the same time related by the reliable patterns of coming and going and other forms of agency; by each other's bodily coherence; and by discernible patterns of arousal over time. When these patterns are repeated in many interactions, self and other are experienced as enduring, and this continuity allows a sense of relationship.

Not only are these qualities represented in children's pretend play, it is significant that the children attempt to preserve the distinct character of the animal model's agency, coherence and affect in these representations. Thus, as with use of language and orientation to social forms, regularities in the children's animal-role pretend play demonstrate how children distinguish various animals from each other and from the human model. Animals offer a range of differences within the basic themes of animacy and intentionality, and young children apply their understandings of these in pretend play.

Functions of animal-role pretend play, and implications for sense of self

It is possible that animal-role pretend play emerged because of the selective advantage it afforded early humans. To the extent it is based on kinesthetic-visual matching, and to the extent that that ability aids empathy and prediction of others' behaviors, it would aid hunting and domestication (Mitchell, 1994b). But the claim that it emerged because of its adaptive value seems questionable. Shepard (1996), attempting to ground human cognition in the interspecies context of our evolution, comes close to this claim in his explanation of the role of such play in development. Anticipating prey's and predator's movements probably added value to the abilities underlying pretend, including kinesthetic-visual matching and higher-order intentionality. But mental abilities like those of humans did not emerge in other species subject to similar selective pressures. Rather, the social context of human evolution seems to hold the key to these abilities. Interaction in early human social groups must have been fluid and ever-changing, transactional and conventional, and must have put a premium on understanding other's motives – tempering selfishness with sympathy and cooperation (Humphrey, 1984). It was pressures within and between such groups that gave an advantage to developmental rates producing dependent infants, large brains, intergenerational learning, language and higher-order intentionality. Then, as Katcher & Wilkins (1993, p. 187) put it, "early humans could use all of their highly developed social intelligence to understand the behavior of animals, vegetation, or any feature of the natural world."

Granting a social crucible for the full flowering of language, metaphor, and pretend, Shepard's (1996) speculations on the roles of animals in thought and the role of pretend in development are suggestive. Animal diversity may have been a template for classification systems, a vocabulary of emotions, and source of the "farrago of selected correspondences" joining personal identity to nature (Shepard, 1996, p. 86). Learning about and pretending to be animals, in his view, are essential developmental bases for symbolism, ritual and "the metaphorical basis of religious conceptions and values" (Shepard, 1996, p. 89).[2] Shepard's thinking may go too far,[3] but it does suggest an important place for animal-role pretend play in identity. We can go further, however, by describing the psychological basis for how early human identity is enriched by make-believe of other animals.

Children's animal-role pretense has been studied little empirically, and when it has, it has been viewed as incidental to accepted goals of development, such as human-role socialization (Corsaro, 1985), or enculturation (Fernandez, 1986). In these cases its function is assumed to be to aid the child in practicing cultural patterns of more mature human behavior. Psychoanalysts since Ferenczi (1913/1916) have given the topic the most attention. Given their assumption of an asocial "animalistic" human body dominated by id-wishes, these theorists typically emphasized how patients are concerned with their own bodily processes and basic needs, and how animal pretend permits a safe distance from which to explore potentially threatening aspects of the self or difficult relationships. For example, Kupfermann (1977) told how one 7-year-old boy identified himself as a cat in response to his mother's lack of ability to meet his needs. The boy instructed the therapist in how to treat "the kitten," thus portraying vividly his experiences and giving him a distance to work through fears of rejection and bodily trauma. "As the cat he was really able to go through all the developmental phases and actively express, face, and resolve the crisis of each stage" (Kupfermann, 1977, p. 384).

Examples that fit this characterization were to be found among my data. Some involved dinosaurs or aggressive animals. Others, baby birds:

> Mrs. Ray talked with Yasmin while pretending to offer her worms and calling her "little bird." Mrs. Ray told her how when she grew up she would fly away and leave the nest. Yasmin was silent. Then Mrs. Ray asked if that made her sad to think of that. She nodded. Mrs. Ray said, "That what's so nice about being a person is that you never have to leave your family, you can keep seeing them as long as you want." This did seem to help Yasmin, but she continued as a bird.

In this example, issues of family ties, autonomy and dependence seem central. But even when such issues were at stake – although they were not most of the time – the patterns we have traced above point to a more pervasive function of animal pretend play in the development of the self. To see this will, however, require a look at the place theorists have given pretend in the development of self.

Since the work of G. H. Mead, it has often been assumed that language is both the foundation of, and the route to, a self-reflective sense of self (Mead, 1934/1974). Mead felt it was the way words let the speaker perceive him/herself from other social points of view – the reversibility language requires – that made it the exclusive route to a truly social sense of self. One implication is that animals, being outside the community of language users, do not play a strong role in crystallizing a sense of self.

Mead was emphatic that pretend, being nonintentional in his view, could not provide the kind of feedback that games with rules did; it could not generate self-awareness. His view, however, has been cogently critiqued by Hanson (1986), who argued that pretend is an intention-dependent act.[4] Her explanation (Hanson, 1986, p. 83) fits all our examples of animal-role pretend:

> The child can be discovering which bits of behavior are most appropriate to the other and which are most characteristic of himself, for he must evidence or emphasize the former and suppress or disguise the latter. A more coherent sense of both himself and the other is thus obtained. Hence, as engaging in this sort of play does further demarcation of the self, it is reasonably cited in an explanation of the development of the self.

Pretend play involves conscious contrast between self and other. This is reinforced by Lewis & Ramsay (1999), who examined the development of intention. Intentions are seen in early pretend play, and the emergence of consciousness is related to the emergence of pretend because the latter "requires conscious intention and self-awareness in that children know that their actions on objects are not real" (Lewis & Ramsay, 1999, p. 85).

Given this interpretation of pretense, animals need not be excluded as "others" in relation to whom the child attains a self-reflective sense of self. Rather than entering into the child's sense of being a human self only by the indirect route of linguistic definition of human versus animal categories, animals' many intriguing differences within animate commonalities are directly available for conscious exploration through pretend. The reflective self-awareness attained in this way would not necessarily be verbal, nor verbalized easily. But it would be revealed in pretend, and would contribute to a clarified and conscious sense of what it is to be human, as well as to an elaborated sense of connection to the animals one has embodied in pretend. In a broader sense, because the basic medium of representation in pretend is bodily translation, and because it is bodily agency, coherence and affect that we share with other species, animal-role pretend play must encourage a sense of continuity between child and animal identities.

It is likely that our abilities for representation and higher-order intentionality are foundations of human pretense. But linguistic representation can turn differences of degree into categorical ones. The metaphoric function of language helps de-literalize such categories, but children, and cultures also, may be vulnerable to affective "capture" or reification of word meanings (de Gramont, 1990). The meanings of animals, and of animal-role pretend, may thus depend on both ecological and psychological factors in every culture. In the case of the human self in Western culture, various socialization messages encourage the distancing of the human self from certain features of animals or our treatment of them. Then, linguistic categories make possible the reified meaning that to be grown-up means to not be an animal (Myers, 1998).

Accordingly, to the extent our theories of the self have stressed language as the primary medium in its development, they have left unexamined the ways children's selves attain their full humanness not just by categorical contrast to other animals, but also by comparison and commonality. Animal-role pretend play contributes to this unique human developmental potential – as yet mostly uncharted – for a sense of connection to other species.

Endnotes

1. Examples of pretend play and imitation are both treated in this chapter. According to Mitchell's analyses, level 4 imitation includes pretense (1987), and pretend play can include imitation (Mitchell, 1990). The key link is intentionality. Pretend play demonstrates intentional simulation; and level 4 imitation requires control over the relationship between model and copy. In this chapter, the imitation examples used were deliberate acts, as evidenced by being directed to a particular other or otherwise marked. In other words, we will look at imitation that serves communicative goals, albeit sometimes literal rather than make-believe ones.
2. Shepard's writings bear on the venerable question of totemism in anthropology. The question entails the metaphoric relations between systems of classification of nature, and of social and psychological categories. The issue of which, if either, domain of thought serves as the basis for generalization is under discussion in developmental psychology (Hirschfeld & Gelman, 1994).
3. Compare the discussion of "biophilic symbolism" in Kahn (1999, pp. 34–36).
4. Such confusions about the nature of pretend and imitation have been clarified by Mitchell's (1987, 1990) comparative-developmental analyses. Mitchell's (1990) analysis of the levels of play distinguished pretend play from other kinds because the design process involved is intentional simulation. Notably, imitation may be critical in attaining awareness of intentionality (Meltzoff & Moore, 1998).

12

Imaginary companions and elaborate fantasy in childhood: discontinuity with nonhuman animals

There are a bewildering number of opinions about what constitutes pretend play and the extent that nonhuman animals are capable of it. Play fighting and object play in mammals and birds strike many observers as having the "acting as if" quality that we associate with pretending. Other researchers are more inclusive, claiming that "in all play, both animal and human, there is an element of 'pretending'" (Aldis, 1975, p. 14). According to Darwin (1871/1896, p. 69; our italics), "even insects play together, as has been described by that excellent observer, P. Huber, who saw ants chasing and *pretending* to bite each other, like so many puppies."

When acts of pretense are assumed to reflect a metacognitive understanding of behavior, claims about pretending in nonhuman animals (e.g., Bateson, 1955/1972) become more controversial (Mitchell, 1991a). For example, Rosenberg (1990) has argued that third-order reasoning about intentionality (i.e., animal A plays with the intention of animal B understanding that A's intention is nonserious) is beyond the intellectual capacity of any nonhuman species. However, according to Bekoff & Allen (1998) such demanding criteria might be said to preclude the possibility of pretending in young children as well! In fact, a growing number of developmentalists are claiming that too much cognitive credit had been attributed to child pretenders. In particular, Lillard (1993b; 1994; 1996; PIAC7) has shown that young children have difficulty when asked questions requiring them to reason about the representational aspects of pretending. This work has led to considerable debate about the extent that children equate pretending with acting in a particular way and fail to understand that acts of pretense are guided by the pretender's mental representation of the pretend entity (e.g., Hickling, Wellman & Gottfried,

1997; Rosen, Schwebel & Singer, 1997; Joseph, 1998; Lillard, 1998b; Woolley, *PIAC8*).

The controversy over the cognitive underpinnings of child pretense could be interpreted as having implications for claims about pretense in other species. After all, the belief that children have a metacognitive understanding of their actions has been a barrier to equating certain types of animal play with pretend play in children. It has been assumed that children, in contrast to chimps, dogs, or ants, understand that pretending involves the projection of an imagined situation onto a real one for a playful rather than a serious purpose. The results of research suggesting that young children have a less sophisticated understanding of pretense than previously supposed could be interpreted as bringing nonhuman animals and children closer together in terms of what is going on when they play. However, there are strong cautions in the literature against this type of conceptual leap (e.g., Mitchell, 1997c; Povinelli, Bering & Giambrone, 2000). In addition, the developmental research in question actually focuses rather specifically on when children develop the capacity to answer explicit questions about the knowledge conditions that must hold for a person to be said to be pretending. It is, to our minds, unlikely that such research has any implication for nonhumans' understanding of pretense.

Regardless of the outcome of the debate about children's representational understanding of pretense, we are convinced that there is something uniquely human about pretend play. Our view reflects a decade of experience studying the creation of imaginary companions (ICs), an elaborate and sustained type of pretend play that is very common in early childhood. Do nonhuman animals ever have ICs? There are scattered observations in cross-fostered great apes involving attention to imaginary objects (Hayes, 1951; de Waal, 1986a), pretend actions (eating, sleeping) (Savage-Rumbaugh & McDonald, 1988), tea-party and makeup play (Patterson, 1978c), and play with dolls (Gardner & Gardner, 1969; Patterson & Linden, 1981; Miles, 1990). The orangutan Chantek used the signs for CAT and DOG in social games of pretense in which he acted as if one of these scary (to him) animals was present (Miles, 1986). Pretend actions in the service of deception (e.g., feigning attention to a distant nonevent to draw others away from some food or a mate that the animal desires for itself) has been observed in chimpanzees raised with conspecifics (for review, see Mitchell, 1993c). However, to the best of our knowledge there are no compelling observations of sustained interactions and attachments between nonhuman animals and imaginary others.

By claiming that the human capacity to pretend is unique and special, we do not mean that animals have no capacity for pretense at all. There is some overlap in the play behaviors of animals and humans, but it is important not to lose sight of the overall distributions that are being compared. The quantitative and qualitative limitations of nonhuman pretense stand in striking contrast to the heights of elaborate fantasy routinely achieved during childhood. This same point is raised in comparisons of human and nonhuman communication. An enculturated chimp once signed "water bird" when referring to a duck, but such rare creative acts by cross-fostered great apes pale in comparison to the everyday creativity in language exhibited by children everywhere (Clark & Hecht, 1983). Similarly, no other species, not even the most intelligent and playful of apes, exhibit pretense to the same degree as young children. In particular, the diversity and creativity evident in children's invention of imaginary companions demands our attention. We suspect that the complex interactions that characterize play with an imaginary companion constitute a form of pretense that even the most anthropomorphizing zoologist would consider to be outside the realm of anything that occurs in nonhuman animals – even when anecdotes and intuitions are included as well as more systematic empirical evidence.

A taxonomy of imaginary companions

Svendsen (1934, p. 988) defined an IC as "an invisible character, named and referred to in conversation with others or played with directly for a period of time, at least several months, having an air of reality for the child but no apparent objective basis." Many researchers have expanded this definition to include personification of objects such as dolls and stuffed animals that are treated as though they have a stable, autonomous personality (Singer & Singer, 1990; Mauro, 1991). Although estimates vary from study to study, imaginary companions are generally believed to be quite common; perhaps as many as 60% of children create them.

For the past decade, we have been interviewing hundreds of children, parents, and other adults about the creation of imaginary companions (ICs). To provide information about the diversity in the forms that they take, we have compiled a list of descriptions from all the studies in which we collected information about ICs. These studies included three different sources of information: child report; parent report; and retrospective reports from adults who had ICs as children. Each source has it strengths and weaknesses (see Taylor, 1999), but taken together they provide a comprehensive picture of ICs. Altogether 341 descriptions were collected: 179 descriptions from the children themselves (Taylor, Cartwright & Carlson, 1993; Taylor & Carlson, 1997; Taylor et al., 1999); 42 descriptions from parents (Taylor et al., 1999; Taylor & Dishion, 1999; Taylor & Luu, 1999); and 120 descriptions from adults recalling their childhood ICs (Kavanaugh & Taylor, 1999; Taylor & Carlson, 1999; Taylor, Hodges & Kohanyi, 1999). The descriptions are not independent observations because many participants (22%) reported more than one imaginary friend. When participants described multiple ICs, they often included ICs of different types and it was not clear to us that there was any basis for selecting or focusing on one IC out of the two or more descriptions. Thus, we decided to include all the descriptions in this analysis. The 341 descriptions were generated by 252 participants: 127 children; 40 parents; and 85 adults (retrospective self-reports). The children ranged in age from 3 to 12 years (56% were 5 years old or younger, 26% were 6 to 8 years old, and 17% were 12 years old). The parents all described the ICs of children under 5 years of age. Many adults did not report exactly how old they were when they played with their ICs, but most of the adult retrospective reports seemed to be of ICs from early childhood.

The descriptions were categorized by two independent coders into the categories listed in Table 12.1. The coders agreed for 87% of the descriptions and disagreements were resolved by discussion. One of the most obvious distinctions to be made was between ICs that were completely invisible and ICs that were based on toys, such as dolls or stuffed animals. Some researchers have excluded toys as ICs, but we think that sometimes children's imagined relationships with stuffed bears or dolls becomes so vivid that it is fair to consider them a type of IC, much like the stuffed tiger "Hobbes" in Bill Watterson's (1990) comic strip *Calvin and Hobbes*.

Altogether we collected 105 descriptions of special toys that seemed to function as ICs rather than as ordinary toys or transitional objects (31% of the IC descriptions). Dolls and teddy bears accounted for 30% and 27% of the toy ICs, respectively. In addition to bears, the stuffed animal category included rabbits, frogs, dogs, monkeys, muppets, a kangaroo, a dinosaur, a hedgehog, a cow, a tiger, a horse, a dolphin, a smurf, a Tasmanian devil, a cat, a donkey, a squirrel, and a moose. The retrospective descriptions collected from adults often provided clear evidence of the intense and sometimes idiosyncratic nature of the relationship that can develop

Table 12.1. *The number of each type of imaginary companion (IC) as a function of data source (the children themselves, their parents, or adults recalling their childhood ICs)*

Type of IC	Child report	Parent report	Retrospective report	Total
Toys				
Doll	28	1	2	31
Stuffed animal	26	9	39	74
Total (toys)	54	10	41	105
Invisible ICs				
Invisible child playmate (ICP)	35	10	18	63
Special ICP	21	6	13	40
Invisible baby (IB)	2	1	4	7
Special IB	3	1	0	4
Invisible older companion (IOC)	11	2	0	13
Special IOC	12	2	2	16
Invisible animal (IA)	17	7	2	26
Special IA	6	2	10	18
Superhero	2	0	5	7
Invisible enemy	4	1	2	7
Ghost	4	0	1	5
Angel	0	0	2	2
Presence	0	0	6	6
Invisible self	2	0	2	4
Other	6	0	12	18
Total invisible ICs	125	32	79	236
Total of all ICs	179	42	120	341

between a child and a special toy. Here is an excerpt from an interview with an adult remembering her childhood 8″ (*c.* 20 cm) sailor doll.

> *Experimenter*: How old is the IC?
> *Participant*: At times she was young, at other times she seemed almost ancient. She was falling apart – almost like she was dying.
> *Experimenter*: What did you like most about her?
> *Participant*: Her courage, strength, tenacity, honesty, will power, spirit – her feelings.
> *Experimenter*: What did you not like about the IC?
> *Participant*: Anger, aggression arguments, controlling her behavior toward the other two ICs. Sometimes I felt awful for the imaginary companion because she had suffered so much (which made her age rapidly). She seemed almost sick with grief . . . she needed to be taken care of.

As can be seen in Table 12.1, there were a variety of categories for invisible ICs. Each of these will be discussed in turn.

An invisible playmate similar to a real child

Although ICs serve many types of functions for young children (see Taylor, 1999), adults often assume that they compensate for a lack of real playmates. Thus, one might expect the most typical IC to be an invisible child who could function in this capacity. In fact, many children do invent invisible child playmates, but a substantial majority (73%) of the invisible characters that children create *cannot* be described in this way. In this category, we included all the descriptions that were consistent with what you might expect if the participant was describing a real-life playmate. In fact, 16% (10 out of 63) of these descriptions were imaginary versions of real friends. Another 16% (10 out of 63) were invisible siblings. Altogether, this category accounted for 27% of the descriptions of invisible ICs (18% of the entire sample of ICs).

> Examples
> "Fred" (report from 12-year-old): "He's funny, has dirty blond hair down to his cheekbones, bluish-green eyes, and long lashes. He's fun to talk to; we always know what the other one is going to say."
> "Freddy French Fries" (report from 12-year-old): "Freddy was mischievous, funny, annoying; always cheers me up, made me laugh."
> "Fake Rachel" (report from 3-year-old): An invisible version of a girl the child once knew named Rachel.

Invisible child playmates with special characteristics (Special ICP)

Imaginary companions often have characteristics that take them out of the realm of what might be expected of a real child playmate. Some have special capabilities such as being able to fly or to perform magic. The category of Special ICP included all the descriptions of invisible children whose special characteristics made them unlike what one would expect for a real child playmate. There were 40 descriptions (17%) of this type.

> Examples
> "Baintor" (report from 4-year-old): A very small invisible boy who is completely white and lives in the white light of a lamp.
> "Simpy" (report from 7-year-old): An invisible 8-year-old girl who is 3 feet tall. She has black hair and eyes and blue skin and likes to wear funny clothes.

"Whistler" (adult report): An elf who was 3 feet tall, had blond hair, and was a trouble maker.
"Alice" (adult report): An invisible girl with special powers who could fight crocodiles, but generally lived under the kitchen table.

Invisible baby

Not all ICs function as playmates. In this sample, 3% (7 out of 236) of the invisible ICs were very young and had to be cared for, taught, or otherwise nurtured. Thus, the child creator takes on the role of parent or teacher, rather than being an equal status playmate. The young age of the IC is a significant part in these descriptions.

Examples
"Little Chop" (report from 5-year-old): An invisible one-year-old baby with black hair and green eyes.
"Amanda" (adult report): An invisible, "sort of a shadow" human baby. "I loved and cared for her and she googled."
"Anka" (parent report): A baby girl who is about 2 years old.

Invisible baby with special characteristics (Special IB)

Of this sample, 2% (4 out of 236 invisible ICs) were babies who had unusual or magical characteristics.

Examples
"Hekka" (parent report): An invisible baby who is very, very tiny.
"Cream" (parent report): A tiny invisible baby who lives on the child's hand.

Invisible older companion (OIC)

When the IC is much older than the child the nature of the relationship often reflects this age discrepancy. The IC serves as an advisor or consultant, rather than a playmate. The advanced age is a significant part of the description of the IC. In this sample, 6% (13 out of 236) of the ICs were considerably older than the child creator.

Examples
"Julia" (report from 7-year-old): An invisible 13-year-old girl, with long blonde hair, green eyes. The child said that the thing she liked best about Julia was that she was older than she was herself.
"Zippy" (report from 6-year-old): An invisible 11-year-old boy who was a foot taller than the child. He looked "like a regular person" with white skin, hair sort of blond, blue eyes and regular clothes.

Invisible older companion with special characteristics (Special OIC)

In this sample, 7% (16 out of 236) of the invisible ICs were described as of advanced age and they had unusual or magical characteristics.

> Examples
> "Skate Board guy" (report from 7-year-old): An invisible boy, about 13 years old, who is 3 feet tall, wears a cool shirt and has a cool skateboard. He is "kind of like liquid." He lives in the child's pocket, but then pours out into a boy shape that does skateboard tricks.
> "Nobby" (report from 6-year-old): An invisible 160-year-old business man who visited with the little boy between business trips to Seattle and Portland.
> "Nothing" (report from 5-year-old): A male IC who is 1000 years old and bigger than the child.

Invisible animal

In this sample, 12% of the invisible ICs were animals instead of being human. The following animal species were included in this category: cow, dog, tiger, dinosaur, mouse, cat, giraffe, horse, pony, lion, elephant, panther, unicorn, and rat.

> Examples
> "Gadget" (report from 4-year-old): An invisible mouse who was very active, mischievous and frisky.
> "Robert" (report from 7-year-old): Robert was an invisible panther who was all black with blue eyes. The child reported that she met Robert in her dreams and that he lived in the jungle.
> "Nobody" (report from 4-year-old): An invisible elephant who lives in Africa.

Invisible animal with special characteristics (Special IA)

When children invent invisible animal friends, the animals are likely to have some features that would not be found in a real-life version of the animal. For example, the invisible animal often has the ability to talk or otherwise communicate with the child. Thus, one might argue that all the invisible animal companions were special in that they were unlike real members of their species. However, some animal friends are further embellished with magical powers (e.g., a cat that flies) or special characteristics (e.g., superior intelligence). In this sample, 8% (18 out of 236) of the invisible ICs were animals that had particularly special characteristics. The species that were represented in this category were: bird, duck, unicorn, mouse, turtle, dog, opossum, monkey, elephant, flea, bear, and dolphin.

> Examples
>
> "Pepper, Crayon, and Golliwod" (adult report): Three invisible "sheas," a type of invisible flea. Pepper was pink with pink hair, Crayon was plaid, and Golliwod was black with black hair. The child carried them around and protected them from the evil planet aliens who were looking for them.
>
> "Dipper" (report from 5-year-old): An invisible flying dolphin who lived on a star, never slept, and was "very very very very fast." He was "about the size of a regular dolphin, but covered with stars and all kinds of shiny stuff."
>
> "Joshua" (report from 4-year-old): An invisible male opossum who is "taller than the ceiling" and "lives in San Francisco." The child complained that Joshua kisses her and she doesn't like it.

Superhero

In this sample, 7 out of 236 invisible ICs were imaginary versions of superheroes, including invisible versions of Aquaman, Superman (twice), Wonder Woman, Wonder Twins, Peter Pan, and Ariel.

Invisible enemy

Not all ICs are friendly or nice; some are predominately mean and frightening to the child – 6 out of 236 (3%) were in this category.

> Examples
>
> "Invisible big blue furry thing" (adult report): It had red eyes, no clothes and was six feet tall. He was really scary.
>
> "Acher" (report from 3-year-old): A 5-year-old invisible boy who "is very bad to me, hits me, kicks me, pulls my shirt, and jumps off my bed." (This child also mentioned a little invisible girl named "Darnit", but did not provide any other information about her.)
>
> "Barnaby" (parent report): An invisible "bad guy" who was scary and had a big mustache. Barnaby used to hide out in the little boy's closet.

Ghost

This sample included 5 ghosts (2% of the invisible ICs).

> Examples
>
> "Mr. Ghost" (report from 4-year-old): An invisible ghost who has white hair, white eyes, and wears no clothes.

"Casper" (report from 7-year-old): An invisible ghost who is about 2 years old and 3 feet tall. He is white with fat fingers, no hair, and green eyes.

Angel

There were two ICs in this sample who were described as angels.

Example
"Pucka" (adult report): Pucka was an invisible guardian angel who talked to the child and listened to her. "It was warm and accepting. I used to have a strong sense of Pucka when I fell asleep at night and a couple of times I caught a glimpse. He glowed and was more animalistic than human."

Undifferentiated presence

Sometimes the IC is not imagined as belonging to a particular species or gender or being a particular age. Instead the IC is experienced as a presence without a particular type of physical instantiation. Six out of 236 were described in this way.

Example
"The Wiz" (adult report): A nature spirit that had no gender and liked to leave gifts.

Self

Some children create another version of themselves, like a double or twin. The child's mirror reflection sometimes plays a role in this type of fantasy. Four out of 236 were in this category.

Examples
"Myself in the mirror" (report from 3-year-old): An invisible person who is based on the child's reflection. She looks exactly like the child and that is what the child likes about her.
"Dewgy" (adult report): An invisible version of the boy, almost like a twin.

Other

ICs can be very idiosyncratic. In fact, 8% (18 of the 236) of the invisible ICs simply did not fit any of the categories above and were not similar to each other.

Examples

"Eyabra", "Akoam", and "Duke" (adult report): Three invisible ICs who worked together as a team. Eyabra was a mask with three faces (one large and two small ones), Akoam was a magic green staff, and Duke was a huge shield made of a turtle's case shell. "Depending on the time of day, a face from the mask would tell me stories while I sat on the shell and the staff kept watch."

"Johann" (adult report): "It was a couch-caster – a little hemisphere of metal that went on the bottom of a couch-leg to keep the carpet from tearing or something. I wore it on my thumb and my thumb and the caster-cap became Johann. I was 3 when it first appeared. The dentist told my parents that I had to quit sucking my thumb, because its was affecting my bite, so I thought of the idea of Johann to keep myself from sucking my thumb. It was male and human, even though its outward visible form was a caster, I still thought of it as human. I would talk to it when alone, and telepathically to it in the company of others. It didn't speak to me as much as it spoke to other people. I felt secure when thinking about or actually in the company of Johann. Everyone knew about Johann. There were even several kidnapping attempts made on Johann." (The coders struggled with the decision of whether Johann should be categorized as an Invisible Other or as a Doll. Neither category seems entirely accurate.)

"Humpty Dumpty's mother" (adult report): An invisible egg that could talk. It has spiky hair, a big round egg-like head and a human body.

"Butcher Shop Guy" (report by 8-year-old): An invisible green Cyclops who was a world traveler and liked to tell the boy, up to the age of 5 years, about his adventures.

Discussion

Creativity is frequently mentioned by authors who wish to draw a distinction between human and animal pretense, but the extent of the creativity, versatility and/or flexibility in young children's play tends to be referred to in passing rather than described in any detail (e.g., P. Mitchell, 1997). In fact, Sutton-Smith (1997, p. 160) points out that, even in research focusing solely on children, there is very little normative work documenting the content and varieties of what he refers to as "the wilder kinds of childish imagination."

Our primary goal in this chapter was to begin to fill this gap by describing the diversity in the types of imaginary companions that children create. We think our analysis provides many striking examples of the creativity that is characteristic of child pretense. Note, however, that by focusing solely on general descriptions and physical characteristics of ICs, we actually are looking at the tip of the iceberg when it comes to diversity. Not only do ICs come in all shapes, sizes, and species, they also seem to fulfill a wide range of both short-term and long-term functions. To date there have been no systematic studies of the functions of ICs, but there are many clues about function provided by *post hoc* examination of how children talk about their ICs and parental anecdotes about them. Aside from providing fun and companionship when no one else is available, they can bear the brunt of a child's anger, be blamed for mishaps, provide a reference point when bargaining with parents (e.g., "Bla Bla doesn't have to finish his dinner, why should I?"), or serve as a vehicle for communicating information that a child is reluctant to say more directly (e.g., "Mr. Nobody is afraid of the new dog next door."). With each of these functions, ICs may serve as a bridge to reality, that is, a means of experimenting with social behaviors or mastering emotions related to real-life objects and events (Partington & Grant, 1984; Bretherton, 1989). More generally, pretend play has been shown to help develop language, social, and imaginative skills that prepare children for later life (Bornstein & O'Reilly, 1993; Sutton-Smith, 1997).

Our reading of the animal literature suggests that a more narrow range of possibilities has been proposed for the functions of play, the most accepted being its role in improving skills that are important in later life, such as hunting and fighting (e.g., Aldis, 1975). However, the empirical evidence for the skill training hypothesis has been mixed (Bekoff & Byers, 1998), and attempts to specify other functions of nonhuman play tend to be problematic and controversial (for a review see Sutton-Smith, 1997; Smith, PIAC9). According to Bekoff & Allen (1998, p. 99), "attempts to define [animal play] functionally face the problem that it is not obvious that play serves any particular function either at the time at which it is performed or later in life." Although Mitchell (1990) points out that animal play cannot be considered functionless because it is *intentional* activity directed toward a specific end, the benefits of animal play have been difficult to specify. Our speculation is that the functions (i.e., the short- and long-term benefits) of animal play are likely to be less diverse, less idiosyncratic to particular individuals, and less fundamental to development than in human children.

Another important aspect of diversity in children's pretend play that was not addressed in our current analysis pertains to culture. In our past research, we have found interesting differences as a function of religion in both the attitudes of adults and the behavior of children with respect to pretense, in general, and imaginary companions, in particular (Carlson, Taylor & Levin, 1998; Taylor & Carlson, 2000). Furthermore, there are cultural differences in play, in terms of both play themes (e.g., imitation of adult roles versus creating exotic scenarios) and adult encouragement (Roopnarine, Johnson & Hooper, 1994; Farver & Shin, 1997; Göncü, 1999). For example, Carlson *et al.* (1998) found that Old Order Mennonite children (similar to the Amish) were less encouraged by teachers to engage in social pretense and their play themes were less "imaginative" than the pretend themes of non-Mennonite Christian children. These types of sociocultural influences suggest discontinuity in the play of human and nonhuman animals. According to Wolf (1984, p. 176), the cultural characteristics of human play establish "if not a hard and fast boundary, at least a steep gradient of change, between the play of humans and the kind of fooling around conducted by even closely-related species."

As for comparisons about the cognitive underpinnings of pretend play, many theorists, notably Symons (1978), have questioned the extent that nonhuman animals experience any connection between a serious activity and the play activity that strikes the human observer as its pretend counterpart (for a discussion of this issue, see Mitchell, 1991a, 1993c). Thus, it is possible that the relation between the "pretend" and real fights of dogs, for example, is entirely in the minds of human observers. For dogs, these activities might be unrelated. Certainly there are enough actual differences in the behaviors that characterize real fighting and playfighting for this to be true (e.g., top speed is used only for real fights, role switching characterizes playfights, real bites are reserved for real fights, different postures, etc.; Aldis, 1975). It might require a human observer to appreciate any similarity between real and play fighting and designate the latter as "pretend."

Fortunately for researchers, during the preschool years children are increasingly able to articulate exactly what they are doing when engaged in acts of pretense. In addition to explicit statements ("Let's pretend you are a monster"), they use distinctive gestures, linguistic registers, and expressions to communicate the nonliteralness of their behavior (Garvey & Kramer, 1989). When children act as if lumps of mud are delicious cookies and entice another child to "eat" them, they indicate they are in a pretense mode via explicit stage direction and knowing smiles. Children expect others to understand the nonliteralness of their actions and are shocked if a play partner appears to have mistaken pretense for reality (e.g., actually bites into "pretend" food) (Golomb & Kuersten, 1996). Although in certain experimental settings, young children have difficulty expressing their knowledge about the representational nature of pretending, it is striking how clear they are about real and pretend scenarios in their own play and how often they strive to keep the record straight. The case of play with ICs is particularly interesting because children show evidence of strong emotional involvement with their friends and, at the same time, they provide a variety of signals communicating their awareness of the friends' fantasy status. The child eyes the experimenter who has been listening closely and taking notes as she describes her imaginary companion and whispers, "You know, it's just pretend."

Part III

Pretense and imagination in primates

13

Pretending in monkeys

Monkeys live in a complex social nexus and rich foraging habitat, and their social and intellectual skills develop in these arenas. Their tool-use and problem solving abilities put them on a different level from most mammalian clades (Beck, 1980; Antinucci & Visalberghi, 1986; Tomasello & Call, 1997; Parker & McKinney, 1999). Young monkeys must learn how to obtain sustenance, as well as how, when, and with whom they should groom, communicate, and play. They must learn to understand facial and vocal communication, recognize and manage their social bonds of kinship and friendship (even if the social ranks are asymmetric), and manoeuver in the field of their dominance relationships (which are asymmetric except for the highest and lowest ranking) (van Hooff, 1967; Stein, 1981; Fedigan, 1982; Gouzoules, Gouzoules & Marler, 1984; Smuts, 1985; Strum, 1987; Whiten & Byrne, 1988; Cheney & Seyfarth, 1990; King, 1994; Zeller, 1994, 1996).

Given the complexity of their world, the psychological complexity of monkeys is to be expected. When I watch free-ranging monkeys using directional gaze to signal an interest in food or approaching danger, or blocking others' views of a yawn (which could be perceived as a threat), I get the strongest impression that they comprehend the effects of their own actions and act intentionally. Take the following observation:

> Three adult female Barbary macaques simultaneously threaten 3-year-old female o1. o1 moves to another young female (the daughter of one of her attackers) who stands in a present-posture–tail stance towards the scene. Upon reaching her, o1 teeth-chatters, hip-touches, and begins grooming her, and the episode is over.

The three adult females had, by their threats, insisted that o1 initiate a friendly approach to the other female. Examples like this abound in observations of monkeys. For example, Fedigan (1982) observed a female Japanese macaque drop and leave a new infant during a group movement. Several other females ran to the mother and threatened her until she went back and retrieved the infant and carried it. She did not abandon it again. Smuts (1985, p. 179) reported a situation in which a female olive baboon manipulated her withdrawal response after a copulation with a consort she had endured all day, so that a consort turnover could occur. "[W]e had the impression that Lysistrata purposefully timed her run-away response in order to increase the chances of a consort turnover."

The elaborateness of monkey communication suggests that monkeys can re-enact their own behavior for particular purposes. When young Barbary macaques are being inspected by adult males they are often held by the hind legs in an inverted position so the adult can see and sniff their genitals easily. When the infants become ambulatory, as they run by an adult, they may extend their closest hind leg straight behind them, in what is called a "leg out" position. The adult may or may not take advantage of this to catch and hold the infant for an inspection. As the infant matures to a juvenile this "leg out" pose becomes transformed to a play invitation, such that when a juvenile runs by another it may perform a "leg out" which often induces play chasing, or a wrestling bout if the second juvenile is quick enough to catch the first. Behavior transformations from function to intention movements to communicative indicators are quite common in monkey communication systems (e.g., ground slaps, head bobs, eyebrow raises, etc., none of which are intrinsically dangerous but all indicating threat).

In this chapter I examine the evidence for monkeys' "reproducing" their own behaviour and that of others. Specifically, I examine monkey behavior for evidence of pretending and its foundations – specifically, self-simulation and imitative matching between their own and others' actions. Some of this evidence is based on my own observations of monkey behavior, some on published reports of behavior. My own observations relevant to pretending derive from almost 30 years of field experience studying communication in free-ranging Barbary macaques (*Macaca sylvanus*) in Gibraltar, long-tailed macaques (*M. fascicularis*) in Borneo and in 10–acre enclosures at Monkey Jungle in Florida (Zeller, 1994, 1996), as well as a short time observing Japanese macaques (*M. fuscata*) at Arashiyama West in Texas. I have also videotaped many primate species in natural habitats in Africa and Central America. Although the examples of pretense are few, they grabbed my attention, and some I even captured on photographs or videotape. Examples relevant to pretending might have been observed more frequently if pretending (rather than communication) had been my focus; such an investigative focus might be fruitful.

Self-pretense and self-simulation

In monkeys, communications are not only behavioral manifestations of an underlying state, but also serve as social modulations of an interior world for external purposes (Zeller, 1998). Such modulations show up in situations of role reversal and role learning. Adult animals may use markers of a younger age in their communications which may reduce manifestations of aggression by others towards them, as when a dominant animal acts in a subordinate fashion, or actually changes rank. The animal's whole activity pattern and demeanor can change, not in an effort to deceive, but in light of a changed situation. When formerly high-ranking males become subordinate in their own group, they act and communicate in quite a different manner than they did before. They know, or can learn, how to change their interaction style to a different pattern. The animals have not changed or forgotten how to act dominant. When a high-ranking baboon or macaque male moves to a new troop, his initial behavior and approach will be quite different than when in the troop he left behind, at least for a short while. He knows how to ingratiate himself into the group by altering the nature of the messages he sends. This behavior may rapidly change if the individual's rank in a new group develops to match the rank he left behind. A transformation from subordinate to dominant animal can be striking – monkeys who only days before acted subordinate now move in and break up fights. Have they suddenly gained strength and fighting skill? More likely they are simply convincing others by their demeanor and actions that they are now invincible. They act capable of overcoming the participants whose fight they are breaking up, whether or not they really are. The others will back down, even if a few days before they might not have done so, because the controller presents the appearance of a high-ranking animal. Such masterful transformations suggest remarkable flexibility in producing images consonant with various roles or contexts.

Gestures are recognized by monkeys as the means by which meanings are communicated (Quiatt, 1984). Why else would Barbary macaques refuse to look at others attempting to direct gestures at them (pers. obs.; also Fedigan, 1982), and try to hide yawns that occur when they are just waking up and have been observed? The meaning expressed in gestures is understood widely enough in primate groups for successful deception (de Waal, 1982; Whiten & Byrne, 1988), as when a monkey's friendly appearing approach changes to aggression when nearby the victim, or when repeated unprovoked fear screams by youngsters get adjacent adults into trouble from high-ranking protectors (as I have seen in Barbary macaques). Such deceptions are reproductions of the monkey's own actions for its own purposes – a form of self-simulation or self-pretense (Jolly, 1988; Mitchell, 1994a).

Macaque play is full of follow the leader, where young in single file run out to the ends of branches and swing or drop to another branch or the ground below. This can occur in ways suggestive of copying or simulation as when I saw several yearling Japanese macaques cross between two trees clinging upside down under a rope stretched between them. Then at the same point, one by one, one after another, each let go with its hands, swung by its feet and finally dropped to the ground. This is similar to the locomotor play I have seen in blue monkeys (*Cercopithecus mitis*) who were playing by trying to walk on a thin wire, which was part of a dilapidated fence. They tried to maintain their balance, but invariably ended by swinging upside down. First the juveniles played this game, and a short time later, subadults and adults repeated it. (This is a videotaped episode from Nairobi, Kenya.)

Some monkey games suggest self-pretense. For example, while being forced off by a play partner, monkeys hang-on tightly to the end of a drooping branch (as if it were a dangerous drop, when it is not). They vigorously and repeatedly play "King of the castle," in which they defend a useless turf, and "I'm going to bite you," in which they feign biting and attempts to bite (Parker & Milbrath, 1994), the latter a common mammalian pattern (Bateson, 1955/1972, 1956; Aldis, 1975). Although one might suggest that monkeys don't have the imagination for such interpretations to be reasonable, some imagination seems present when monkeys locomote while their eyes are covered (Gautier-Hion, 1971a; de Waal, 1986b; Burton, 1992; see Russon, Vasey & Gauthier, *PIAC17*). Also deceptive feigning of other sorts is common in play, such as when a monkey looks away, but then grabs at another or catches it by the tail. Similar to this feigning of disinterest in play, foraging monkeys may move their hands around locating fallen fruit by feel, but not looking down. This reduces visual cues which would lead others to investigate the area. Deceptive feigning of disinterest is also reported by Smuts who observed it between males and females who are trying to unobtrusively reduce the distance between each other. They may sit, self groom, look down, quickly look at the other and slide a little closer, but carefully not catch the other's eye. When one finally does look up while the other is looking, exaggerated "come hither" looks and approach grunts ensue. Days of patient manoeuvering may pay off in an approach being accepted by the other individual (Smuts, 1985, p. 5).

A more complex example of self-pretending I observed was a situation in which two M. *fascicularis* females were sitting in nervous proximity. One hit an infant between them, but was already grimacing, anticipating the threat of the other. Once the interaction had begun it very quickly became a grooming episode which allowed the two females to continue in relaxed proximity (Zeller, 1992). From video replay and slow motion analysis of interactive gaze direction and gesture, it seemed quite clear that the attack on the infant was pretense, a social ploy, to begin a grooming session. The infant did not respond as though it had been attacked and its mother, who was in close proximity hardly looked up, which strongly suggests that the "aggression" was for appearances only, even though the female's hand actually contacted the infant hard enough to reorient its body. This episode was not deception; there was no evidence that any individual misunderstood what was going on. The action provided an appearance of reality although the participants knew it was not reality. This became clear to me when the videotaped interaction was replayed in slow motion (Zeller, 1992). The first female who initiated the interaction had not previously been able to catch the gaze of the other who was feeding and avoiding looking up. When the infant vocalized on being hit, the second female threatened the first and visual contact was established. There was a series of alternating gazes and a slow approach by the first female to the side of the second. After repeated teeth chattering, the second female presented for grooming.

Monkeys can also act playfully with an aggressive intent. Breuggeman (1973, 1978) suggests that some mothers play with their infants to distract them from seeking the nipple which may serve as a mild reprimand. This may be the case, particularly when a yearling tries to suckle if the mother loses her next infant, or if the juvenile tries to sneak a drink. Breuggeman (1978, p. 185) refers to play in the context of what appears to be parental behavior as a "veiled threat." While this may be true, it also may be the case that distracting the juvenile without being aggressive may be easier on the mother especially if she is carrying a young infant. The mother is still simulating friendly social behaviour because the aim of the play is not play, but distraction of the youngster from the nipple. Breuggeman (1978) discusses several potential functions for play, especially when older animals are involved. Play can be fairly aggressive, especially when two or three youngsters unite to chase or wrestle with one recipient, and play chasing and biting transforms to fighting and screams. However, even if such play results in aggression, the same youngsters seem willing to play again quite soon, either that day or the next, so the aggression is still framed in the context of play. This differs from more adult forms of aggression which frequently functions to teach a lesson of nonparticipation or evoke a reluctance to interact socially. Thus, play-based aggression can still be considered to have a more simulated quality than is the case with other agonistic situations. Adults have also been seen to play briefly after agonistic episodes which may allow them to interact in a more positive manner before the encounter is terminated, thus improving the sociability of subsequent encounters. Social play with a third party can also ensue after an aggressive approach by another, almost as a parallel to redirected aggression, but serving more calming and affiliative social ends (Breuggeman, 1978).

Monkeys sometimes act deceptively or camouflage their actions, and at times these actions suggest pretense (Whiten & Byrne, 1988; Byrne & Whiten, 1990, 1992). These can be "game" episodes or episodes in which they benefit from such misrepresentation. Using a vocalization to indicate danger and thus diverting an attack on oneself as every one scrambles to safety is one example (Cheney & Seyfarth, 1990). The call is the same as would be used in a real predator situation, but its function is different. The sender is simulating what his action would be in a different circumstance, but to gain the goal required at this moment. One example I observed involved an old, very peripheral female Barbary macaque Wm. At the end of one field season, she and I were sitting quietly in the bushes within a foot of each other and out of sight of the rest of the group. We were sharing a small pile of biscuits which were sitting on the ground between us. First she had one, then I did, then she did, in a very relaxed atmosphere. Suddenly she grabbed all the biscuits remaining and stuffed them in her mouth, chewing and swallowing in a hurry. Within a few seconds the head male burst through the bushes to see what was going on. However, all he saw was the two of us sitting in the sun. Wm had gotten rid of the evidence of the illicit treat we had been sharing. She was deceiving him by pretending that nothing had been happening, in spite of the fact that he could probably have smelled the cookies. As it was, the evidence was gone so he did not attack either of us. (On a previous occasion he had ambushed me from a roof and stolen the food I had been saving for Wm, who as an old peripheral female did not get a large share of the provisioned food. On another occasion, he had grabbed my belt and tried to snatch food from my hands.) It was very clear to me that she got the food out of sight this time in order to keep us out of trouble with the greedy leader. She manipulated the appearance of the situation to make it seem different from what it actually was, but her pretense was in acting as if nothing was going on. Wm did not respond to the head male's close rapid approach as she normatively would, with a hasty retreat. She sat still and looked at him, as he looked at both of us. This was unusual behavior for her in such a situation of close approach by him.

Pretense behavior at several levels has been recorded in Byrne & Whiten's (1990) data base compiled from records contributed by many primatologists. In particular, directing threats to nonexistent objects can be used by individuals to redirect attention away from themselves which frequently terminates an aggressive episode. This is recorded in a number of macaque species in the data base. In a particularly complex example in rhesus (*M. mulatta*), reported by Sade (1965, cited in Byrne & Whiten, 1990, p. 28), 1956–Male-1, who attacked a female, was being chased by the ranking troop male, Old Male-A.

> Suddenly, out from under another bush 1956–Male-1's parent, Old Female-1, came running on her hind legs carrying her five-day-old infant. She ran to stand by her son's side, and together they made violent threat gestures and vocalisations at a part of the area empty of all monkeys and observers, and away from the dominant male. Old Male-A stopped chasing to look at what they seemed to be threatening, then chased 1956-Male-1 again, who again threatened loudly away from Old Male-A. Old Male-A sat down three feet away peering again in the direction in which 1956-Male-1 was threatening.

In this case the Old Female reinforced 1956–Male-1's efforts to convince the ranking male that there was a nonexistent danger. The episode ended with mother and son walking away and grooming while Old Male-A was left "peering at nothing" (Byrne & Whiten, 1990, p. 28). Sade ends this record by commenting, "Parents often defend their offspring from attacks of monkeys more dominant than themselves by diverting the attention of the attacker in just the manner described" (Sade, 1965, cited in Byrne & Whiten, 1990, p. 28).

In terms of a different function the Byrne & Whiten compendium also records several cases of individuals hiding from others in the group, either in play, or to maintain possession of resources, such as food, or uninterrupted copulation opportunities. The hiding in play was reported in hybrid baboons who alternately hid from each other in a play session. While hiding they maintained a "characteristic facial and postural pattern: they stand crouched and motionless, and attentive (visually and auditorily) to the appearance of the other partner" (Colmenares, in Byrne & Whiten, 1990, p. 34). Several episodes were reported of female hamadryas moving slowly away from their unit male until they were partly concealed by rocks or shrubs, and were able to groom or even copulate with other males hiding there (Byrne & Whiten, 1990). Sometimes the females caught the observer's attention because they moved so slowly and cautiously, looking to see if the unit male was watching, and sometimes it was the young recipient male whose cautious behaviour, while moving into the concealed position, focused the observer's attention on the interaction.

Pretense and matching

I have observed many instances of one monkey's behavior matching that of another monkey, but the matching often seems incidental to the behavior. For example, both young and subadult blue monkeys I observed show the pattern of rushing up trees one after another and jumping out of them, or of trying to walk on a wire fence and dropping off at the same place halfway along, but the similarity between these actions may simply be the consequence of playing chase or following the purported "model" during play. Juvenile and even adult vervets and blue monkeys reproduce the highly stylized bouncing locomotion of infants, but this may be a reproduction of their own earlier behaviors – a self-simulation, rather than a matching to another.

Another more convincing example of matching I observed involved a spectacular jump onto a log emerging from water and branch shake frequently performed by the highest ranking male *M. fascicularis* in a 10-acre enclosure at Monkey Jungle. Younger subordinate males reproduced his actions in his absence. At times the highest ranking male would chase a younger one off "his" branch when he came back, and shake it himself, allowing the observer a good opportunity to see the similarity between the two behaviours. A skeptic might regard this example as social facilitation, but the similarities in timing and location seemed unlikely to result by chance. Also, in this case the young animals were performing the action first, often a day after the last time the high-ranking male had done it. The high-ranked male most frequently performed the activity, so while it is true that his performance right after the young males may have been caused by social facilitation, that argument does not hold up for the younger males. Most of the time, with a whole 10–acre forest around them, the subadults did not branch shake unless an avian predator was sighted. As recounted above, upside down rope crossing by Japanese macaque juveniles is another example of matching another's behaviour by dangling upside-down from a rope after watching another macaque do the same.

In my extensive research on components utilized in facial gestures of macaques, the most striking discovery was the similarity between adult animals of the same kin lines. This was clearest in the Barbary macaque group where my knowledge of kin relationships was most secure. The background behind differential use of components was not genetic, since most of the animals used the various components at some time or other, but the pattern of the use and relative frequency varied from one matrikin line to the next, and the one non-kin individual (a migrant subadult male) had a different pattern than the others (Zeller, 1992, 1996). The process of learning facial gestures is fairly long-term and since the patterns of offspring and mothers are similar it seems reasonable to argue that the young learned from their mothers. The simulation of the mother's communicative pattern by all her young was quite evident in their choices of the components used in various gestures and relative frequency of these. Even after the mothers had died, adult female sisters continued to use the pattern learned from her and passed this pattern to their offspring. The similarity of pattern use between matrilinealy related animals allowed distinctions by kin line to be recognized. The capacity to copy what another animal is doing was clearly demonstrated.

The research literature provides a few instances of simulation by monkeys. One juvenile vervet replicated in part another vervet's use of a human arm as a swing, but was unable to reproduce a crucial part:

> Debbie ... has invented a swing game, employing humans ... First, she climbs up the human, and pushes one arm, so that the perceptive human will extend the arm. Next she sits, then slides backwards so that her knees grip the arm, and she crosses her arms across her chest, gripping her shoulders with opposite hands. The human is supposed

> to sway the arm, so that Debbie gets to swing. Donald tries. He gets as
> far as climbing the human, and is helped into prodding the arm into
> place. He gets into the right leg position, but he can't deal with both
> arms crossing so one crosses, the other hangs down. The swing-
> through is therefore imbalanced and Donald gets irritated, scratches
> and runs off.
>
> (BURTON, 1992, p. 41)

Over the course of 1.5 years, four captive rhesus macaques took on another's distinctive tic of touching its forehead with its hand (Rivers *et al.*, 1983, p. 8). Another Barbary macaque imitated gait and eye-covering of another:

> A sub-adult stands looking down the road. He looks down the road,
> then up, then places his right hand over his eyes, and jauntily ambles
> in a curious three-legged stride down the road. A juvenile has been
> watching from the vantage point of the wall. Once the sub-adult is
> gone, the juvenile takes his place at the top of the road; covers his eyes,
> and in that odd gait, follows the leader.
>
> (BURTON, 1992, p. 39)

One rhesus macaque matched her mother's positions while holding an infant, an example which has been interpreted as pretending because of the apparent use of a coconut shell to represent the infant (Breuggeman, 1973, p. 196, cited in Mitchell, 1987):

> A 2-year-old female, 7N, moved after her mother, [AC], while [AC]
> carried 7N's infant brother in the ventral position. 7N clasped a piece
> of coconut shell with one hand to her ventrum. When AC stopped, she
> lay on her side and rested one hand on her infant's back. Just a few feet
> away, 7N adopted the exact posture of AC, while still holding the
> coconut shell to her ventrum.

A similar situation occurred at Monkey Jungle, Florida, among the long-tailed macaques. An older female, Je, had a male infant early in the birth season. The infant was already showing the beginning of natal coat change (about 20 weeks) when another primiparous female Su was very near parturition. Je stayed with Su the two days before the birth. Right after (the neonate was still wet), Je got her now large infant, who was perfectly capable of independent locomotion, returned to Su's side and held him close to her ventrum, directing his head to the nipple. He struggled a bit but she held him firmly in her arms in the sucking position or carried him ventrally all that day and the next as she stayed beside Su and her new infant. Su held her infant close and did not leave Je's vicinity. It seems to me that Je was simulating correct infant care and carrying which Su matched. Je was simulating her own behavior with a young infant in acting as if her own was younger than he actually was. In a sense she was manipulating the appearance of a situation to provide a model for Su. You could not really call this teaching because she did not direct Su's attention other than by sitting in close proximity and she did not act to modify Su's behavior, but the timing and intensity as well as the unusual nature of this behavior were quite striking. This was particularly so because the group of *M. fascicularis* were in the middle of the birth season and Je had not shown the same type of behavior before with other mother/neonate pairs even when the other mothers were primiparous. She did not behave this way with any other female either that year, the year before, or the year after when Su had another infant. My interpretation of this behavior rested on the possibility that Su was Je's daughter, which was certainly possible in terms of age and relative rank.

Another example of simulation in object play behavior I have observed in *M. fascicularis* involves a pattern of behavior with stick and stone being transferred to a new object. Many times these macaques manipulate a stick, holding it between their teeth and the ground vertically. They then rub a stone up and down the stick, which makes a small noise. One day, a gate lock had fallen into the enclosure and one male was videotaped holding the stick in position and rubbing the lock up and down. When he left the lock, another young male immediately approached, picked it up, broke a stick to the right length and repeated the lock rubbing. When he finished, a female came immediately and did the same. This all occurred in half-an-hour, all recorded on video. The similarities in body posture and lock manipulation were noticeable. They also spent some time inserting sticks into holes on the padlock such as the keyhole and the hole for the lock shaft. These animals had seen keys used in these locks all their lives. In this episode, I am really only arguing for simulation in the repeated lock rubbing up and down the stick, but poking of small twigs into the keyhole might be considered to arise from observation of humans, although it might also just be exploratory activity.

In an example of vocal imitation I observed, the model was the peripheral Barbary macaque Wm, who functioned as a sentry in a relatively small group of 22 animals. Barbary macaques tend to be very quiet animals, but when Wm noticed an external threat to the group or aggressed toward a group member, she used a threat with an added inhale/exhale vocalization (she sounded like an asthma sufferer). This behavior functioned to direct attention to her on the periphery when she threatened. This behavior developed a year after I had begun observing this group. A year later, three juveniles in the group were also using this inhale/exhale vocalization as part of threats although, since they were central in the group, this was not necessary to direct attention to their facial gestures. Their mothers did not use this component in their threat gestures, and the youngsters had only begun to use it in their third year. After about six months they gradually stopped these vocalizations. This matching may have had some underlying functional component, but it seemed to be a nonfunctional simulation of Wm's behavior in a similar situation.

Although these examples suggest that a few individual monkeys may have the ability to simulate others' actions and sounds, evidence against a species-wide capacity for bodily imitation is the failure of a well-directed attempt to train motoric imitation in a long-tailed macaque (Mitchell & Anderson, 1993) and the general failure of monkeys to show salient imitations of distinctive behaviors and sounds (Fragaszy & Visalberghi, 1990; Whiten & Ham, 1992; Byrne & Russon, 1998).

Unequivocal examples of extended pretending are not easy to observe. As evidence, I looked for instances of imitation which suggested role play, since (as I noted above) monkeys are so talented at enacting diverse role-related behaviors. Youngsters may watch others sitting in trees performing the sentry role and perform this behavior, including making the warning barks. Play mothering develops into alloparenting where youngsters (of both sexes in Barbary macaques) learn to hold, carry, groom, and play with infants. But whether any modeling of another's actions is required to produce similar actions is unclear.

In one case I observed, modeling others' actions seemed evident. In a free-ranging group of Barbary macaques, a family of youngsters aged 3, 2, and 1 year, had been orphaned. They continued as a group, all sleeping together, grooming and foraging, with the 3-year-old female o1, dorsal-carrying her 1-year-old sister when necessary. The 3-year-old remained small that year and the next year her age-mates became fertile and had infants, but she did not. The others were ventral-carrying their infants and receiving a lot of attention from the head male, common in Barbary macaques (who show a lot of male care). o1 began to carry her now 2-year-old younger sister ventrally as if she were an infant, cuddled her, groomed her, held her to the ventrum and lip-smacked to her in the pattern of the other young females with their infants. This type of care was noticeably different from the helping behavior of the previous year and much less age appropriate for the "infant." It very much seemed to me that o1 was role playing as a mother, pretending that she had an infant. This behavior went on for a number of weeks. (o1 eventually did have a baby the next year and cared for it very well.)

Conclusion

Overall, the foundations for pretending in monkeys show up in the sensitive use of role-related behavior and self-simulation (particularly in deception). There is some evidence for simulation of others, self-pretense, and pretense occurring in play, aggression, resource acquisition, redirection of attention and concealing of resources. In some cases this involves matching the bodily behavior of others, while in other cases it reveals an understanding of how an individual's own appearance and activity will be perceived by others. Only a few instances suggestive of pretending based on others' actions have been found. Their occurrence may be influenced by a variety of factors: individual intelligence; context; social position; and possibly enculturation.

14

Pretending primates: play and simulation in the evolution of primate societies

The simulative modality

Play, imitation, and pretense can all be explained in terms of a theoretical construct called *the simulative modality* (Bruner, 1972; Reynolds, 1976; Mitchell, 1994a). All these phenomena imply a separation between the form of behavior, its normal motivation, and its typical consequences in the wider world. The concept of the simulative modality, first scientifically proposed in the work of Gregory Bateson (1955/1972), is now poised to become a major theoretical framework, encompassing data from field primatology, human evolution, and cognitive psychology.

The strong form of the theory presented here consists of four major theoretical propositions: (1) the simulative modality consists of neurohumoral states that evolve in conjunction with the cerebral cortex of the primate brain; (2) social interaction stimulates the simulative modality and induces a mental model of one's own body and the bodies of others called the cognitive body; (3) the major psychological differences among primate species are due to the types of subcortical systems that are integrated into the simulative modality and by differentiation of a species-specific cognitive body; and (4) each type of cognitive body is expressed through a characteristic form of social organization.

Although the dominant theories of primate evolution treat symbolic processing as a laboratory curiosity, the theory of the simulative modality indicates that it is a normal and essential aspect of simian social organization, required for the development of status hierarchies, enduring mother–child dyads, and normal sexual relationships among adults. In this chapter I argue that symbolic communication, especially play and pretense, are essential and wide-spread features of nonhuman primate society, setting the stage for further developments in the hominid lineage.

A sense of neurohumors

Ever since the nineteenth century, evolutionists have emphasized the continuity in the anatomy of the brains of apes and humans, pointing out that there are no structures in the latter that are not found in the former (Huxley, 1863). The similarity of human and simian brains is supportive of Darwinian evolution, but it is hard to reconcile with the great disparity of human behavior in such areas as language, technology, and art. As neuroanatomy has been supplemented by electrical and chemical techniques, science has accumulated a sophisticated body of knowledge on the biology of brain states, hormonal modulation of brain development, the role of neurotransmitters in nerve conduction, and the modes of operation of the major tranquilizers and hallucinogenic drugs (Cooper, Roth, & Bloom, 1996; Stahl, 1998). Taken together, these findings show that the chemicals that bathe the brain are potent agents of neural function, and that even small differences in chemical sensitivity or the ratio of complementary transmitters can have big effects on behavior and state of mind. In many cases, the neurohumors can be increased or reduced by relatively small shifts in the amount of precursor compounds or of the enzymes that catalyze their synthesis, enabling small genetic or environmental changes to move a neural system toward a new state of equilibrium. This provides a physical model for understanding how even innate anatomical structures can function differently depending on the social and behavioral context.

In parallel with these developments in neuroscience, psychologists have shown that non-utilitarian states of consciousness involving the whole brain, such as play, dreaming, and "flow," contribute to creativity, learning, and a sense of well-being in humans (Tart, 1972; Csikszentmihalyi, 1991, 1997; Fromberg & Bergen, 1998)

Simian society

Much of the evidence for the simulative modality comes from field primatology – the study of the way of life of other species of primates based on observational studies of the animals in their native habitat. Field primatology demonstrates that there is no close correspondence between the anatomy of a primate species and the type of society it develops (DeVore, 1965; Smuts *et al.*, 1987; Dolhinow & Fuentes, 1999). Among the apes, the

Pongidae, gorillas live in groups consisting of several mature females with one breeding male (Schaller, 1963). The common chimpanzee has a central hierarchy of adult males that overlaps a number of family units each consisting of a mother with her dependent offspring (Goodall, 1971). Bonobo chimpanzees form close attachments among females while male hierarchies are poorly defined (Badrian & Badrian, 1984; de Waal & Lanting, 1997). Gibbons live in life-long mated pairs that aggressively defend a territory against other pairs (Carpenter, 1940), and orangutans are usually solitary (Rodman & Cant, 1984). It is also known that these variations are not explained in any direct way by the species' habitat or ecological adaptation. Some arboreal species, such as the squirrel monkey (Rosenblum & Cooper, 1968), live in large troops containing adults of both sexes, a type of society more common among such terrestrial species as macaques and anubis baboons (Altmann & Altmann, 1973). On the other hand, a close relative of the anubis baboon, the hamadryas, lives in aggregations of nuclear family groups, each led by a single adult male (Kummer, 1968).

Underlying the diversity of social organization are distinct dimensions of social life that are universal among primate species, even though the forms they take are as variable as the societies themselves. One ubiquitous feature is the intense, mutual social interaction between mother and child, the *maternal dimension*. Co-existing with this is an even more complex interrelationship among attention-getting, approach-avoidance behavior, aggressive display, and social centrality – often called *agonistic* behavior, after the Greek word for "contest."

The maternal dimension

The mother–infant dyad in primates is a relationship involving both mutual attention and asymmetrical care-giving (Altmann, 1998; Bowlby, 2000). Infant primates are totally dependent on their mothers for transportation and food. Once immature primates are old enough to move about by themselves, they play in close proximity to their mother and follow her when she moves. A mother also attends to her baby, often picking it up and clasping the infant to her chest when the troop is about to march. The flow of care from mother to child coexists with a mutuality of action. One of the most striking aspects of the maternal relationship, and the one which differentiates it most from the agonistic, is eye contact and relaxed face-to-face interaction. In agonistic relationships, asymmetrical eye and head movements signal the relative ranks of the participants. Subordinates look away when dominants stare, and direct eye contact is avoided. In the mother–infant dyad, in contrast, both mother and child can look at one another. As the child matures, mutual gaze is replaced in many species by mutual grooming, in which each animal combs the fur of the other, either simultaneously or by taking turns.

The agonistic dimension

The maternal dimension of primate behavior co-exists with the agonistic dimension. In rhesus monkeys, for example, a species I observed for many years, the so-called dominant animals are always more interesting to other monkeys than peripheral members of the troop. Troop members keep an eye on the high-status individuals, moving when they move, and they compete with one another to groom the dominant males and hold the babies of dominant females. Although only a minority of primate species form social hierarchies, in societies with more than one adult male there is usually something akin to what has been called *attention structure* (Chance & Jolly, 1970) – the tendency for members of a group to attend to one individual more than another, giving rise to a social center and a social periphery. The attention structure is conveyed both to human observers and the simians themselves through behavioral cues: by stares and glances; by avoiding eye contact; and through attention-getting behaviors, such as postures, movements, and vocalizations. Male rhesus monkeys vociferously shake branches, gorillas thump their chests, and gibbons sing in a dawn chorus (Altmann, 1967).

The two basic dimensions of primate social life, the maternal and the agonistic, are treated very differently in discussions of primate society and social theory. People readily empathize with the mother–child dyad of nonhuman primates, such as the intimate family gatherings captured in van Lawick's footage of chimpanzees; but agonistic social centrality is a less intuitive and sympathetic concept, not because it is unfamiliar, but because in Western society it impinges on concepts heavily invested with a negative ideology. Most theoretical discussions of social rank are derived from the intellectual tradition known as social theory (Béteille, 1994), which was developed in the eighteenth and nineteenth centuries by people interested in either justifying the imperial class system of Europe or in violently overthrowing it; and both approaches begin with the premise that violence and property are the essential factors in the historical development of rank-related behavior and social hierarchy.

But among nonhuman primates rank-related behavior has only a tangential relation to the distribution of resources and the control of wealth – if only because there is no wealth. Unlike humans, monkeys and apes have no cooperative production, no trade, no capital accumulation – each animal forages for itself. The only exception to this generalization is the limited sharing of certain rare but highly prized foods among chimpanzees (Teleki, 1973; Kuroda, 1984). Consequently, when a nonhuman primate successfully challenges a higher-ranking individual, the usual result is not a re-allocation of property but an increase in social attention and deferential behavior paid to the victor by other members of the troop. The extent to which an individual is the center of group attention, not just in agonistic encounters but in positive contexts as well, is a measure of the animal's *social centrality*. In many species, high-ranking individuals receive a disproportionate share of the grooming, getting more grooming than they give in return. Thus, nonhuman primate social hierarchies are consistent with the allocation of *social* and *psychological* rewards, but this qualification is cold comfort to Marxist, capitalist, and physicalist theories of society, all of which assert the primacy of the material over social and symbolic factors. From a primatological point of view, social rank is not a way of allocating bananas and genes but of expressing social centrality.

In nonhuman species with the most developed social hierarchies, such as macaques, violence is not the normal means of expressing and establishing social rank (Rawlins & Kessler, 1986). Provided the animals have not been raised under socially deprived conditions or trained by people to be fighters, violence is most likely to occur as a consequence of a deliberate challenge to a higher-ranking animal by a subordinate individual or by a newcomer to the group. The dominant animal reacts aggressively when the challenger does not express the appropriate gestures of appeasement and avoidance. But violence is not the sole nor even the most significant determinant of social rank; and it is not co-extensive with social centrality.

As with humans, aggression among nonhuman primates has a strong political component, in that successful challengers have allies that back them up in encounters with higher-ranking individuals. More recent observational studies have underscored this political dimension. For example, research on chimpanzees indicates that the devious pursuit of self-interest, what primatologists have come to call Machiavellian intelligence, is as important as threat and retreat in the everyday competitive behavior of this species (de Waal, 1982; Whiten & Byrne, 1997). Even in such overtly aggressive primate species as macaques and anubis baboons, where bites and chases are common, and linear hierarchies organize much of social life, social status is as much inherited from one's mother through shared experience as it is achieved through agonistic encounters (Lindburg, 1980; Smuts, 1985).

But the strongest argument against classical social theory, whether of the left or right, is the fact that rank-related behavior in primates is usually the *opposite* of fear and aggression. The normal response of a frightened animal is to run away from the threat, while the goal of aggression among animals, as in the defense of young, of mates, and of oneself, is repelling the intruder. Yet high-ranking primates, whether male or female, are more *attractive* to others, and serve as a focus of group attention. Although the submissive behavior of subordinates, such as grimacing and looking away, is often interpreted as "fearful," this analysis begs the question of why "fear-inducing" individuals should be so attractive in the first place. The concept of the simulative modality makes a significant contribution to our understanding of this seeming anomaly.

Symbolic indicators of rank

Social hierarchy in nonhuman primates is a *transformation of the behavior patterns of fear and aggression by the simulative modality*. In real fear, the subordinate runs away; in real aggression, the dominant attacks. But the social primates have evolved a system in which the behaviors normally linked to fear and aggression function symbolically. As Bateson indicates in his concept of metacommunication in animals, the "fear grimace" of the subordinate is not "expressing fear" but intentionally conveying the *negation* of threat. "I'm no threat to you," the subordinate is saying as it snuggles closer to the dominant animal, "see how scared I am" (Bateson, 1955/1972). In this theoretical interpretation, the fear grimace is a kind of pretense, a simulation of an innate behavior pattern.

The simulative modality, by definition, is the separation of behavior patterns from their usual motivational and consummatory correlates, allowing social learning to be interposed between the instigation to the behavior and its actual expression. In nonhuman primates, it is usually the behavior patterns of fear and aggression that are simulated because it is these motivational systems that are potentially the most disruptive of social organization. In rough-and-tumble play, young animals grapple with each other while pretending to bite. Their mouths are open, their canines exposed, and their muzzles in intimate proximity to each other's necks – but the biting does not take place. In approach-avoidance play, they take turns chasing and fleeing, but there is no final victory to either party. Certainly, both types of play are highly developed in primate species that grow up to form social hierarchies, but playful animals are not "practicing aggression," as utilitarians have hypothesized (e.g., Symons, 1978). They are developing neural structures by means of which the innate behavior patterns of fear and aggression can be *re-deployed in a symbolic manner*.

Significantly, the absence of social rank in rank-oriented primates does not produce social equality but pathological fear and aggression. In the 1960s the psychologist Harry Harlow separated infant rhesus monkeys from their mothers at birth and raised them in social isolation on various inanimate devices that simulated aspects of maternal behavior (Harlow, 1971). Not surprisingly, such monkeys grew up devoid of the social skills needed for interacting normally with other monkeys, but one of the most interesting results, unpredicted by any theory of development or dominance then prominent, was the effect on fear and aggression. If paired with a normal animal, the socially isolated monkeys curl up in the corner in abject terror, and they are prone to self-inflicted biting and mutilation. In normal monkeys, the high-ranking individual hardly ever bites – he or she stares, exposes a canine, approaches in a deliberate manner, maybe even lunges. Similarly, the low-ranking animal almost never curls into a ball screaming. It pretends not to notice, looks away, moves aside, crouches down, runs a little faster. In Harlow's monkeys, the whole communicative middle range of normal agonistic behavior is missing.

This finding is consistent with the hypothetical role of the simulative modality in the development of rank-related behavior. The normal monkey develops in the course of rough-and-tumble and approach-avoidance play a finely tuned range of social responses that simulate many of the behavioral characteristics of fear and aggression (Symons, 1978). But because they lack the painful consequences of real fear and aggression, they actually function to increase affiliation and solidarity. In Harlow's monkeys, in contrast, there is no modulation, only full-throttle fear or aggression, which in turn facilitates further social isolation and withdrawal.

The behavior patterns of the primate brain, constrained as they are by the primate body, are very conservative, changing little from one species to another; and such basic patterns as running, throwing, screaming, and grimacing are easily recognizable to simians and ourselves. So similar are behavior patterns among closely related species that classical ethologists emphasized the innateness of behavior, regarding it as comparable to the anatomical traits long used by taxonomists to classify organisms. However, the concept of the neurohumoral basis of the simulative mode substantially alters this theoretical formulation. With neurohumoral modulation, the states of mind in which the behavior is embedded can alter dramatically, making the ethologist's concept of homologous behavior largely irrelevant to an understanding of intention, motivation, and social context. Once a grimace can be feigned and avoidance simulated, the activity is no longer "fearful behavior," even though its neuromuscular component may be identical to fear-induced behavior in primitive quadrupeds. For this reason, the social hierarchies of monkeys and apes, far from being evidence of the inevitability of fear and violence, are better explained as a symbolic dimension of social life, the agonistic, that *simulates* the social signals of fear and aggression to facilitate social solidarity and physical proximity.

The cognitive body

The play behavior of primates facilitates the development of a cortical representation of the body – the *cognitive body*. The cognitive body develops during social interaction and by exposure to special-sensory images that can be easily mapped to the shape of the physical body. When a rhesus monkey sees its reflection in a mirror, it does what it normally does with unfamiliar animals of the same species – it threatens it – and research indicates that it continues to see the mirror image as another monkey, even though it may learn to ignore it. Many great apes, however, soon learn that the reflection in the mirror is not another ape but is systematically related to their own body movements (de Veer & van den Bos, 1999). Although usually interpreted in terms of abstract cognitive and perceptual abilities, it is not an accident that self-recognition is found in the great apes, which are the nonhuman primates most dependent on learning from the behavior of others.

In imitative learning the observer must be able to map an image of the other's body, made known primarily through the visual channel, onto the appropriate parts, positions, and movements of one's own body, which are known usually through the senses of touch and proprioception. To do what you are doing, I need to make a body shape that is superimposable on the image of your body as it appears to me – what I term *intersimian mapping* (what others have called "kinesthetic-visual matching", see Mitchell, 1993a, 1994a). If the observer is face-to-face with the mentor, as would be expected in the mutual gaze of mother–child intimacy, *the actions of the mentor are necessarily a mirror-image reversal of the observer's own body image* – that is, the image of my left hand is superimposable over the image of your right hand and vice versa. Indeed, experiments suggest that humans perceive their own mirror images in terms of images of an external "other," giving rise to the paradox of mirror image reversal (Gregory, 1987). Thus, the great apes can solve problems with man-made mirrors because the normal, every-day social relations of imitative learning require it.

Imitative learning also requires a neural infrastructure for social intelligence that is much more developed in anthropoid apes than in monkeys. Observational studies indicate that apes attend to the reward value of actions by and to others, perhaps less by direct observation of physical results than by sensitivity to emotional communication by animals with whom they have a close social relationship. After all, the impetus for a chimpanzee to imitate ladder-building is the perception that its builder escaped from the zoo. Without the ability to translate the consequences of behavior for others into anticipated consequences to oneself, imitative behavior would not be practical. The well-documented Machiavellian intelligence of chimpanzees is consistent with this ability, and suggests that intersimian whole-body mapping and intersimian reward assessment emerge together to create a pongid style of life.

The theory of the simulative modality explains many of the differences between humans and apes in terms of changes in the content of the cognitive body. In humans, the cognitive body can form detailed representations of small group interactions, such that individuals can organize their own behavior in reference to anticipated actions by others. For example, in the course of several years of research using videotape to record the play activities of preschool children, I recorded many examples of pretend play in which every-day implements were transformed into imagined objects (Reynolds, 1982). On one occasion three boys digging in the sand pit with plastic spades used them as if they were rifles, raising them to their shoulders and pretending to hunt lions. The word *imitation* suggests a leader and a follower, a stimulus and a response; but frame-by-frame analysis of the video shows that all the spades begin their upward movement in the same frame; and they continue to rise together in synchrony until they come to rest on the shoulders of the boys. Initially I expected such collective actions to be triggered by a behavioral cue given by one of the participants, to which others then respond, but in many instances there is either no precipitating cue or it travels from the instigator to the followers in less than 1/24 of a second, the temporal resolution of the video. Extensive psychological research with infants and mothers shows similar mutuality (Nadel & Butterworth, 1999). I conclude that much of normal imitative play in children, provided they have played together before, appears to be simultaneous in its expression; and for this I coin the word *synchromimesis* (synch-ro-mi-ME-sis), which means "imitation at the same time."

Synchromimesis is not to be confused with the diurnal rhythm of activity, in which all members of a group perform similar activities in synchrony, such as story time and nap time. In synchromimesis the activity takes place within a face-to-face group; it involves simultaneous actions appropriate to a shared frame of reference; and the frame of reference may itself be imaginary – in pretend lion hunting, there are no lions present.

The phenomenon of synchromimesis in humans works in conjunction with another process that occurs in ape tool-use only episodically if at all, namely *role complementarity*. Although simians often perform technical skills that presume a good understanding of cause and effect, as when a chimpanzee leans a log against a wall to make a ladder, ape tool-use is an individualistic activity. In fact, *no* nonhuman primate society ever studied by primatologists has cooperative work groups involving physical objects, such that two or more individuals must work together in order to achieve the result (Reynolds, 1991, 1993a). That is to say, the most elementary and pervasive technical activities of the human species do not occur in nonhuman primate society at all. With the exception of a few experimental situations (Crawford, 1937; Savage-Rumbaugh, Shanker & Taylor, 1998), one simian does not hand a tool to another. One simian does not saw the planks while another nails them on. One simian does not lift one side of a heavy trunk while his partner lifts the other.

The fundamental building block of human technology is not tool use but the *cooperative task group* (Reynolds, 1993b). The latter presumes both synchromimesis (e.g., all the sailors rowing the boat in unison) and reciprocal complementary roles (e.g., the boatswain steering while the sailors row). The precursors of synchromimesis and reciprocal complementarity are found in the play and mutual grooming of nonhuman primates. Reciprocal complementary roles are a normal part of approach-avoidance play, when animals take turns chasing each other, as well as in reciprocal grooming, when two animals alternate between grooming and being groomed. The social precursor of synchromimesis is the simultaneous expression of similar behavior by multiple individuals who are interacting with one another. This occurs in simultaneous grooming, in the grapples of rough-and-tumble play, and in round-robin play chases of three or more simians, in which each animal chases the one in front around a fixed object such as a tree.

Thus the absence of technology and economy in nonhuman primates is not primarily a question of low intelligence and deficiencies in physical skill. Rather *simians lack the social relations of cooperative production*, specifically the ability to take on complementary roles in a technical activity and to perform the reciprocal action when required. As with the other aspects of human behavior, tool use becomes technology and grooming becomes economy when the object manipulation skills of simians are incorporated into a more differentiated cognitive body – one capable of pretend social situations that include neural representations of the roles of others as well as oneself, and which can be expressed through the role switching of social play.

A new approach to social theory

The simulation of the basic motivational systems of mammals – fear, aggression, sex, maternal care, and so on – within the simulative modality explains why there is so little correspondence between the anatomy of nonhuman primates and the societies they develop. The social relations do not develop from the ancestral substratum of mammalian behavior – from real fear, real sex, and real aggression – *but from the symbolic redeployment of innate behavior patterns within socially induced play states*. This development also explains the phylogenetic changes in the hormonal control of behavior. Whereas in animals like insects there is a fairly direct relationship between the presence of hormones and pheromones and the expression of contextually appropriate behavior, in playful animals neurohumoral states uncouple innate behaviors from the ancestral vertebrate systems of motivation and control. The neural maturation of primate behavior remains endocrine-dependent, but the contexts of behavioral expression are no longer defined by the circulating levels of hormones.

The detailed application of the theory of the simulative modality to primate behavior is shown in Table 14.1 as four grades or types of functionality: monkey C_1; pongid P_1; pongid P_2; and hominid H_1. A grade

Table 14.1. *The relation of the simulative modality to the cognitive body, social organization and brain development*

Grade	Cognitive body	Emergent social organization	Brain development
Monkey C1	Simulated fear and aggression Facial and voice recognition Voluntary control of facial expression Delayed onset of sexual maturation Hand integrated into oral grooming	Large troop size Status hierarchies Peer play groups Groups of mothers with children Grooming groups	Integration of sensorimotor cortex of the hand with other special sensory areas Integration of special sensory cortical areas with limbic areas
Pongid P1	More delayed onset of sexual maturation Inferring reward value of action by others Intersimian mapping of whole-body actions Configuring objects to the whole-body image	Life-long duration of mother–child dyad Machiavellian strategies of status competition Social learning of whole-body survival skills Polypods (object constructions supported by gravity)	Integration of visual, kinesthetic, and somesthetic cortex with occipito-parietal association areas
Pongid P2	Partial voluntary control of sexual expression Integration of foods into sexual expression	Social relations based on simulated sexual receptivity and expression Male contribution of food to consorting female	Corticalization of reward systems Subcortical motivation of cortically organized behavior
Hominid H1	Tonic lithogrip Lithogrip replaces the hand in grooming and the mouth in agonistic competition	Long-distance carrying of physical objects for agonistic and exchange purposes	Simulated limbic states control cortically organized hand/object schemas

expresses the incorporation of a mammalian survival function into the simulative modality, and this is made manifest by three simultaneous developments: (1) the expansion of neocortical tissue involved with the function; (2) the development of new cognitive capacities; and (3) a new form of social organization that presupposes these capacities. Although the grades are arranged in linear order, they are not to be construed as an evolutionary tree. Nor is the list exhaustive – there are three additional hominid grades not shown here.

Studies of the most highly social species of Old World monkeys, especially macaques and baboons, suggest a level of psychosocial function called C1 (for the Cercopithecinae). In the monkey of the C1 type, the motivational systems of fear and aggression are transformed by the simulative modality into symbolic indicators of rank that function to bring animals together instead of driving them apart. On a cognitive level this transformation is hypothesized to require facial recognition, voluntary control of agonistic posture and facial expression, and contextualization of the social signals of fear and aggression.

In the development of the pongid P1 grade, the ranking systems of monkeys remain a possibility, and actually occur among male chimpanzees; but the distinctive psychosocial characteristics of the anthropoid apes are their larger brains and increased integration of the visual, motor, and somesthetic systems into the simulative modality (Parker & Gibson, 1979; Gibson & Petersen, 1991; Mitchell, 1994a). This enables apes to learn whole-body skills through imitative learning and to map their bodies to mirror images. The apes have also increased the length of childhood, delaying puberty until the animal is 10 years old or more (see Roberts & Krause, PIAC19).

Furthermore, as indicated by research on the bonobo (Thompson-Handler, Malenky & Badrian, 1984; de Waal & Lanting, 1997), there is a second grade of behavioral development summarized in Table 14.1 as pongid P2. Apes of this type combine a longer period of sexual receptivity with voluntary control of sexual expression, leading to closer integration of adult males with adult females, expressed by increased grooming and the sharing of food between adults.

As theorists have pointed out, the P2 grade adumbrates important features of human social life. In the hominid lifestyle, shown in Table 14.1 as H1, there is: (1) an extension of nurturant behavior derived from the mother–child dyad into the relations of adult males with adult females; (2) the loss of periodic sexual receptivity by females, enabling voluntary control of sexual signals; and (3) the gathering of food and other objects by both males and females in order to give it to someone else. The effect of the postulated changes is a material economy in which objects move back and forth between the adult male and female cohorts, while adults develop cross-sex exchange relationships that are independent of estrous-related consort pairing.

In H1 hominids, the P2 grade is extended by the development of a cognitive structure called *tonic lithogrip* – the prolonged holding and manipulation of objects (Reynolds, 1982; Wilson, 1998). Far from being a trivial function, videotape studies of adult humans show that the hand assumes the appropriate shape *before* it makes contact with an object, indicating that the sensory properties of the perceived object are controlling the muscles of the hand. This is a cortical function requiring visual recognition of the shapes of external objects, kinesthetic memory of the associated hand shapes, and mapping of object schemas to the cognitive body.

In addition, the lithogrip combines with the social scenario of grooming to enable economic activity in a strict sense of the term. Although contemporary chimpanzees (Goodall, 1986) carry food and objects about – with their feet as well as their hands – they only rarely hand objects to others or willingly share what they have. Like many other primate species, however, chimpanzees commonly groom one another with their hands and reciprocate grooming they have received. H1 hominids build on this social context by supplementing hands-on grooming with the hand-off of desired objects to others.

Conclusion

Evolutionary changes in primate social organization presuppose the cognitive functions of pretense and simulation. The subcortical systems of mammals become represented on a cortical level, which in turn stimulates social play using neocortically modulated behavior patterns. When the scenarios of social play continue into adulthood and are transmitted by social learning, they become in effect the adult patterns of social interaction on which biological adaptation is based.

15

Representational capacities for pretense with scale models and photographs in chimpanzees (*Pan troglodytes*)

Examining symbolic, pretend play in a nonverbal individual, especially one who is not human, is inevitably fraught with difficulties in identification and interpretation. For example, a young wild chimpanzee maneuvered a small stick in a manner similar to the ant-fishing technique seen in older individuals, but there were no ants present. Goodall (1986) interpreted this action as possible evidence that the chimpanzee was imagining or pretending that there were ants present, although other interpretations are possible. Call & Tomasello (1996; Tomasello & Call, 1997), by contrast, suggest that such behavior may be more parsimoniously explained as simple manipulative play with sticks, and thus attribution of mental imagery and pretense is not necessary in the interpretation.

Still, there are accounts of captive members of all four great ape species producing pretend play. For example, they have been observed participating in pretend play with dolls, engaging in behaviors such as tickling and feeding (Hayes, 1951; Gardner & Gardner, 1978; Tomasello & Call, 1997). Controversy remains as to whether these behaviors truly constitute symbolic play. Arguments against the existence of symbolic object play in the great apes have been supported by a study conducted by Premack & Premack (1983) which suggested that young chimpanzees were unable to recognize the correspondence between a pretend room and its real-world referent (Call & Tomasello, 1996; Tomasello & Call, 1997). Chimpanzees watched as a small object was hidden within a model of a room. They were then given access to the full-size room, but were unable to find the real object hidden in the analogous location. Thus, the chimpanzees did not appear to recognize the similarity between the model and its referent. A series of studies described in this chapter represent a re-examination of chimpanzee scale model comprehension, using methodologies patterned closely after those of DeLoache (1987, 1991, 1995) used with young children.

Scale model comprehension by children

DeLoache (1987, 1991, 1995) found a striking difference between 2.5-year-old and 3-year-old children in the ability to understand the representational nature of a scale model. In her paradigm, children were tested to see if they recognized that a scale model of a room could be a source of information about the full-size room it represented. Children initially watched as an experimenter hid a miniature toy in the model while explaining to the child that a real, life-size toy was hidden in the "same place in the big room" (DeLoache, 1991, p. 740). The child was next allowed to enter the real room and search for the toy. As a way of measuring the child's memory for the original hiding site within the model, the child was then brought back to the model and asked to find the hidden miniature toy. Three-year-old children readily retrieved the large toy from the room after observing the hiding event in the model, and also readily remembered the original hiding site in the model. Young children (2.5 years of age), however, did not show the same level of success. Nevertheless, their performance on the memory test indicated that their failure in finding the full-size toy was not due to forgetting the original hiding site in the model, since they reliably found the miniature toy during the second retrieval test.

Therefore, younger children were unable to recognize that the model could serve as a source of information about the location of the real toy. Their inability was made even more interesting given that each child received an extensive orientation phase during which the correspondence between the sites in the model and room was explained in detail. DeLoache (1987, 1991, 1995) explained the weaker performance of younger children based on her dual representation hypothesis. That is, younger children exhibited an inability to form a "dual orientation" to the model. In order to solve the scale model task, the child had to represent the model as both a tangible, real object, and secondly, as a representation of the real room.

Support for DeLoache's dual orientation hypothesis came from trials done with the same children during which the model was replaced with photographs of the hiding sites (DeLoache, 1987). An experimenter pointed to the photograph of the site in the room where the toy was hidden, and the children were then asked to go to the room and search for the hidden toy. Both age groups were successful with this version of the task. By using photographs, the same children who had been unable to solve the scale model task could reliably find the toy in the real room. Thus, in situations in which the 2.5-year-olds had to form a dual orientation (scale model task), their performance was poor, whereas photographs, which did not require a dual orientation, were readily interpreted representationally. In contrast, 3-year-olds were prepared to recognize the model as both symbol and object, and could respond to the model as easily as photographs.

DeLoache and her colleagues have since found new evidence suggesting that the younger children were having difficulty comprehending the symbolic relation between the model and room (DeLoache, Miller & Rosengren, 1997). They examined children's performance when a symbolic understanding was not necessary to find successfully an object hidden in the room. In this version of the experiment, the children were shown a "shrinking/enlarging" machine which they were told could make a room small or large. The children watched a miniature toy being hidden in a scale model. After leaving the area to "let the machine work," they came back to find a larger room where the model had been. They were then asked to find the hidden toy. For these children, the model and room were one and the same, and thus the relationship between them was seen as literal, not symbolic. Therefore, solving the task would not require the child to form a dual orientation to the model. The 2.5-year-olds under these test conditions showed better performance than with the typical scale model task, thus supporting the dual orientation hypothesis.

Scale models and pretend play

DeLoache (1990) extended the dual orientation hypothesis to other aspects of pretend play. Object substitution (e.g., using a banana as a telephone; see Leslie, 1987) involves a type of dual orientation in that a child may realize the real use and nature of the object (food), but still treat it as a representation for another object. Given that 2.5-year-old children participate in object substitution, it seemed puzzling that this same age group could not succeed with the scale model task. DeLoache, however, explained that the model had high "referential specificity" in that it had a specific, real world referent, and children must use the information gained from it in a specific manner. Symbolic play, in contrast, usually involves an object that represents a generic object. For example, a block being used as a car is usually not referring to a specific car. Thus, the model task may involve a different level of pretense, as the children must map the model onto a highly specific referent.

Scale model comprehension by chimpanzees

Given the difficulty in definitively demonstrating symbolic play and scale model comprehension in chimpanzees, as well as the well-documented studies of similar skills in children by DeLoache and her colleagues, we re-examined chimpanzees' understanding of the representational nature of a scale model in a set of experiments patterned closely after DeLoache's innovative paradigms (Kuhlmeier, Boysen & Mukobi, 1999). The results of the three studies described below, contrary to the work of Premack & Premack (1983), suggest that chimpanzees are able to understand the symbolic nature of a scale model. Of the three studies we have completed, during the first, we created a scale model of an indoor room that was very familiar to two chimpanzee subjects. Both subjects were able to interact safely with one of the authors (S.B.) outside of their home cage. In the second experiment, a different scale model, which depicted the chimpanzees' outdoor enclosure, was used, allowing seven adult chimpanzees to be tested. The third study was designed to examine whether the animals were solving the task by mapping the correspondence between individual objects in the model and room, or whether they were recognizing the more complex spatial/relational relationship between the scale model and its referent.

Indoor scale model
Initial task

Two chimpanzees (*Pan troglodytes*), housed at the Ohio State University Chimpanzee Center, USA, participated in the first of the three studies (Kuhlmeier *et al.*, 1999). Sheba (adult female, 15 years old) and Bobby (young adult male, 10 years old) were tested in a carpeted playroom familiar to both. Throughout testing, the room contained four furnishings: a blue metal cabinet; an artificial tree; a large blue plastic tub; and a brightly colored fabric chair. The scale model, which was one-seventh the size of the room, was placed in the hallway directly outside of the room.

The model was perceptually very similar to the room and contained miniature versions of the furniture, carpet, and other permanent features of the room. The hidden items included an aluminum can of soda and a miniature version of the can.

Prior to any formal testing, each chimpanzee completed a three-step orientation phase, not unlike that described for the children in the DeLoache (1987) study. First, the model was placed in the center of the room. The experimenter then displayed each miniature item next to the analogous full-size item, and verbally and gesturally drew the chimpanzee's attention to them. Next, the model was moved to an adjacent hallway, directly outside of the room. As the chimpanzee watched, the experimenter hid the miniature can of soda behind the tree in the scale model. The experimenter and the chimpanzee then entered the full-size room, and the experimenter "hid" the real soda can behind the tree as the chimpanzee watched. Both then exited the room. At this point, the chimpanzee was allowed to re-enter the room and was encouraged to locate the "hidden" soda. A second orientation trial was then completed. However, this time the chimpanzees only witnessed the hiding of the miniature soda in the model, and were not allowed to watch as the experimenter hid the real soda can in the room. Both chimpanzees were immediately successful in locating the hidden can in the room.

The testing procedure also closely followed DeLoache's, and consisted of three phases:

(1) Hiding event: The chimpanzee watched as the experimenter placed a miniature soda can in one of four hiding places (each site was used twice) in the model (Figure 15.1). The experimenter then displayed the real soda and moved into the real room to hide it, out of the chimpanzee's view. The experimenter returned to the hall, where the chimp was waiting, after hiding the can.

(2) Retrieval 1: Upon her return, the experimenter reminded the chimpanzee (through pointing) where the miniature soda had been hidden, and then the chimpanzee was allowed access to the room to search for the analogous site (Figure 15.2). The experimenter remained in the hallway.

(3) Retrieval 2: To test for the chimpanzee's ability to remember the site in the model, the subject was led back to the model, and encouraged to indicate where the object had been hidden. During both Retrieval 1 and 2, only retrievals that occurred on the first search attempt were considered correct.

Representational capacities for pretense in chimpanzees 215

Figure 15.1. The Hiding Event. The chimpanzee watched as an experimenter hid a miniature soda can in the scale model.

Given both chimpanzees' successful performances in the last phase of the orientation, we were surprised to find that only Sheba was able to find the real soda at a level statistically above chance, (7 out of 8, 87.5%, correct responses)[1] (Figure 15.3). Sheba entered the room and moved directly to the hiding site during most trials. Her performance suggested that she was able to use the model as a source of information for the location of the hidden can in the real room. Bobby, however, appeared to approach the task quite differently. On all but one of the trials, he entered the room and immediately searched the blue tub. If unsuccessful, Bobby would routinely search the other sites in the room until finding the soda can.

At this point we attempted to simplify the task by only presenting the miniature item associated with the hiding site; the entire model was not present. Sheba again performed at a level significantly above chance (5 out of 8, 62.5%, correct responses). However, even in this simplified version, Bobby continued to show perseverative responses. It appeared that Bobby failed to use any knowledge gained from the model and the

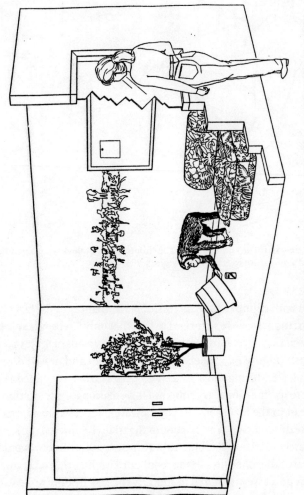

Figure 15.2. The Retrieval. The chimpanzee was allowed access to the full-size room to find the hidden soda can.

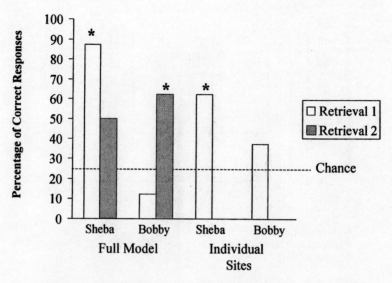

Figure 15.3. Percentage of correct responses during the two conditions of the indoor scale model task. *$P < 0.05$.

sites themselves, and instead, moved from location to location, using a fairly rigid search pattern.

Photograph task

We next examined whether Sheba and Bobby could use photographs to locate a desired object hidden in a room. Of particular interest was whether Bobby, who had failed previous tests with the scale model, might be able to perform better with a photographic representation of the room, and thus show evidence of the "picture superiority" effect reported by DeLoache (1987) for the younger children who performed poorly on the model task but were successful when photographs were used instead of the model.

The paradigm remained similar to the model task, with the same full-size room and hiding sites used. However, the model was now replaced with color photographs of the four hiding sites, mounted on the wall outside of the room. The Hiding event was slightly modified from the scale model task. The chimpanzee watched as the experimenter displayed the real soda can and pointed to the photograph of the correct hiding site. Retrieval 1 then occurred in the full-size room as in the previous scale model task. However, since no miniature can was hidden, there was no Retrieval 2 during the photograph task.

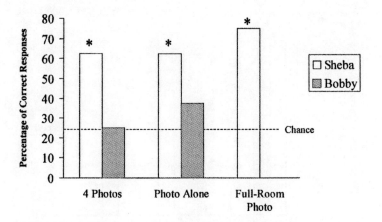

Figure 15.4. Percentage of correct responses during the three conditions of the photograph task. *$P<0.05$.

Again, Sheba was the only subject who found the can at a level above chance (5 out of 8, 62.5%, correct responses) (see Figure 15.4). Bobby responded in the same perseverative manner as he did with the model. We then modified the photograph task in two ways to see if Bobby's performance might improve. The first manipulation was an attempt to minimize any interference that may have been occurring because of the availability of all four photographs at once. Consequently, we conducted eight test trials with only one photograph of the hiding site present on each trial. During another series of trials, we replaced the photographs of the individual sites with a panoramic photograph of the entire room, which showed all the hiding sites. In both instances, however, Bobby's performance remained poor, while Sheba continued to be successful (individual photographs: 5 out of 8, 62.5%, correct responses; panoramic photograph: 6 out of 8, 75%, correct responses) (see Figure 15.4).

The indoor scale model and photograph tasks suggested that a chimpanzee could understand the relationship between a scale model and photographs and the larger space they represented. The data also suggested that Sheba was able to recognize not only that the room and model were related, but that the elements of one were analogous to the elements of the other. Thus, events that occurred in one represented events that should occur in the other. It is unclear, however, whether Bobby's poor performance was due to an inability to form the type of dual orientation that DeLoache (1987) proposed was necessary for success with the scale model task or his inability to inhibit a rigid motor pattern, or some combination of both. If Bobby's difficulty was simply an inability to represent the model as both object and symbol, as has been suggested for the younger subjects in DeLoache's studies, his performance should have improved when the photographs, which could not be readily perceived as objects, were used. Instead of showing a "picture superiority" effect, however, Bobby continued to show perseverative responding. It remained possible that Bobby was unable to form a dual orientation to the model (in which case, his success on the orientation trial would be attributed to chance) and simply guessed the hiding site, subsequently developing a rigid search pattern. Alternatively (though less parsimoniously), Bobby may have understood the representational nature of the model, but was unable to inhibit the expression of an inflexible search pattern, and consequently was unable to demonstrate his understanding of the model's relationship with the real room.

Outdoor scale model
Initial task

The intriguing results of the indoor model study led to some unanswered questions. Was Sheba's superior performance due to age, gender, both, or some other factor(s)? By creating a scale model of one of the outdoor enclosures, we were able to test all seven of the adult chimpanzees housed at the Center (Table 15.1) (Kuhlmeier *et al.*, 1999). All but two of these chimpanzees (Digger and Abby) had extensive previous experience on other cognitive tasks (e.g., Boysen & Berntson, 1989; Thompson, Oden & Boysen, 1997; Boysen, Mukobi & Berntson, 1999). The model was one-seventh the size of the enclosure, and similar to the indoor scale model, both the enclosure and model contained four hiding sites. These sites were represented by large outdoor toys familiar to all the chimpanzees and included a rubber tire, a large plastic barrel, a large plastic alligator-shaped teeter-totter, and a plastic sandbox. Plastic bottles filled with fruit juice and a miniature bottle were used as hiding items.

The testing procedure was similar to the previous study. The chimpanzees were tested individually and watched as the miniature bottle was hidden in the model. The animals were then allowed to search the enclosure to find the real juice. However, because of the structural features of the enclosure, it was not possible to require a second retrieval with the

Table 15.1. *Chimpanzee subjects participating in scale model tasks*

Chimpanzee	Age (years)	Sex
Abby	23	Female
Bobby	10	Male
Darrell	17	Male
Digger	8	Male
Kermit	17	Male
Sarah	38	Female
Sheba	16	Female

model. That is, the subjects did not return to the model a second time, in order to show that they remembered the location of the miniature bottle's original hiding site.

The data suggested that the chimpanzees were able to use the scale model as a source of information as to the location of the juice bottle in the enclosure. At the group level, the chimpanzees found the hidden juice bottle at a level above chance[2] (sign test, $P<0.05$). Three of the seven chimpanzees completed the task at levels significantly above chance (for all $P<0.05$; Sheba: 11 out of 20 (55%) correct responses; Sarah: 13 out of 20 (65%) correct responses; Darrell: 11 out of 20 (55%) correct responses) (see Figure 15.5). These chimpanzees, after witnessing the hiding event in the model, entered the enclosure, went directly to the correct site, and retrieved the hidden juice. Thus, the results replicated Sheba's performance with the indoor model, and demonstrated that other chimpanzees could also recognize the model as both an object and a representation.

Analyses of the choice patterns of the unsuccessful subjects yielded especially interesting results. All four animals who failed the task showed a significant preference to visit a particular site first (the tire in the left front of the enclosure) on each trial, regardless of whether that was the correct site. The successful chimps did not show this spatial bias. Second site choices were also evaluated. One subject, Abby, chose the correct site in her second search attempt significantly above chance (12 out of 14 (85.7%) correct responses, $P<0.05$). She would first search the tire, then move directly to the correct site and retrieve the juice. The other three unsuccessful subjects, after their initial incorrect choice, typically went to the adjacent site, and, if third choices were then made, they were sites that were in the same clockwise direction. Thus, these chimps appeared to be using the search strategy of checking each site, moving in a clockwise rotation around the room.

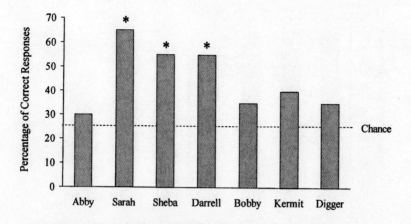

Figure 15.5. Percentage of correct responses during the first condition of the outdoor scale model task. *$P<0.05$.

Moving sites task
In an effort to minimize the stereotyped search patterns displayed by the unsuccessful animals, we modified the paradigm slightly. In this version of the task, the toys in the enclosure, and their miniature counterparts in the model, were moved into new positions between each trial. Both adult females were again successful, replicating their previous results, and Abby, the third female, now also performed at levels significantly above chance (for all females $P<0.05$: 5 out of 8 (62.5%) correct responses) (Figure 15.6). Abby's success was particularly interesting given her performance when the sites were kept in the same positions in the first task. Moving the sites appeared to disrupt her routine of visiting the front, left position before going to the correct site.

The performance of the other subjects, all males, was not facilitated by the revised task. In fact, Darrell's performance weakened to below chance levels (although both Darrell and Kermit chose the correct site on 50% of trials and approached statistical significance at $P=0.10$). Digger and Bobby again were unsuccessful, and had the lowest performance of all the animals. Again, we examined the unsuccessful chimpanzees' response patterns. Similar to their performance when the sites were fixed, the unsuccessful animals showed a significant preference for the front, left position, regardless of the site item. Also similar was their frequency of visiting the adjacent site as their second choice, and then continuing the search in a clockwise direction. Thus, except for one female (Abby),

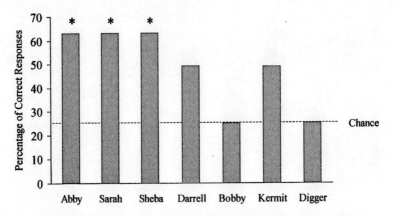

Figure 15.6. Percentage of correct responses during the second condition of the outdoor scale model task in which the hiding site positions were moved on each trial. *$P<0.05$.

moving the sites between each trial did not affect the strong search patterns used by the male chimpanzees.

Preliminary conclusions

Both the indoor and outdoor scale model studies have shown that chimpanzees have the representational capacity for some features of pretense with both scale models and photographs. They were able to approach the model as something other than an object in their environment and were able to utilize it as a source of information related to the location of a hidden item. These results replicate and extend the innovative studies of similar skills in children by DeLoache (1987; 1991) and her colleagues to a nonverbal, nonhuman species that had previously been thought incapable of comprehending the complex symbolic nature of a scale model (Premack & Premack, 1983).

However, it should be noted that scale model comprehension is not an easy task for all chimpanzees, as evidenced by the failure of some of our animals to complete successfully the task. The outdoor version of the scale model task allowed us to examine the error patterns demonstrated by the unsuccessful chimpanzees. The response patterns of the animals who failed to perform above chance were quite different from those of the 2.5-year-old children who were similarly unable to complete DeLoache's task. It has proven informative in previous work to examine the failure of chimpanzees to successfully learn a task, "especially if the determinants and mechanisms of that failure can be identified" (Boysen et al., 1996, p. 76), and again we were able to explore the constraints that some chimpanzees faced when asked to demonstrate an understanding of a complex representational concept.

The type of rigid search patterns shown by the chimps had not been reported for young children. DeLoache (1991) reported that unsuccessful children most frequently used a strategy of searching the site where the toy had been hidden on the previous trial. Our unsuccessful subjects, however, methodically searched the room, and, in the outdoor version, moved clockwise from site to site. The perseverative searching patterns survived many manipulations of the original task, including when sites were presented individually, or two types of photographs were used (Bobby, indoor scale model task), and when sites were moved between trials (all unsuccessful chimps, outdoor scale model task).

While Digger and Bobby's performance is difficult to interpret, they may simply be unable to understand the symbolic function of the model. The performance of Abby, Darrell, and Kermit lead to questions of what information these subjects gained as they observed the hiding event in the model. Abby's correct second choices when the sites were fixed, and subsequent success when sites were moved, suggested that she was able to understand the correspondence between the model and room. However, until the sites were moved for each trial, she was unable to disengage from repeatedly searching the same site. Darrell's performance when the sites were fixed suggested that he was also able to use the model as a source of information. However, when the sites were moved trial by trial, his responses deteriorated to an inefficient search method which did not permit him to respond optimally, and his performance decreased to just below chance. The other adult male, Kermit, performed slightly better than the adolescent males when the sites were fixed, but he remained at only 40% correct. When the sites were moved, however, Kermit's performance increased to 50% correct, a level equivalent to Darrell's. As with Darrell, Kermit may have been able to gain some information from the model, but his inefficient and intrusive search pattern overrode any significant understanding of the model's function.

Object and spatial/relational similarity

Given the overall success of the chimpanzee group in the scale model task, our third study examined what aspects of the model the chimpanzees were understanding and representing in order to generalize from the model to its real-world referent (Kuhlmeier & Boysen, 2002). Controversy exists in the human literature as to whether children solve the scale model task by simply recognizing the similarity between individual objects in the model and room (Blades & Spencer, 1994), or whether they recognize the more complex spatial/relational correspondence (DeLoache, 1995). To explore these issues, we tested the chimpanzees in two experiments to investigate the contribution of relational similarity and object correspondence on scale model comprehension. In the first experiment, success with the scale model task relied principally on recognition of the spatial, relational similarity between the model and the outdoor enclosure. The second experiment examined the influence of object and spatial similarity by systematically controlling position (spatial) cues and the color/shape of the sites (object cues).

Identical sites

Similar to the approach of Blades & Spencer (1994), we removed the individual object cues from the hiding sites by using four sites that were identical. However, different from Blades & Spencer (1994) who studied children using two unique and two identical hiding sites, in our chimpanzee version we used four identical red plastic tubs as our hiding sites. Success on the task required recognition of the relational similarity between the model and its corresponding full-size space, since other perceptual cues were controlled.

Procedures remained unchanged from the original scale model task (Kuhlmeier et al., 1999), and the same seven chimpanzees were tested.[3] At the individual level, three chimpanzees immediately searched the correct hiding site in the enclosure at levels approaching significance ($P = 0.11$; 50% correct for each). For the group overall, the animals' performance was above chance level (sign test, $P < 0.05$) (Figure 15.7). Thus, these results suggest that chimpanzees are able to recognize the relational similarity between a scale model and its full-size referent, although this similarity is not as readily recognized in the absence of other contributing cues such as color and shape. The task was difficult for the chimpanzees, since performance was not as robust as the original scale model study (Kuhlmeier et al., 1999) in which individual object cues as well as relational/positional cues were available.

Controlling position and color/shape cues

The next experiment was designed to evaluate the influence of spatial and object cues on scale model comprehension, and to examine if either

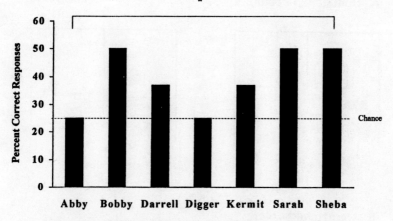

Figure 15.7. Percentage of correct responses during the identical sites condition. *$P<0.05$.

object or spatial cues took priority for the chimpanzees' success (Kuhlmeier & Boysen, 2002). Similar to the original scale model studies (Kuhlmeier et al., 1999), the hiding sites in the model and enclosure were unique in color and shape, but now the subjects were tested under three new conditions (Figure 15.8). In the unmatched positions condition, the position of the sites differed between the scale model and the outdoor enclosure on each trial. Consequently, individual color and shape cues of the objects were present, but spatial cues were discordant. In the unmatched colors condition, the colors of the individual sites were incongruent between the model and the enclosure, yet spatial cues and shape cues were consistent. And, finally, in the unmatched shapes condition, the shapes of the different sites were dissimilar between the model and the enclosure, and thus, only spatial and color cues were available.

The unmatched positions condition was the only condition during which the chimpanzees' performance as a group was above chance (sign test, $P<0.05$) (see Figure 15.9). Thus, with only color and shape cues present, the chimpanzees were able to use the model as a source of information about the location of the hidden item in the full-size enclosure. They did not appear to require that the positions of the sites be analogous between the model and enclosure. However, their performance was not as strong as during the original scale model task when spatial and object cues were both congruent between the two spaces. In both the unmatched colors and unmatched shapes conditions, when object cues were not

A. Unmatched Positions

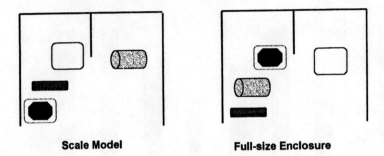

B. Unmatched Colors

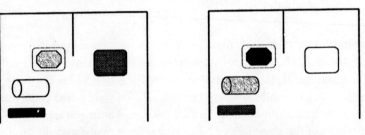

C. Unmatched Shape

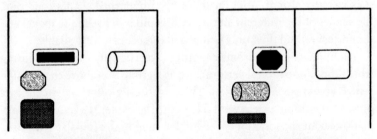

Figure 15.8. Examples of hiding site arrangements for the model and full-size enclosure for each condition of the Unmatched conditions. The dashed line represents a hypothetical hiding situation in the model and enclosure for each condition.

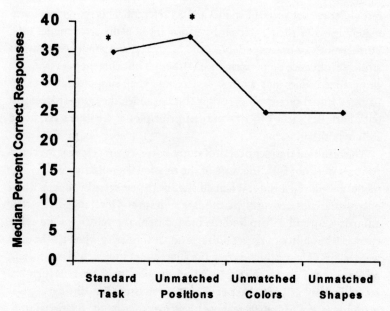

Figure 15.9. Percentage of correct responses during the Unmatched conditions. *$P<0.05$.

present, all the animals exhibited the previously observed perseverative, clockwise search behavior, as seen in the original study (Kuhlmeier et al., 1999). They did not use the position of the hiding site (and one of the object cues such as color or shape) as a cue for the location of the hidden juice.

Discussion

The data from the unmatched conditions suggests that chimpanzees received some limited information from the model despite the absence of spatial/relational cues. Therefore, they may have relied solely on the object correspondences between the model and the enclosure. When the object cues (color and shape) were not present, the chimpanzees were unable to solve the scale model task on the basis of spatial cues alone. These data suggest that object cues may be a more salient source of information than the relational cues for the chimpanzees.

DeLoache (1990; 1995) has developed a similar theory to explain children's understanding of scale models, and this theory can be directly related to Gentner's (1988) relational shift hypothesis. According to DeLoache, children initially recognize both the relational and object correspondences between the model and its referent. This recognition ultimately leads to their successful performance on the scale model task. Furthermore, with experience, the object correspondences have a stronger influence on performance. Although our current data from the chimpanzees does not address the factors that supported the chimpanzees' initial success on the original scale model, they are in agreement with both DeLoache's theory and the principles of Gentner's relational shift hypothesis.

The results of the identical sites study, however, suggest that the chimpanzees were not fully unaware of the relational similarity between the model and the enclosure. When all sites were perceptually identical and only spatial cues were present, the data at the group level suggest that the animals were still able to find the hidden item at levels that were above chance. The ability to respond to relational similarity when all the sites were identical, yet failure during the subsequent unmatched conditions, may have been due to the sparse visual complexity of the four identical red tubs. Similarly, Gentner *et al.* (1995) reported that during a cross-mapping task, children were more likely to recognize the relational similarity between stimuli when the stimuli were not visually complex. Under these conditions, the children did not respond solely on the basis of perceptual correspondence, and instead recognized the relational similarity. The more complex and perceptually rich objects provided "a more tempting competitor to the relational mapping than the sparse object match" (Gentner *et al.*, 1995, p. 278).

Although relational similarity may not be as easily recognizable for chimpanzees as children, these data reflect the capacity for such mapping. However, the salience of the object cues can apparently interfere with recognition of relational similarity, especially when objects are perceptually unique in color and shape. Nonetheless, the chimpanzees demonstrated the capacity for complex mapping, as well as the susceptibility to influences that can impede similar mapping in young children.

End notes

1. For all analyses at the level of the individual, statistical significance testing was performed with the binomial test, with an alpha level set at 0.05.
2. Statistical analyses at the group level were completed using a one sample sign test, with an alpha level set at 0.05.
3. However, due to one chimpanzee's (Sarah) reluctance to maneuver on the ground during this condition, the hiding sites were hung from the enclosure's mesh ceiling.

ANNE E. RUSSON

16

Pretending in free-ranging rehabilitant orangutans

This chapter examines four possible cases of pretending in rehabilitant orangutans. Only about a dozen other cases are known, so these cases are worth dissecting because they are so rare. Seven incidents have been reported for Chantek, an enculturated language-trained orangutan, spanning pretending to feed a toy animal at 1 year 8 months, deceptive self-pretense at 1 year 10 months (acting as if needing to urinate in order to stay in the bathroom) and pretend play at 4 years 10 months (clay man symbolic play) (Miles, Mitchell & Harper, 1996). Two incidents reported as deceptions may also qualify as pretending: a wild adult male lulled a competitor into dropping his vigilance by pretending to lose interest in combat; and an adolescent female rehabilitant fooled a human by pretending to lose interest in forbidden goods (Byrne & Whiten, 1990). Some imitative behavior in free-ranging rehabilitants resembles other-pretend, or pretending at behaviors observed in others, at representational levels (Blake, 2000; McCune & Agayoff, PIAC3). Examples are: nonfunctional tooth-brushing, and siphoning fuel from an empty fuel drum (Russon & Galdikas, 1993, 1995; Russon, 1996).

These new cases offer a chance to explore whether characteristics linked with human pretending, especially complex ones like metarepresentation, feature in great ape pretending. All four cases show complex social maneuvering, so they offer a rare window on complex social cognition in a species best known for its minimalist approach to sociality. Three cases also qualify as deception, so they offer a window on the relationship between pretending and deception in one of the great apes. Two cases involve non-enculturated orangutans, so they show the sophistication possible without intensive human tutelage.

The four cases were observed incidentally in the course of my systematic studies of other facets of behavior in free-ranging rehabilitant orangutans. Two cases in rehabilitants at Camp Leakey, Tanjung Puting National Park, Central Indonesian Borneo, were identified after about 1000 hours of systematic observation. Two cases in forest-living rehabilitants in Sungai Wain Protection Forest, East Indonesian Borneo, were identified after about 500 hours of observation. Each case is described then discussed with respect to whether pretending occurred, what cognitive processes seem to have been involved, and what role enculturation may have played.

Princess

Princess was an adult female (about 16–18 years old) who had been under rehabilitation at Camp Leakey since early infancy. I did not observe this incident directly but it was reported to me by several reliable witnesses, independently, in a highly consistent fashion. This, and the incident's consistency with other rehabilitant orangutan behavior, make it worth examining. Princess had been trained in sign language as a young infant at Camp Leakey (Shapiro & Galdikas, 1995) so she was highly human-oriented and could be considered enculturated. Her human orientation plus her gentle temperament made her a perennial visitor favorite.

Princess was socializing with human visitors on their bunkhouse porch one afternoon. After a while she left in typical orangutan style, climbing up the bunkhouse wall; she climbed the end wall to the peak of the roof. Atypically, she stayed up there for several minutes then returned to the porch instead of continuing on her way. Within minutes of her return, humans were commenting on how especially sweet she was. Even more atypically, she stayed on the porch with humans instead of going for provisions at 16:00 and she went to sleep at nightfall, around 18:30, on the doorstep instead of in a nest. Near 20:00, humans had to step over the sleeping Princess when they left the bunkhouse for dinner and locked the door.

Two hours later, when humans returned to the bunkhouse, they found Princess gone but also a door that would not unlock. They looked inside by flashlight and saw an inner bar lock slid closed and Princess on the floor. Princess, on request, unlocked the door. Inside, humans found that she had gone upstairs and rummaged through the food in their suitcases. While inspecting the damage they spotted torn screening that must have allowed her to enter. It was at the top of the end wall at the roof peak, right where she had paused when she had left the porch that afternoon.

Princess probably noticed the break-in spot in the afternoon, when she went to the roof peak, but knew she had little hope of entering with humans present because they regularly foiled such attempts. She began behaving atypically right after pausing at the break-in spot. She feigned indifference to this spot and even moved away from it. She became noticeably more friendly to humans at the bunkhouse, probably to create a false image of her reasons for staying; this likely increased her chances of monitoring the scene without detection. She waited until she knew humans would be inattentive to the break in – the nightly dinner hour. Other great apes have been known to find food then wait until no one is around to get it, sometimes for several hours and sometimes moving away from it (Byrne & Whiten, 1990).

Unyuk

Unyuk was another adult female rehabilitant at Camp Leakey (about 18–20 years old), one less human-oriented than Princess. She spent much of her time in the forest and could survive independently there. She also frequented the camp but was particularly feisty so few humans wanted her around. Once I followed her accompanied by a very experienced forest assistant, Ucing. While I videotaped Unyuk, Ucing cut a walking stick from a branch with his new Swiss Army knife, an unusual gift. He sawed off its ends with the knife's saw then scraped off its bark with a cutting blade. Unyuk watched intently then copied his sawing and scraping using one stick against another.

Probably because of Unyuk's interest, Ucing opened the knife's scissors and pretended to cut her hair, snipping the air near her forehead. She grimaced and winced but whenever he paused, she gestured to continue (e.g., squeaked at him, grabbed a fistful of her hair and held it towards him). Each time, he resumed snipping near the spot she indicated and she resumed watching intently. Suddenly, after 14 minutes, she appeared to lose interest. Later, from the video, it was clear that she abruptly stopped tracking the knife's movement – her face remained near the knife but her eyes were trained beyond it. Also, when she first looked away from the knife she did a classic double-take – she glanced beyond Ucing, glanced back at the knife, then, with a start, looked back again beyond Ucing (her head jerked forward slightly) and stared fixedly at whatever was there for about 30 seconds.

Beyond Ucing, where Unyuk's eyes were trained, were our backpacks.

She and other orangutans regularly stole backpacks for any food inside. Unyuk made no move towards our packs, however, and her face remained close to the knife as if she were still watching it. Unyuk glanced back at the knife only once, however, when it came close to her face. Then, abruptly, she sat up and resumed scissors play with Ucing. She let him hold her wrist and watched him snip near the hair on her forearm. When he paused snipping she shifted position as if for comfort (her shift moved her closer to Ucing but also to one side, so she was no longer directly opposite him). Then she offered her forearm again and Ucing resumed snipping. Unyuk remained very still and attentive until Ucing snipped near her head – then she winced and he paused. The first time he paused she invited him to resume by grabbing fistful of hair on her head, squeaking, and leaning towards him, so he resumed snipping. The next time he paused she shifted her position, again moving closer and sideways. Her second shift put her beside Ucing instead of facing him, so his body no longer blocked her access to the packs. Within seconds she lunged for the packs. We managed to guard the packs, but barely. Ucing had expected nothing, although he'd known Unyuk for about six years.

Unyuk enacted behavior inconsistent with her motives in order to distract Ucing, probably because he regularly foiled her direct attempts at theft. She feigned indifference to what she really wanted, the packs, as Princess and as other orangutans have done (Byrne & Whiten, 1990). Also, she became unusually "nice" to Ucing, as she and other rehabilitants did when bent on deception (A. Russon, pers. obs.), and she actively engaged Ucing in grooming-like interactions, presumably to distract his attention while she surreptitiously set up better conditions for nabbing the packs.

Paul and Bento

Paul (about 6 years old) and Bento (about 7 years old) were juvenile males in Sungai Wain Forest, already semi-independent (supplementary provisions were offered daily at a feeding site but human interaction was strictly curtailed). Both ranged around the feeding site so they met often. Bento, an inveterate moocher, regularly plagued Paul. He sometimes followed Paul for whole days on end and stole much of his food. So Paul began hiding from Bento within weeks of his own arrival in the forest in 1995. By 1996, Paul had refined his techniques for escaping Bento's harassment.

It was typical, then, that one day in 1996 Bento passed close by Paul as

Paul was eating termites on the ground, so Paul shifted furtively to a secluded spot and monitored Bento's movements surreptitiously. Bento made no move in Paul's direction, however, apparently more interested in obtaining the heart, or meristematic tissue, of a nearby forest coconut palm (*Borassodendron borneensis*, locally "bandang"). He climbed into the palm's crown about 3 m above ground, pulled out the newest leaf from the crown's center, bit off the heart matter adhering to new leaf's base, and ate it. Bandang hearts were among Paul's preferred foods but he had not yet learned how to obtain them, so he often scrounged others' leftovers. He rarely scrounged them from Bento, however, because of the potential for harassment. This time Bento shifted his position in the palm's crown so that his view of Paul, below, was blocked. Paul then approached Bento's palm, crouched low, walking very cautiously, and coming only close enough to reach leftover bits of Bento's new leaf that had fallen to the ground. Equally cautiously, he picked up the leftovers and crept back to his secluded spot, turned his back to Bento, and ate them – frequently glancing over his shoulder towards Bento. Bento did nothing; either he had not detected Paul's movements, or he did not care.

Bento finished eating and left the palm, so the remaining leftovers became available. Instead of staying hidden until Bento left, as usual, Paul ran directly in front of Bento to the feeding site (about 30 m away) then ate leftover papaya right in front of Bento's eyes. The result was predictable. Bento tried to steal Paul's papaya, staff fended Bento off so Paul could eat, and Bento finally gave up and left. Once alone, Paul poked around the papaya remains only briefly. Then he raced directly back to Bento's palm, climbed into its crown, and ate the leftovers lodged there.

Paul appeared to conceal his interest in the leftovers from Bento, to act as his own decoy to lead Bento away from food he wanted, and to manipulate the well-worn script of the others' typical behavior to his own advantage (Bento's mooching, human protection). Luring competitors or guards away from desired items has been reported in children 19–24 months old (Chevalier-Skolnikoff, 1986) and in Chantek at 30 months old (Miles *et al.*, 1996).

Panjul and Aming

Aming, an adolescent male about 9 years old, was relocated to a new site in Sungai Wain Forest. I observed his first meeting with the dominant male at the new site, Panjul, a subadult about 11 years old. When Panjul first sighted Aming, and six more times over the next three hours, Panjul approached Aming. Aming always fled nervously but Panjul gradually came closer. Panjul's seventh attempt brought him close enough to touch Aming's mouth. Although Panjul was careful and nonaggressive, Aming withdrew nervously.

Panjul approached within 3 m twice more but instead of touching Aming, he turned away and began eating leaves. Both times Aming stayed and watched. The second time Panjul tried coming closer, Aming grabbed a bit of liana in his mouth and stayed for about five minutes, then squeaked and withdrew up a tree. Panjul left, then Aming came to the ground and ate soil. Panjul soon returned, approached within 3 m of Aming, then again began eating termites. Aming, in full view of Panjul and watching Panjul closely, reclined on a log with his genital area exposed towards Panjul and pulled a leaf from a small nearby *Licuala* sp. palm over his head. Panjul continued eating termites as if indifferent to Aming's presence. Aming began eating pith from a mature petiole of the *Licuala* palm. This was peculiar. No Sungai Wain orangutans had ever been seen eating *Licuala* pith although they commonly eat pith from *Borassodendron borneensis* palms and Aming obtained *Licuala* pith using the technique for *B. borneensis* pith. Panjul watched Aming eat. Aming became increasingly agitated and finally jumped away from Panjul.

Five minutes later, Panjul approached within 2 m of Aming then ate leftover pith from Aming's *Licuala* palm, turning his back to Aming while eating. Aming watched very closely, then again reclined and exposed his genitals to Panjul. Panjul immediately approached, very slowly, but Aming jumped and climbed away. Panjul returned to eating leftover pith and reclined on his back, then moved away and ate rattan. Aming watched from behind, about 5 m away in a tree. After a few minutes he ate some fruit then started to climb down. At the noise, Panjul approached again. Aming climbed up again, then ate *B. borneensis* pith.

Panjul approached very slowly, squeaking repeatedly, then touched Aming. Aming withdrew, then reciprocated briefly, then retreated again. Panjul turned away and ate Aming's leftover *B. borneensis* pith. Aming watched, reached towards Panjul, but retreated within two minutes. Panjul returned to eating *B. borneensis* pith then approached again and this time inspected Aming, making rhythmic rumbling-bubbling vocalizations like those that males make while copulating. Aming squeaked throughout and finally climbed away. Panjul tried again and the two grappled briefly. Aming retreated again. After two more unsuccessful attempts at contact, Panjul left.

The key act seemed to be eating before a companion's eyes. Four similar orangutan encounters I have detected support this interpretation – a dominant male (4/4) repeatedly approached a subordinate companion (2/4–male, 1–female, 1–human) in a distinctly nonaggressive manner but the subordinate consistently withdrew. After several failed overtures the dominant ate before the subordinate's eyes. This eating visibly calmed subordinates, 3/4 of whom also then ate and allowed the dominant to make contact. The single reported deception in wild orangutans also suggests this component – an aggressor male ate termites when near and probably visible to his competitor, just before deceptively withdrawing from him (Byrne & Whiten, 1990). The literature suggests that such eating is ritualized or symbolic, an encoded signal intended to communicate a message rather than a functional act aimed at providing nourishment, and that the message concerns conflict management. That Panjul's and Aming's eating was ritualized is supported by their taking an item that no one in their community was known to eat at the time. Primates have used ritualized eating to serve notice of their interactive intentions in the near future, wild mountain gorillas and orangutans included; in some species it can indicate benign intentions and counter the avoidance induced by expectations of conflict (e.g., Schaller, 1963; Schürmann, 1982; Silk, 1996; Cords & Killen, 1998). In gorillas, conspecifics appreciate both the communicative intent and the message (Schaller, 1963). Panjul's and Aming's ritualized eating also suggests mutual feigned interest in a common object, a tactic observed in chimpanzee conflict resolution. It has been likened to a "collective lie" that helps break tension and bring adversaries back together, with one deceiving and the other acting as if deceived (de Waal, 1986a).

Discussion

Recent research indicates that pretending occurs in various forms and levels, some governed by simple cognitive mechanisms and others by complex ones (e.g., Harris & Kavanaugh, 1993; Lillard, 1993a; Mitchell, 1994a; McCune, 1995; McCune & Agayoff, PIAC3). McCune, for instance, has distinguished a series of forms that emerge in developmental sequence in children's play, grading from presymbolic to symbolic in the cognitive levels involved. There is, then, no simple answer to the question of whether these orangutan cases represent pretending – they may qualify by some definitions but not others. I therefore assessed the four orangutan cases from several perspectives.

Simple pretending

Simple pretending is defined as deliberately behaving as if something were real when it is not, behaving as if doing one thing while really doing something else, or projecting a supposed situation onto an actual one (e.g., Lillard, 1993a; Mitchell, 1994a). All four orangutan cases qualify as simple pretending. All the orangutan actors behaved as if honestly involved in one activity (Princess visiting humans, Unyuk studying knife use, Paul foraging for papaya, Panjul and Aming eating) when really acting to further another (break into the bunkhouse, steal packs, obtain leftovers, meet a competitor).

Deception and pretending

Deception may be a rudimentary form of pretending found in many non-human species that represents an evolutionary precursor to the more complex pretending seen in humans (Mitchell, 1994a). Like pretending, deception involves deliberately behaving as if something were the case when in fact it is not. Actors who deceive enact normal actions in such a way that partners are likely to misinterpret their meaning, to the actors' advantage (Mitchell, 1986; Byrne & Whiten, 1990; Lillard, 1993a). Deception likely involves strategically varying shared scripts, where a script is a representation of the sequence of elements that constitute an event, so it would consist of minor variations to the behaviors normally enacted in a script or deliberate enactments of scripts under atypical circumstances (Mitchell, 1999a).

Princess's, Unyuk's, and Paul's cases qualify as deception because each of the three orangutans led their partners to misinterpret a situation by enacting normal behavior in an abnormal way, to their own benefit. Tactically, all three concealed their interests by creating affiliative images (Princess, Unyuk), distracting partners (Unyuk, Paul), and using humans as social tools (Paul). Strategically, all three appeared to manipulate two scripts – inhibiting one, deliberately playing out another. Paul inhibited the script of taking leftovers from Bento and activated another well-worn script, eating provisions with Bento and staff present; Unyuk inhibited her theft script and promoted a grooming-like script; Princess inhibited a break-in script and promoted a visit-humans script. Panjul's case does not qualify as deception because Panjul did not lead Aming to misinterpret a situation, but to reinterpret it correctly. Otherwise Panjul's behavior resembles deception; he benefited from his ruse by enacting normal behavior in an abnormal way and he used a typical deceptive tactic, creating an affiliative image.

All three deceptions involved self-pretending, i.e., voluntarily simulating one's own past actions out of context, which constitutes a rudimentary, presymbolic form of pretending (Piaget, 1945/1962; Mitchell, 1993a, 1994a; McCune, 1995). All three deceptions also involved more complex processes. They show one feature suggestive of Mitchell's (1986) level 4 deceptions, planful or intentional deceptions that reference the whole layout of the field: Princess deceived a human group for at least five hours as part of a plan for breaking into the bunkhouse; Unyuk feigned a complex interactive role, flexibly keeping Ucing distracted while positioning herself to steal his goods; Paul orchestrated a three-way competition for leftover papaya when the papaya was distant and invisible and the ruse had to delude humans as well as Bento. Panjul's ritualized eating suggests symbolic communication about future acts. These cases may then qualify as complex pretending.

Complex pretending

Definitions for complex pretending impose a variable set of additional criteria, primarily cognitive ones. (1) The actor must entertain two representations of a single situation, one literal/veridical/real and the other nonliteral/distorted/pretend/imaginary, simultaneously, without confusion, and deliberately (Piaget, 1945/1962; Werner & Kaplan, 1963; Leslie, 1987; Harris & Kavanaugh, 1993; Lillard, 1993a; Mitchell, 1994a). (2) The actor must mark the pretend representation as different from a literal one (e.g., Piaget, 1945/1962; Werner & Kaplan, 1963; Leslie, 1987; Harris & Kavanaugh, 1993). (3) For some, pretend representations must be deliberately imaginary/unrealistic in that they flout realistic constraints of the situation (e.g., Leslie, 1987; Whiten & Byrne, 1991). (4) For some, pretend representations must be symbolic or metarepresentational (Leslie, 1987; Whiten & Byrne, 1991). Several indices have been proposed for detecting symbolic pretending, including object substitution, attribution of pretend properties, multiple representations of the same object, and imaginary objects (e.g., Leslie, 1987; Lillard, 1993a).

All four orangutan cases meet the criterion of deliberately holding two distinct representations of one situation, simultaneously and without confusion. In all three deceptions, each actor's behavior was initially honest (2/3, friendly interaction; 1/3, foraging) but became dishonest when he or she changed aims (3/3, stealing food). From the moment each orangutan's aims changed, each simultaneously simulated innocent behavior and covertly advanced the ulterior aims. The two representations must have been held without confusion because the deceptions were not detected – i.e., the actors' behaviors were not themselves confused. That holding two representations was deliberate is evident in signs of voluntary simulation in their feigned behavior, such as exaggerations or distortions of normal behavior (e.g., being too nice, not going for provisions, and eating papaya before Bento's eyes) and careful control of actions towards covert aims, especially delaying timing and positioning until conditions were ripe. The final case, Panjul's and Aming's, involved two representations of the act of eating food – the literal one and the communicative one.

The orangutans must have marked pretend representations as different from literal ones because they handled them differently. They expressed and exaggerated behavior consistent with their pretend representation, the deceptive or the symbolic one, but inhibited and constrained behavior consistent with their ulterior one.

Deliberately flouting reality when enacting the imagined representation did not characterize these orangutans' deceptions – this would have given the ruse away. In fact this would seem to apply to all deceptions, which aim to simulate reality, so by this criterion no deceptions would qualify as pretending. Panjul and Aming flouted reality if they actually consumed *Licuala* pith in their ritualized eating because in reality they did not eat this item. My evidence is, however, inconclusive on this point – both appeared to really eat but the amount was abnormally small and the item was not a known food. At this time, no Sungai Wain orangutans had ever been observed eating *Licuala* pith in over 1000 hours of systematic observation collected over two consecutive years, although it was readily available and orangutans ate other parts of the same palm (H. Peters, 1995; Russon, 1998). The first observations of normally eating *Licuala* pith occurred two full years after this incident and involved two different orangutans.

These cases show various signs of symbolic pretending. Unyuk's deception resembles McCune's combinatorial pretending, portraying a variety of differentiated event components in sequence, which qualifies as symbolic pretending. It involved a variety of simple actions (e.g., feigned watching, presenting arm then head for clipping, positional adjustments, vocalizations, gestures) performed to suit the flow of a complex and novel interaction. Some of Unyuk's actions were elicited by Ucing's overtures but others were initiated independently. Unyuk enacted some actions so as to sustain the interaction and advance her covert aims simultaneously; her positional shifts, for instance, suited the interaction but also moved her into place for snatching the packs. Some of her actions must have been almost if not completely novel – this could well have been her first close encounter with a Swiss Army knife and she would have to have modified any simple actions on the spot to make them suit both representations of the situation simultaneously. It is then improbable that she replicated her own past behavior in the sense of simply reproducing previously performed actions or even action sequences. What Unyuk simulated was more akin to an interactive role than individual behaviors. Although she simulated past behaviors, she had to alter and deploy them in an original fashion to sustain the deception and simultaneously advance her ulterior aims. Human children similarly elaborate symbolic gestures in longer scenes from about 18 months of age (Blake, 2000).

Panjul's and Aming's ritualized eating may represent even more sophisticated pretending because it seemed to be governed by internal mental processes rather than perceptual features of real entities in the environment. It could qualify as McCune's hierarchical pretending, which closely resembles Leslie's metarepresentational pretending (Blake, 2000). Their ritualized eating appeared to be instigated by social circumstances (close proximity under conditions of social tension), not hunger or food availability (both ignored many readily available foods and when they did eat, if they did, they picked an item that was not a normal food). Its significance lay in the social message conveyed, not the ingestion of food. Panjul's ritualized eating appeared to signify friendly intentions, i.e., future behavior, and some consider that agreements concerning future behavior are intrinsically symbolic (Deacon, 1997).

Other details further suggest symbolic pretending. Princess and Unyuk held multiple representations of a single entity, their human partner(s), because both simultaneously treated them as attractive companions and interfering guards. Panjul and Aming may have attributed pretend properties to *Licuala* pith because they treated it as edible when they normally treated it as inedible.

Conclusion

Like other great apes, these orangutans behaved in ways that readily qualify as simple pretending. More interesting are indications that they were pretending at more sophisticated levels, up to and including symbolic pretending.

Claims have been made that pretending and other high-level abilities are limited to great apes who have been symbol-trained (Savage-Rumbaugh & McDonald, 1988) or enculturated, i.e., immersed in human culture (e.g., Tomasello, Savage-Rumbaugh & Kruger, 1993). Evidence of high-level deception in great apes already belies both claims (Whiten & Byrne, 1991). These orangutan cases further challenge them. Three of these four orangutan cases, including the two most complex (Unyuk, Panjul/Aming), were engineered by orangutans who could not be considered enculturated or language-trained. Two cases also involved orangutan–orangutan interactions, feral settings, and feral orangutan issues. In sum, these data indicate that symbolic abilities, including pretending, can and do develop in great apes without extensive human tutelage and they play an important role in feral great ape life.

Acknowledgments

The research on which this chapter is based was sponsored by Indonesia's Nature Conservation Agency (PPA), in the Ministry of Forestry and Estate Crops (MOFEC). Work at Camp Leakey was sponsored by the Orangutan Research and Conservation Project. Work in Sungai Wain Forest was sponsored by the Wanariset Orangutan Reintroduction Project and the Tropenbos Foundation of the Netherlands. Funds supporting the research were provided by Canada's National Sciences and Engineering Council (NSERC), and by York University and Glendon College in Toronto. I am very grateful to Dr. B. M. F. Galdikas and Dr. W. T. M. Smits for making this work possible. Thanks also to Pak Adi Susilo and Nita Boestani, for their great help in facilitating the work, and to Ucing, Dolin, and Adriansyah, for their invaluable assistance in the forest.

17

Seeing with the mind's eye: eye-covering play in orangutans and Japanese macaques

Introduction

Eye-covering play, deliberately closing or covering one's eyes during a play sequence, has been documented in various nonhuman primates. Some observers have suggested that eye-covering play involves pretending – acting *as if* one can't see, doesn't exist, or exists in some altered form (Cunningham, 1921; Hahn, 1982; de Waal, 1986b, 1989). Only one systematic study has been made of eye-covering play in nonhuman primates (Thierry, 1984) and no systematic attempts have been made to investigate the cognitive processes involved. We systematically studied eye-covering play in captive orangutans (*Pongo pygmaeus*) and Japanese macaques (*Macaca fuscata*) with the aim of assessing its cognitive implications in relation to pretending and imagination.

Cunningham (1921) was perhaps the first to describe eye-covering play in a nonhuman primate – in an immature lowland gorilla (*Gorilla gorilla*) who would shut his eyes tightly then run around knocking into furniture, which was interpreted as pretending to be blind. Eye-covering play has since been reported in all the great apes. Their play often takes the form of stumbling or groping about while "blind" (Hoyt, 1941; Harrisson, 1961, 1962; Lang, 1963; Gautier-Hion, 1971b; Goodall, 1971; Rensch, 1972; de Waal, 1986b, 1989). Occasionally they use objects to cover their eyes (Harrisson, 1962; Lang, 1963; Goodall, 1971; Rensch, 1972; de Waal, 1986b, 1989) and eye-covering play sometimes occurs as a social game (Hoyt, 1941; Harrisson, 1962). For example, Harrisson (1962) described an elaborate version between two juvenile rehabilitant male orangutans, a "hankie contest," in which one covered his head and eyes with a handkerchief, the other raced forward to tear the hankie off and make the "blind" one "see," then they changed roles and started again.

Eye-covering play also occurs in other nonhuman primates, including gibbons (Hahn, 1982), several cercopithecine monkeys (guenons – *Cercopithecus nictitans, C. pogonias, C. neglectus*; mangabeys – *Cercocebus albigenal, C. galeritus*; talapoins – *Miopithecus talapoin*; rhesus macaques – *Macaca mulatta*; olive baboons – *Papio anubis*) (Gautier-Hion, 1971b; Thierry, 1984) and one New World monkey (brown capuchins – *Cebus apella*) (P. Vasey, pers. obs.). The gibbon, a pet, circled the room "in a drunken stagger" with his arms spread out and his eyes shut tightly (Hahn, 1982). Captive talapoins moved about blind in a "drunken" manner, peeking if they struck an object or just prior to jumping from one substrate to another, then re-closing their eyes and carrying on (Gautier-Hion, 1971b). Captive rhesus macaques walked or jumped around their cage with their eyes covered, occasionally groping with the free hand; they sometimes paused to peek then re-covered their eyes and resumed locomoting (Thierry, 1984).

Interpretations of eye-covering play have concentrated on pretending but simpler alternatives have also been proposed. Eye-covering play may be a voluntary means of self-stimulation (Gautier-Hion, 1971b; Thierry, 1984) or a form of autotelic play (Mitchell, 1990). Autotelic play involves actions, like somersaults, performed for their own sake (*sensu* Suits, 1977) or to amuse the actor (Mitchell, 1990) rather than to serve other ends; it involves attempting to recreate previous actions through repetition. Mitchell (1990) interpreted bonobos' eye-covering play as "third-level learned autotelic play." Level 3 play is typically a form of self-imitation but it can involve repetition of another's actions (Mitchell, 1990). The autotelic play explanation suggests that more complex mental processes than self-stimulation may underlie eye-covering play, at least in great apes.

Whether eye-covering play by nonhuman primates involves pretending hinges on what pretending means. Its essence is deliberately behaving as if something were real when it is not, behaving as if doing one thing while in fact doing something else, or projecting an imaginary situation onto an actual one (e.g., Lillard, 1993a; Mitchell, 1994a). It requires simultaneously entertaining *two* cognitive representations of a situation, one real and one imaginary, deliberately and without confusion, as well as recognizing and "marking" the imaginary representation as different from the real one (e.g., Piaget, 1945/62; Werner & Kaplan, 1963; Leslie, 1987; Whiten & Byrne, 1991; Harris & Kavanaugh, 1993; Lillard, 1993a; Mitchell, 1994a). Some see symbolic or metarepresentational cognition as a defining feature, where metarepresentation means re-representing existing representations (e.g., Leslie, 1987), but others recognize a graded range of simpler forms governed by pre-symbolic cognition (e.g., McCune, 1995).

Cognitive criteria are of interest because they may distinguish primate species in their capacity for pretending. The simplest metarepresentations, for instance, are second-order representations that re-represent first-order, i.e., sensory-motor or "real," representations of the world. Second-order representations may represent great apes' peak cognitive achievements and distinguish their cognition from that of other nonhuman primates (e.g., Matsuzawa, 1991; Whiten & Byrne, 1991; Gibson, 1993; Langer, 1993, 1996; Mitchell, 1994a; Byrne, 1995; Parker & McKinney, 1999; Russon, Bard, & Parker, 1996). A few cases of playing with non-existent objects in language-trained great apes (e.g., pulling an invisible pull toy) appear to satisfy criteria for metarepresentational level pretending (Hayes, 1951; Savage-Rumbaugh & McDonald, 1988; Whiten & Byrne, 1991; Miles, Mitchell & Harper, 1996). We know of no accepted evidence for pretending at this level in other nonhuman primates.

We studied eye-covering play in Japanese macaques and orangutans, two primates known for spontaneous eye-covering, to assess whether their eye-covering play constitutes pretense and what cognitive processes may be involved.

Methods

Subjects and settings

The Japanese macaques were housed at the Université de Montréal's Laboratory of Behavioural Primatology, Canada, in a mixed-sex group of 23 adults (18 females, five males) and 14 immatures (seven females, seven males) (for details see Vasey, 1998). Their enclosure was furnished with climbing and swinging devices plus loose enrichment items like coconut shells and pails.

The orangutans were housed at the Metropolitan Toronto Zoo, Canada. In 1995, the group comprised three adults (two females, one male), three adolescents (two males, one female), one juvenile (male), and one infant (female). The same individuals were present in 1998. Their enclosure was furnished with permanent climbing structures, changing arrays of ropes, nets, logs, and behavioral enrichment items like boxes, clothes, and plastic scoops.

Sampling and data collection

We defined *eye-covering play* as an actor deliberately covering his or her eyes as part of a play sequence, where *play* was intentional activity performed for its own sake, for amusement or to simulate another end-directed activity for benign ends (Mitchell, 1990). In eye-covering play, *events* were periods when eyes were continuously covered and *bouts* spanned several temporally contiguous eye-covering events that were interspersed with brief peeking (i.e., uncovering the eyes for a few seconds) but interrelated (i.e., involved similar activities). For each event we recorded actor (name/age/sex), component behaviors, and type of play (stationary/locomotory/positional, arboreal/terrestrial, social/nonsocial). When we recognized bouts we noted their occurrence and the behavior during interspersed eyes-uncovered periods.

For Japanese macaques, we recorded all spontaneous eye-covering play that occurred during observational sessions for other projects, between March 1993 and March 1995 (1264 hours). Eye-covering was rare, so more systematic procedures were not feasible. For orangutans, we recorded eye-covering play in two-hour observational sessions via two schedules: (1) weekly sessions, October 1995 to March 1996 (50 hours), balancing subject focus to obtain six or more events per subject (Hickerson, 1996); and (2) biweekly sessions, videotaped, February 1998 to July 1998 (15 hours), focusing on the two orangutans who performed eye-covering play during that period (females Ramai and Sekali, aged 12 and 5 years respectively). Eye-covering was facilitated by providing the orangutans three to six items of the sort they used spontaneously to cover their eyes (e.g., cloth, containers) at the start of each session.

Results

Overview

In the Japanese macaques, eye-covering play was rare ($N = 75$ events/ 69 bouts by 13/37 individuals – 3.1 bouts/month; mean = 5.8 events/individual, range = 1–14), brief (mean = 4.4 seconds, range = 1–20 seconds), and sporadic. All events were performed by juveniles 1 to 3.5 years old (82.7%) or adolescents 3.5 to 4.5 years old (17.3%). The rate of eye-covering play was higher for males ($N = 3$ males, mean = 7 events/male) than females ($N = 10$ females, mean = 5.4 events/female). Events were solitary more often than social (84% vs. 16%) and slightly more often terrestrial than arboreal (56%

vs. 44%). Types of eye-covering play included simple locomotion (walk/run), positional behavior (spin, hang from swing, somersault), object manipulation (rub surfaces, manual exploration) and sitting (alone and inactive).

Eye-covering play was relatively frequent in the orangutans, although it was facilitated (1995/6: 100 events/50 hours; 1998: Ramai – 38 events/15 hours, 2.7 events/hour, Sekali – 33 events/15 hours, 2.2 events/hour). Based on the 1998 sample, events were brief (mean = 12 seconds, range 1–45) but many occurred within longer bouts (mean = 41.2 seconds, range 2–199). In 1995, all orangutans indulged in eye-covering play but juveniles and adolescents did so more than other age classes (Hickerson, 1996). Events were solitary more often than social (1995/6: 79% vs. 21% (Hickerson, 1996); 1998: 97% vs. 3%) and in 1998, exclusively arboreal. Orangutan eye-covering play included arboreal locomotion, positional behavior (spinning, hanging), wrestling, chasing, object manipulation, sitting (alone or with a companion), and one social game like Harrisson's (1962) "hankie contest."

Eye-covering candidates for pretending

Because eye-covering play did not appear to reflect a unified purpose or cognitive apparatus for either orangutans or Japanese macaques, we examined only complex events – those involving either elaborate behavior or "cheating," i.e., deliberately using vision or touch to obtain extra information in the midst of an eye-covering bout. The term "cheating" suits because both species sought extra sensory information only when they could not continue their activity without breaking their apparent "rule" for the game, directing behavior "blind" and from memory. We recorded two forms of cheating used by both species, often after bumping into something or missing an item they were grabbing for: *peeking* (deliberately uncovering eyes for a few seconds to look around then re-covering them) and *groping* (manually searching for an item by feeling around for it or fumbling manipulation). Cheating confirmed that our subjects' vision was substantially reduced when their eyes were covered, even if they were not completely blind. It also provided indices of the cognitive processes used during eye-covering play – it offered evidence on the use and enhancement of representations (e.g., when actions were adjusted according to sensory information obtained by cheating) and, via its timing and consequences, some evidence of using higher-level cognitive processes.

Japanese macaques

Sixteen events (21.3%) involved peeking and/or groping. All were performed by females (N = 8); most occurred in non-social contexts (87.5%); more occurred arboreally than terrestrially (68.7% vs. 31.3%) and while locomoting than stationary (75% vs. 25%). Except for once, locomotion was along continuous structures. The exception occurred when a female covered her head with a pail and ran towards a fence; when near the fence, she paused, groped for the fence, grasped it, and climbed a few feet before uncovering her eyes.

Groping for objects (e.g., swings, platform edges) or individuals nearby occurred in nine eye-covering play events, performed by seven different females. Actors did not sweep their hands widely through space as if they had no sense of what was there; instead, their groping seemed directed. In five of these cases, actors positioned themselves before covering their eyes so that they were in proximity to and oriented towards the object or individual for which they subsequently groped; then, remaining stationary, they covered their eyes and groped in a directed manner. In all five instances, groping appeared clearly to be the raison d'être for the eye-covering play. One female, for instance, positioned herself beside a tire-swing before covering her eyes with a deflated plastic ball then groped for the swing with both hands in a well-directed manner. On another occasion a nulliparous female made several attempts with her eyes open to touch her newborn nephew which failed because the infant's mother moved to avoid contact. Eventually she tried a new tactic, first covering her eyes and then reaching once more towards the infant, but the mother shifted her position once again to avoid contact. In the other four cases, groping served as one element within a larger behavioral strategy. Actors covered their eyes then groped for nearby objects (e.g., fences, walls, edges of platforms) while locomoting, apparently trying to avoid traveling errors. Twice they slowed down as they neared barriers, moved more cautiously, then groped for the barrier with an outstretched arm. In the other two instances, they groped along the edge of a continuous arboreal structure while locomoting on it.

Peeking occurred in four eye-covering play bouts performed by four females, all while locomoting across continuous substrates. In three cases, they appeared to peek to prevent errors. One covered her face with a coconut shell that blocked all but her peripheral vision then locomoted on an arboreal platform while repeatedly turning her head to the side. This way, she could peek in the direction she was moving and correct her position. In the fourth case, the actor appeared to be checking her position in space. She covered her eyes, spun around rapidly in circles, stopped, uncovered her eyes, looked around, recovered her eyes, and spun around for a few more seconds.

None of the macaques cheated to correct errors. On four occasions a female made a mistake while traveling blind by bumping into objects like walls, swinging doors, or poles. Each time, they simply uncovered their eyes and stopped locomoting.

Orangutans

Twenty-four (34%) events in the 1998 sample involved blind travel (Ramai-14, Sekali-10). One event in the 1995–96 sample was a social game that resembled Harrisson's hankie contest (Hickerson, 1996).

In the social game, Ramai and Juara, the adolescent female and male, repeatedly put a shirt over each other's head and eyes. The seeing partner then backed up a few meters and charged the blind one, then the two wrestled vigorously until one grabbed the shirt and put it over the other's head. The game lasted over 15 minutes.

In blind travel, Ramai and Sekali typically clambered through climbing structures with their eyes covered, including transferring from one arboreal structure to another. Ramai once traveled blind for 15 seconds through a tangle of bars, poles, and ropes along a route involving three transfers between discontinuous structures. The only hint of difficulty was an occasional brief hesitation. Despite the complex trajectories, they made only minor errors, such as fumbling while grabbing a support. They never had to backtrack because of arriving at an impassable or awkward gap or withdraw because of grabbing an inappropriate support.

Ramai and Sekali groped in 9/24 and 5/24 blind travel events respectively. As in the macaques, their groping was aimed and showed they had a good sense of what was there. Their groping was always a component of a broader behavioral strategy, not the main focus of eye-covering play. Systematically, both groped as they neared discontinuous or unstable points such as ropes, curved poles, junctions (e.g., log to platform), edges, or obstructions (e.g., a rope across a pole). They did *not* grope when transferring from one straight pole to another or traveling along continuous or stable supports (e.g., walking over a flat open surface, climbing up/down one pole or rope). They groped to prevent minor errors (Ramai 5/9, Sekali 1/5) – commonly by slowing down as they neared a discontinuity, reaching directly but slowly towards the next support object, then cautiously grabbing hold. Once Ramai walked blind along a log with a rope dangling across it. As she neared the rope, but before bumping it, she paused her stride; *then* she reached towards the rope. Only *after* her hand contacted the rope did she grope a little to grab it, and immediately *after* grabbing it she resumed her stride. Both also groped to correct errors (Ramai 5/9, Sekali 4/5). Their reach sometimes fell just short of or beyond the best point for grabbing or touching an item so they fumbled or missed it; then they groped, improved reaching details, and contacted the item successfully. Once, as Ramai climbed a straight pole with her eyes covered, her uppermost hand bumped a rope and a log attached to the pole as she reached higher. With that hand she felt around these obstructions for the pole, grabbed the pole just above the rope (point "X"), then resumed climbing. As her uppermost hand let go of "X" to reach even higher, she placed one foot exactly to "X."

Ramai and Sekali peeked in 8/24 and 4/24 of these eye-covering events respectively. They peeked just *before* attempting a difficult transfer in 6/8 and 2/4 cases respectively, i.e., they anticipated an upcoming difficulty while blind, obtained fresh visual information, re-covered their eyes, then made the transfer without error. Sekali peeked just *after* groping in 2/4 of the events in which she peeked. Hickerson (1996) also noted that when peeking, orangutans sometimes looked at a specific location in the enclosure then repositioned and/or reoriented themselves (e.g., shifted to the other side of a bar). When they re-covered their eyes and resumed traveling, they traveled until they reached the location they had just looked/turned towards. Ramai and Sekali showed the same pattern in 1998.

Discussion

Our aim was to determine whether pretending occurred in eye-covering play. If pretending means acting upon mental images deliberately inconsistent with reality, the macaques and orangutans were probably not pretending in most of their eye-covering play. In traveling blind, it would have been dangerously foolhardy to act upon imaginary features of the climbing space. Orangutans are extremely cautious in moving arboreally so their indulging in flights of fantasy is doubly unlikely. Eye-covering play could not have involved pretending about a self that cannot see or exists in another state (e.g., feeling vs. seeing) because by covering their eyes, they *really* could not see (as indicated by cheating) and *really* existed in another state. Hickerson (1996) suggested that in the social game she observed, the blind partner may have been pretending to be caught off guard by the other's play charge despite expecting it – but by eye-covering, the blind partner was *in reality* off guard about *when exactly* the charge would come, not just pretending to be. Eye-covering play could have involved pretending not to exist when it reduced stimulation or contact, as when children think they can not be seen if they can not see themselves, but most of our eye-covering play involved stimulation seeking. As such, the point is that actors *do* exist, not that they do not.

The alternatives that have been proposed have some explanatory value. Generating sensory stimulation (Gautier-Hion, 1971b; Thierry, 1984) characterizes many of our events (e.g., hanging and spinning) but not all. Sometimes our subjects' eye-covering play reduced sensory stimulation, e.g., orangutans sat and "hid" under cover or macaques covered their eyes then attempted to touch a newborn infant despite the mother's prior attempts to prevent such contact. Autotelic play can account for the orangutans' acrobatic eye-covering play (e.g., hang from a structure then flail or twist about, squirm inside a T-shirt) and much of the Japanese macaques' eye-covering play (e.g., spin, hang, somersault, rub surfaces, most blind walking and running), but not for all the blind travel. Both species seemed to move about blind for its own sake in some cases, but with the goal of reaching a planned destination in others. Our analyses of the goal-directed blind travel did not convince us that all cases served amusement; some suggested simulating end-directed activity for benign ends. Further, our blind travel did not show the deliberate replication of actions associated with Mitchell's (1990) level 3 learned autotelic play – it replicated actions that actors already knew (e.g., walk, clamber), but no differently than normal sighted locomotion does. Neither of these alternatives, then, self-stimulation or autotelic play, seems to span the range of cognitive processes involved in these primates' eye-covering play.

Considering pretending from a cognitive perspective, we assessed whether eye-covering play involved: (1) two representations of the situation held simultaneously and without confusion; (2) an imagined representation which was marked as such and which directed the eye-covering play; and (3) metarepresentation.

(1) Both Japanese macaque and orangutan subjects showed signs of holding two mental images of their space during their eye-covering play. They likely had well-established first-order, "real" representations of their quarters from their extensive living experience with enclosures and

climbing structures that were permanent, plus movable items that were commonly available for months. They also took in new sensory–motor information from the immediate experience of moving about blind. Whether this produced a new first-order representation, a perceptual representation, or was simply assimilated to the stored representation is unclear. What is clear, however, because both species made errors, is that the stored and the immediate images were discrepant. Both species recognized the discrepancy (they cheated to avoid making errors, at points of reduced predictability), and even seemed to enjoy experiencing or playing with it (they elected to travel blind). The orangutans were perhaps more able to manipulate the discrepancy voluntarily than the Japanese macaques (they corrected their errors on the spot through deliberate cheating while the Japanese macaques did not). For both species, at least at error points, a stored and an immediate image must then have co-existed simultaneously and without confusion.

(2) Both species' stored representations could be considered "imagined" in that they were retrieved from memory instead of generated from immediate experience. We found no signs, however, in either species of imagined representations in the sense of deliberately flouting rules of the real world. Our "imagined" representations must have been marked as distinct from immediate experiential images because both species could, and did, voluntarily choose which to use during bouts of blind travel. Both species opted for an imagined representation to direct their blind travel – they followed a self-imposed rule of using an image held in the mind to guide their movements, to the point of deliberately excluding visual input that would normally enhance or refresh that image. They also voluntarily chose if and when to use experientially based images (i.e., peeking or groping) and they appeared to check the match between these two images (e.g., Japanese macaques peeked after locomoting blind as if to check their position in space, orangutans often tore off their cover as they reached the end of their trajectory). That the distinction was effective is shown by the fact that both species were reasonably good at deciding when to choose the imagined representation over the normal experiential image to handle problem conditions. Both locomoted and anticipated the placement of nearby items with their eyes held covered, even though the orangutans were much more proficient at this than the Japanese macaques.

(3) We found indicators that may shed light on whether eye-covering

play was governed by metarepresentations; these indicators suggested differences between the two species. The orangutans seemed to plan specific destinations and travel routes in advance but the macaques did not, unless groping at a wall or fence constitutes a travel goal and moving a few meters in a straight line constitutes a route. Wild orangutans are known to plan travel routes (MacKinnon, 1974; Galdikas & Vasey, 1992), so this is feasible. Our orangutans sometimes looked intently at and reoriented towards a distant location before covering their eyes then traveled blind to that location. While traveling blind they anticipated problems and they made few overt route changes (e.g., never backtracked, rarely paused to peek then reorient), few routing errors, and only minor mistakes over long, complex trajectories. This suggests that pre-existing representations of *routes* guided their trajectories, not immediate experiential cues from the space. The macaques positioned themselves relative to a target object/individual before eye-covering and groping from a stationary position, which suggests some form of planning or attempt to "see ahead" but not planning routes. We found several associated patterns. The orangutans' trajectories were elaborate whereas the Japanese macaques' were much simpler. The orangutans peeked and groped to handle immediate problems within complex eye-covering activities, whereas the Japanese macaques often peeked or groped as an end in itself. When travel problems occurred the orangutans corrected them immediately, "on-line," whereas the Japanese macaques did not, and instead stopped dead. The travel problems the macaques tackled blind were much simpler than the orangutans', however, so their solutions could not reveal the same order of complexity.

Planning travel routes is a non-trivial cognitive task that probably requires more than first-order representations of spatial features. In the forest, ex-captive orangutans make frequent routing errors in their early days of traveling arboreally – with their eyes wide open (A. Russon, pers. obs.) – and in the zoo, climbing structures were complex enough to afford varied routes to most destinations. Our orangutans' routes involved crossing discontinuities in the climbing structures and changing direction, so their representations must have involved mapping a sequence of movement vectors that could be connected onto the space, from the starting location through the climbing structures to the chosen destination. To be usable, these vector sequences must have been defined in terms of the spatial relations between climbing devices, calibrated against the self –

i.e., the actor's size and capabilities determine whether devices are interconnected and which gaps can be spanned. Spatial reasoning that operates in terms of object–object relations is considered to entail second-order representation, i.e., rudimentary metarepresentation (Case, 1985; Langer, 1996; Byrne & Russon, 1998; Russon et al., 1998), and the self-concept implied in orangutans' arboreal locomotion suggests second-order representation (e.g., Povinelli & Cant, 1995). Some evidence supports this interpretation, over the possibility that routes were defined by first-order representations generated from past travel experiences. The orangutans could *handle errors on-line* then resume travel along the original trajectory, they used peeking and groping as *tactics in the service of* elaborate eye-covering activities, and they could *voluntarily alter* their cheating tactics (they sometimes peeked and groped in one bout). These features of the orangutans' blind travel point to hierarchization, the generative cognitive process considered responsible for metarepresentation; the Japanese macaque pattern more closely resembles the linear, chained organization characteristic of first-order representations and association-based cognitive processes (e.g., Langer, 1996; Byrne & Russon, 1998; Russon, 1998).

Other patterns in the two species' eye-covering play suggest similar cognitive differences. (1) More of the orangutans' than the Japanese macaques' blind travel was arboreal and through discontinuous structures, so the routing and locomotor problems the orangutans posed themselves were more difficult than those tackled by the Japanese macaques. (2) The Japanese macaques' eye-covering events and bouts were shorter than the orangutans', suggesting they were unable to sustain the mental images without frequently refreshing them with new experiential input. (3) The orangutans' blind travel was more clearly governed strategically by the stored representation. In particular, the way the orangutans peeked and groped to avoid errors, before reaching difficult locations, suggests that they used the stored representation to "see ahead" to the upcoming difficulty. The Japanese macaques also peeked to avoid errors but rarely (in three cases only) and over very short distances, in a straight line, on arboreal structures. They even seemed to have difficulties following a straight line with their eyes covered, and would drift to the edge of the structure. They peeked after coming near enough to the edge to feel it with their hands. They used experiential cues to detect difficulties at an early stage whereas the orangutans could predict difficulties mentally, ahead of time.

Conclusion

Our analyses suggest Japanese macaques and orangutans were not pretending in their eye-covering play. We found wide variation in their play, however, which suggests that several cognitive processes and motives can underlie eye-covering play, not a single one. This implies that rejecting any single case or type of eye-covering play as pretending, or showing that other processes may have been used, does not eliminate the possibility that pretending occurred in others. It also raises the issue of what other cognitive exercises than pretending might be involved.

Both species seemed to be playing with discrepancies between different mental images of the world, both appeared to appreciate that their stored representation could differ from the image emerging from their experience of traveling blind, and both had some voluntary control over which image to use. This could be one way in which individuals discover the possibility of pretending – by experiencing the fallibility of any representation when tested against the real world. In the macaques, it is likely that all representations were first-order while in the orangutans, discrepancies between first- and second-order representations appear to have been at play.

Pretending is only one of several ways of playing with normally occurring discrepancies among multiple representations of a situation. Taking second-order representations as an example, second-order representations of a situation are derived from first-order representations so the two are *linked* but *distinct* and actors come to recognize that second-order representations are *not* tied to real-world constraints in the same way as their founding first-order representations. In pretending, actors voluntarily play with that distinction in a particular way – they elaborate and behave in line with a second-order representation so as to deliberately flout some of the rules of the first-order representation, the situation "as it really is" (Leslie, 1987; Whiten & Byrne, 1991). A second way of playing with the same discrepancies is to reduce them, to come closer in the mind to meeting real constraints. The difference in such games could be likened to that seen between representational and abstract painters, or between scientific and literary writers – both sides generate images inspired by "reality" and both experience discrepancies between what they create and the original reality. One side aims to come closer to the original and the other to escape it, but both could be seen as engaging in the same basic game of manipulating discrepancies between representations. This second game is the one that both orangutans and macaques appeared to be playing.

These orangutans and Japanese macaques did not engage in pretending because they did not aim to flout the constraints of their first-order representations. Both species did, however, engage in a cognitive game similar to pretending – playing with the discrepancies among multiple mental images of a situation. Their eye-covering play seemed to involve some element of imagination in that their actions were guided by images held in memory, so it could be considered another form of imaginative play. The orangutans, but probably not the Japanese macaques, appeared to be functioning at the same cognitive level involved in pretending, so they were engaging in an equally sophisticated form of imaginative play. Our subjects may have been playing with their sense of self even if they were not pretending about it, or with the awareness that when they can not see with their eyes, they can see with their minds.

Acknowledgments

P. L. V. would like to thank Bernard Chapais, Annie Gautier-Hion, Gary Linn, Richard Pawsey and James G. Pfaus. A. E. R. would like to thank Pam Hickerson, Colleen O'Connell, Sheryl Parks, and Juan Carlos Gómez. The research was funded by a post-doctoral grant to P. L. V. from FCAR, Quebec.

JUAN CARLOS GÓMEZ AND BEATRIZ MARTÍN-ANDRADE

18

Possible precursors of pretend play in nonpretend actions of captive gorillas (*Gorilla gorilla*)

This chapter discusses cases of spontaneous behaviors resembling human pretend play in hand-reared lowland gorillas studied longitudinally over a period of several years. In their early years these gorillas had regular contact with humans and human objects. We discuss instances of behaviors that apparently meet some criteria of pretense (such as object substitution or role enactment). Our argument is that there is little evidence that these apparently pretend behaviors are of the same kind as the pretense shown by human children. However, they might reflect the operation of some basic cognitive mechanisms that played a role in the evolution of pretense in the human lineage. The spontaneous presence of these precursors of pretense may also explain why more complex cases of possible pretense can be induced in apes subject to formal procedures of symbolic training.

The notion of pretend play

Pretend play, also known as symbolic or make-believe play, is a characteristic behavior pattern of the human species that appears in children during the second year of life (Piaget, 1945; Leslie, 1987; other chapters in this volume). It consists of the ability to "act as if" something is the case when it actually is not. For example, acting as if one were a monkey (if one actually is a human, that is) or as if a wooden block were a camera and one were taking pictures (perhaps of an imaginary monkey!). Pretend actions usually display only a fraction or an otherwise modified version of their serious counterparts. Thus, a child pretending to be a monkey does not need literally to carry out the actions of a monkey – she may just make a few stereotypic movements that may not even correspond to any real monkey behavior. What matters is her intention of acting as a monkey. This is one of the reasons why pretense is also called "symbolic play." The actions are carried out with the intention that they stand for different actions, objects, or situations that actually are not there.

This on-line symbolic intent is also what distinguishes pretense from other forms of play, such as playfighting, common to all mammals including humans (Fagen, 1981). Playfights look in many respects like real fights, as if evolutionarily they derived from a modification of aggressive behaviours. However, neither animals nor human children (when they engage in this kind of play) intend their playful wrestling to stand for a real fight – they are simply engaging in rough-and-tumble play for its own sake (see Mitchell, 1990 for notion of schematic play). In contrast, the gist of pretense is not only its playful mood, but the intended symbolic substitution – an on-line, intentional representation of what is not the case at the moment.

There exist different theoretical accounts of pretend play in humans. We will briefly discuss two with special evolutionary significance. In Piaget's (1945) classical account, pretend play emerges during the second year of life as a particular manifestation of a more general ability to use symbols – the semiotic function. This ability is also manifest in other behaviors such as deferred imitation or language. Like other symbolic behaviors, pretense appears gradually in development, as infants take control of their "schemas" (in Piagetian theory, the abstract structure underlying actions) and produce them in a decontextualized way, with increasing independence from the environment. In his longitudinal account of the emergence of pretense in children, Piaget describes many intermediate forms of behavior which are not yet pretend actions, but "precursors" of it.

In contrast to this classic conception, in Leslie's (1987) popular and controversial account, pretense is not part of a more general symbolic ability, but one of the first manifestations of a domain-specific cognitive mechanism – metarepresentation – which allows children to deal with mental representations as different from the things they represent. It is the same mechanism at the core of mindreading functions, especially false-belief understanding. Since in Leslie's account this mechanism is innate and modular, its manifestation in pretend play would be an all-or-nothing issue, without intermediate forms of behavior. Other recent views of pretense, although rejecting its links with metarepresentation

(Perner, 1991; Jarrold *et al.*, 1994), do not address the issue of its developmental precursors.

Piaget's view, with its ontogenetic sequence of precursors, appears to be in principle better fit for an evolutionary perspective on pretense. However, we will argue that both Piaget's and Leslie's conceptions can be integrated into an evolutionary framework.

Our criteria for pretense

According to Leslie pretense actions must display at least one of the following features: (1) an object is used as if it were a different object, i.e., behavior patterns that are typically applied to a particular kind of object are applied to an obviously different one; (2) an action typically performed with an object is carried out in vacuo, i.e., as if the object was present; or (3) an action is carried out with the appropriate object but as if this had a property that it actually does not have at the time (e.g., drinking from an empty glass). Of course, a single act of pretense can combine several of these features.

Detecting actions apparently fulfilling any of the above criteria does not guarantee that we have found pretend play. For this to occur it is essential that the above actions do not happen by mistake or by chance (e.g., drinking from an empty glass because one wrongly thinks it contains water); they must be intended by the actor. However, intentionality by itself may not guarantee that a given action involves pretending – the action must be intended as a symbolic action, i.e., intended not only as it is but as if it were a different action (Perner, 1991). It is possible to list objective criteria of intentionality at large, such as repetitiveness of the action, availability of information that would allow rectification of a mistake, or signs of goal-directedness, such as adjustments in relation to the possible goal (Bruner, 1981); but criteria for symbolic intent are more difficult to find, especially when one is dealing with nonverbal organisms.

A solution in the human literature to the problem of identifying symbolic intent is the production of some concurrent signal: smiles or laughter; descriptions of the ongoing actions with labels corresponding to the pretended action; or explicit verbal declarations that the ongoing action is "pretense;" "make believe;" etc. However, such signals are by no means necessary components of pretense in humans. They usually appear in socialized versions of pretend play, i.e., when children pretend together or are aware of an audience. But we can still attribute pretense to children who are acting without any of the above indications.

In any case, most of these criteria of symbolic intent cannot be used with our gorillas (with symbolically trained apes, the criterion of verbal labeling could be used – e.g., Patterson & Linden, 1981; Jensvold & Fouts, 1993; Matevia, Patterson & Hillix, PIAC21). The only applicable criterion would be "smiling" and "laughter," if we assume that primate play faces are comparable to human smiles and their play vocalizations are forms of laughter (see van Hooff, 1972, 1973). In fact, even with human children, especially in their earliest stages of play development, it is frequently difficult to be sure that a particular action is intended as pretense (Huttenlocher & Higgins, 1978). An attempt to overcome this problem is to focus upon the structure of the pretend action and the context in which it is produced. This is usually subject to some sort of schematization that renders it structurally different from its serious counterpart. In the human literature, authors (Dunn & Wooding, 1977; Fein & Apfel, 1979b) have suggested the following criteria to identify "nonliteral" actions independent of any verbal descriptions by the subjects:

(1) familiar activities are carried out without the necessary materials or outside their usual context;
(2) actions are enacted incompletely without attaining their usual results;
(3) inanimate objects are treated as animate;
(4) an object (or a gesture) is replaced by another; and/or
(5) the child "impersonates" another person or an object (e.g., acting as if being a firefighter or as a train).

These are the criteria that we will use to analyze possible cases of pretense in our gorillas.

Subjects and methods of the study

The subjects were five wildborn captive gorillas with estimated ages ranging from 8 to 12 months at the beginning of the study. They were initially hand-reared by humans in a zoo nursery environment. One of them (Muni) was hand-reared as an "only" gorilla baby without any contact with other gorillas for over one year. Then three more gorillas arrived – Bioko, Nadia and Guinea – that were kept together as a group into which Muni also was eventually integrated. The fifth subject – Arila – arrived much later and was never fully integrated into the main group.

Muni had extensive contact with humans and human objects during her first two years of captivity. When the four gorillas were integrated into a single group, the amount of human and object contact decreased, but the four had almost daily access (averaging about 10 hours a week) to some objects and humans. Straw, ropes, tires, some plastic containers and boxes were part of their daily environment. For further details about the hand-rearing conditions of the subjects see Gómez (1992, 1999).

The observations reported in this chapter come from an extensive database about the behavioral and cognitive development of these gorillas compiled during several years of study. The data consist of handwritten descriptions, audiotape transcriptions, and videotapes of both spontaneous behaviors and reactions to experimental and semi-experimental situations. Notes were taken on the spot or audiotaped and later transcribed by one or two observers. The procedure is similar to that used by Piaget in his baby studies and the "diary method" used in longitudinal studies of linguistic and communicative development (Braunwald & Brislin, 1979). We report on data from the first five years of the database regarding the production of behaviors resembling "pretend actions" according to the criteria discussed above.

"Pretend" actions in gorillas

Several behaviors apparently met some of our criteria for pretense. Most cases occurred when the subjects were at least between 2.5 and 3 years old. At younger ages, only sporadic examples of possible pretense were recorded. For ease of analysis, we have grouped our observations into four sections: object substitution; context substitution; imaginary properties; and role enactment.

Object substitution

Playing mothers with pretend babies
Two of our female gorillas – Nadia and especially Muni – displayed behaviors resembling maternal behavior patterns applied to objects such as stones, shoes, balls, pieces of cloth, and even a rubber doll. These patterns usually involved placing the objects between arm/armpit region and sides/chest, or on the neck or the top of the back area while standing on all fours, usually followed by walking while keeping the object in balance, but readjusting it with the hand if necessary. A third pattern

involved holding the object against her chest region with the arm folded while walking on three legs. No playfaces or play vocalizations were recorded during the performance of these patterns.

These actions are reminiscent of maternal behavior patterns characteristic of gorillas, especially primiparous gorillas dealing with their first offspring (Maple & Hoff, 1982). Indeed when Muni and Nadia grew older and had babies, some of their initial maternal behaviors resembled the patterns we had observed years before with the substitute objects. This could appear to be a perfect example of pretense in gorillas. However, both females had been captured in the wild when they were very young (Muni probably at around 4–6 months; Nadia at around 8 months) and had been reared by humans and among other young gorillas. During most of their infancy, therefore, they lacked models of gorilla maternal behavior that they could reproduce in a pretend mode. Moreover, most of the observations were made when the gorillas were about 2.5 years old.

A more plausible explanation is that they were displaying "instinctive" behavior patterns designed to start maternal behavior in female gorillas. These behavior patterns were activated by the above objects, as later they were activated by real babies. The reactions of the babies (suckling, grasping, whining, etc.) would favor the differentiation of these initial patterns into proper maternal behaviors. This process of progressive attunement of initially primitive maternal behaviors appears to happen also in wild gorillas (Maple & Hoff, 1982).

In the above examples, therefore, the gorillas may be accidentally applying patterns that are evolutionarily designed to deal with gorilla babies to substitute objects. If the observed activities are play at all, they would be instances of what Mitchell (1990) calls "schematic play." We have no evidence of "as if" or symbolic intent, i.e., that they are representing the objects as babies. Quite on the contrary, we suggest that it is more likely that, in the first moments of their first maternity, Nadia and Muni were representing their babies as just one more object to which those patterns were applied. Only later, as a consequence of their interactive experience with the babies, they would develop some notion of "gorilla babies" and a special set of differentiated patterns to deal with them. A more convincing case for symbolic play would be made if these sorts of proto-maternal behaviors were observed in individuals shortly after they had witnessed actual maternal behaviors by a mother in their group. Videotaped data of Muni and Nadia after the withdrawal of their babies for hand-rearing (since their maternal behaviors were judged insufficient to guarantee successful mother-rearing) remain to be analyzed in detail, but at the moment of videotaping no obvious maternal-like behaviors applied to "pretend babies" were apparent.

Playing ball with apples

Something closer to intended object substitution happened with Nadia at about 2.5 years old. She had developed an elaborate game that involved throwing tennis balls into the air or against a wall, catching them, and throwing them again. The balls were eventually worn out and disappeared. Six months later, after watching a human throwing an apple against one of the glass panels of their cage, Nadia started to manipulate the apple as she had previously manipulated tennis balls. Apples were familiar, everyday things (a regular component of their diets) whose proper function until that moment had been to be eaten. From this moment onwards, apples also became a sort of substitute ball for games similar to those played previously with genuine tennis balls.

Was Nadia pretending that the apples were tennis balls? Our interpretation is that she was just applying a recently acquired manipulative pattern to a similar, round object of about the same size as a tennis ball. The point of the throwing game can be achieved with the physical properties of the apple itself, without having to "imagine" it as a tennis ball.

In this example, there is certainly an initial element of object substitution (suggested by the demonstration of a human), that could have been accompanied by some sort of activation of the representation of the absent tennis ball while the apples were watched and manipulated; but the manipulation of the apples appears to proceed for its own intrinsic interest.

Occasionally, Nadia came to produce some "apple substitutions" as part of her throwing games. For example, in a social version of the game in which she would exchange apple throws with a human, Nadia, having no apples at hand, once used a straw stick, throwing it in exactly the same way, with the same playful facial expression and vocalization with which she threw apples. Was she pretending that the stick was an apple? Our answer is again "No." The gist of the game lay in the action of exchanging something – she was just opportunistically substituting a different throwable object.

There is, again, the possibility that when Nadia was throwing the stick, somewhere in her cognitive system the representation of the apple – the usual object in this context – was simultaneously activated. This could amount to a representation of "I use stick instead of apple." This is different from the "I use stick as if apple" representation that, according to authors like Leslie (1987) or Perner (1991) underlies human pretense. However, representations of the type "stick instead of apple" could constitute evolutionary "precursors" of the cognitive structures that support pretense in humans. We return to this point later.

Other possible instances of object substitution observed in our gorillas (rough-and-tumble play with straw bundles when a play partner was unavailable, "tracing" movements with a stick instead of a piece of chalk –therefore, producing no marks – offering pebbles instead of leaves to humans) are amenable to similar nonpretense explanations.

Context substitution

Building "sofas" and "beds" out of context
Rough-and-tumble play among our gorillas was punctuated by resting periods during which the play partners would remain watchful of each other until resuming their playbout. During these "vigilance" periods gorillas might engage in some sort of distracting manipulation; Bioko's favorite "diversion" was "nest" construction – he would sit on a bunch of straw and start arranging it around his body. However, his manipulation of straw was accompanied by visual vigilance of the partner and was readily interrupted to resume play, sometimes throwing the straw onto his partner.

The "pretend" element of this action would lie in the behavioral decontextualization of nest construction. Every evening our gorillas made a straw "bed" or sleeping nest (as gorillas do in the wild; van Hooff, 1972; Sabater Pi, 1984). Nest construction was relatively frequent also during daytime, as a way of making "sofas" rather than "beds" – i.e., nests that were used for sitting, while eating, playing with some object, grooming, or resting. However, this "sofa" building activity, when launched during play vigilances, was subordinated to play. The "sofa" was incomplete, rarely used as such, constantly arranged and rearranged, and readily interrupted, even instantly destroyed to perform a play invitation with the straw onto the play partner. Were the gorillas engaging in pretend nest construction during these vigilance periods?

Against this "pretense" interpretation is the fact that the material, the physical manipulations, and the general structure of the nest corresponded to those of actual nests. The only element of schematization was the lack of completion, the readiness to interrupt the activity and "recycle" the nest, and perhaps a higher rate of rearranging manipulations at least in Bioko. There is also some decontextualization (or one should rather say "intercontextualization" of social play and "sofa" building), but the gorillas did engage in the activity itself, not in an activity that was a symbol of nest building; however, the activity was decoupled from its usual context and consequences, and interestingly it was re-coupled to a playful behavioral context and therefore presumably to a playful mental context. These decontextualized actions, subordinated to other playful activities, may again be precursors of human pretense. But they are not yet pretense themselves.

Pretending to play

Play can occasionally be used by primates as a strategy to handle potential conflicts and aggression (Breuggeman, 1978). It is as if animals are pretending to play when actually they are engaging in serious behaviors. Possible cases of this strategic use of play were observed in our gorillas (see Mitchell, 1991b, for similar observations). For example, at the beginning of her integration with the younger gorillas, Muni addressed numerous play invitations to them. These were initially not responded to at all or responded to with aggression. The male, Bioko, would occasionally respond to Muni's insistence by chasing her away in a clearly aggressive attitude. However, Muni would run away from Bioko displaying a play face and behaving as if it were a playful chase. The impression was that she was trying to transform Bioko's aggression into a playbout (the threat vocalizations and the overtly aggressive maneuvers of Bioko render implausible the possibility of a genuine mistake by Muni). Similar "re-direction" behaviors were observed when humans reprimanded Muni (e.g., for having stolen an object) or after she was frustrated in some request.

Was Muni pretending to play? Our suggestion is that she is engaging in actual play as an attempt to transform another's aggression into play. Her behavior was not symbolic, but directly instrumental in achieving a goal. However, she was demonstrating a flexible ability to evoke behavior patterns out of context and perhaps some rudiments of deception (Whiten & Byrne, 1988; Mitchell, 1991b). These again may be precursors of genuine pretense.

Playing with nonexistent properties

Examples of possible use of completely nonexistent objects were never observed. However, some actions were recorded that, although carried out with an appropriate object, appeared to assume a property or situation that actually did not exist at the time

Cleaning what is already clean

Seeing humans do some sort of cleaning would frequently evoke from the gorillas an "imitation" of this activity with the same or similar implements on the same or similar surfaces that were not necessarily dirty. Once in their repertoire, it was common to see spontaneous cases of "cleaning" without any real dirt to be cleaned.

A favorite variety was the "cleaning" of garments worn by humans (who frequently had to engage in real cleaning throughout the day). The visual attention of the gorillas was very intense upon their own actions and the surfaces they were brushing. For the observers, there were no easily appreciable changes produced on the clothes by this activity. It seemed to be carried out for the sheer pleasure of the scrubbing and brushing. No playfaces or vocalizations were produced.

There seems to be something especially attractive for gorillas in any activity that involves wiping with something upon a substrate. All our subjects did it more or less frequently, from very early (beginning of second year of life), and similar reports exist for other hand-reared gorillas (Perinat & Dalmáu, 1989). A possible explanation of the prevalence of these patterns may lie in their similarity to some components of mildly aggressive displays that consist of sliding objects (e.g., straw) on the floor while locomoting. These were frequently observed in our gorillas from the age of three years in a serious, mildly aggressive context. Perhaps the cleaning actions of humans were activating early, incomplete manifestations of these instinctive patterns, without any attempt by the gorillas at simulating cleaning actions (see maternal-like behaviors already discussed).

As to the cleaning of human garments, there is nothing in our observations suggesting that the gorillas were imagining nonexistent stains – the exercise of the activity itself, and perhaps some inconspicuous effects upon the garments, appeared to be their sole motivation.

Drinking from empty cups

The drinking schema (turning a container over one's mouth) was frequently applied to a variety of actually empty objects from an early age by all our gorillas. One possible reason to do this could be to test if there is anything in the container. However, on many occasions the drinking schema was repeatedly applied to the same obviously empty container.

The action, then, appears to be neither a mistake nor a probe – it is intentional.

However, there is no indication that gorillas were pretending that there was liquid in the container – no playfaces or vocalizations, no mouth movements simulating ingestion. Actually, perhaps it is misleading to call this a "drinking" schema; a better characterization could be a "turning-over-mouth" schema. The frequency of this action could be due to gorillas' extensive use of their lips for tactual exploration of objects.

What these observations reflect is, again, a well-developed ability to abstract salient action schemas from their usual functions and environmental supports, without necessarily implying a concurrent ability to imagine what remains absent.

Role enactment

Playing experimenters

During training sessions in a discrimination learning procedure (match-to-sample), the reinforcements were administered to the gorillas with a spoon that the experimenter charged with jam in a jar and then put in the mouth of the gorilla. In a particular trial, other spoons were used as discriminanda. Before starting the trial, the experimenter showed the spoons to the gorilla (Muni) and mock-fed her with them. After answering correctly, instead of consuming the reinforcement offered by the experimenter with his spoon, Muni took one of the sample spoons, introduced it in the jam, stirred it, and directed it to the mouth of the experimenter. Muni repeated the "feeding" of the experimenter several times, but without recharging the spoon in the container.

This example involves an inversion of the training roles – the gorilla "rewards" the experimenter – and feeding actions without any actual food. There appear to be, therefore, two possible elements of pretense: adopting a pretend role; and acting with a pretend property.

As to the first, Muni does not appear to be simulating that she is the "trainer" – her activity is limited to the delivery of food and the introduction of the spoon in the human's mouth, without simulating any of the other training operations. As to the second, what others do with their mouths was the focus of intense attention among our gorillas. They frequently put food, leaves, or straw in the humans' mouths and intently watched any ensuing chewing or mouthing. Thus, Muni did not necessarily pretend that there was food in the spoon, either then or previously, when she accepted being mock-fed by the trainer (here she probably was just enjoying the feeling of the cold, smooth surface of the spoon in her mouth). However, there remains the interesting reversal of the normal agent–patient structure in the context of training. This can be a close precursor of some components of genuine pretense, but need not involve any "as if [an experimenter]" intent – it would be enough that Muni engaged in an "acting instead of [the experimenter]." Admittedly, if a similar behavior were reported from a child, it would be difficult to resist the interpretation that he was engaging in pretense. In fact, perhaps many early behaviors that are classified as pretense in young children are not really very different from what Muni was doing in this rare observation – i.e., perhaps they are complex instances of "acting instead," rather than "acting as if."

Another example from the same context of discrimination training consisted of Muni taking three colored circles and positioning them in front of herself in exactly the same triangular configuration of the match-to-sample trials. This appears to be a case of reproducing a familiar state of affairs (the training configuration of circles) without pretending that she was "training" herself or the experimenter (there was no follow-on action, like "responding" by choosing a circle). However, this ability to take up segments of observed routines and reproduce them out of context again may be a precursor component for genuine pretense and another instantiation of an "acting instead" ability.

Other potential pretenses

Some other examples were observed of actions apparently containing some elements of pretense: highly schematized chest-beating movements; sexual games with schematic elements of adult mating patterns; and moving around with eyes covered. The latter has been discussed by some authors as a possible case of pretense (see Russon, Vasey & Gauthier, PIAC17, for a critical review). We believe that it could rather reflect a playful exploration of displacement patterns in which the normal coordination between vision and movement is altered. In one of our subjects (Muni) we observed a whole family of "vision games," ranging from blind locomotion with eyes closed and/or head covered with clothes or boxes to moving around holding a mirror against her face or even, in one occasion, moving around while pressing an eyeball with her finger. What these games appear to have in common is an exploration of vision and its relationship with locomotion and prehension – a remarkable achievement in itself, but one that need not imply pretense. We suggest, therefore, that the same nonpretend explanations that we have discussed in this chapter can be applied to these examples.

Conclusions

We have described different examples of behaviors produced by gorillas showing some of the features that characterize pretense in human children. We have suggested, however, that none of our examples constitutes genuine pretend play. After analyzing in detail the observations, we propose that, although reminiscent of some features of pretense, they can be explained as resulting from the operation of nonpretend mechanisms, such as the ability to decontextualize actions, and to store action representations as abstract schemas decoupling them from their contents and usual supports. These might be necessary, but insufficient prerequisites of pretense, i.e., they might be only part of the cognitive machinery used by human children when they engage in symbolic play. The specialized system (or combination of systems) responsible for pretense in humans does not exist in apes. What apes show is some reflections of possible precursors – or rather co-cursors – of pretense in human evolution. These precursors are not specialized for pretending, but for flexible object manipulation and possibly for the generation of flexible representations of actions. These imperfect examples of pretense in apes are offshoots of cognitive mechanisms that are fulfilling other functions; but they are at the same time samples of the sort of phenotypic outcomes that could have favored the emergence of more specialized pretend mechanisms in our hominid ancestors.

The spontaneous presence of these precursors of pretense in apes might also explain why more complex cases of pretend-like behaviors can be induced in apes subject to formal procedures of symbolic training. A protracted and intensive training in symbol use may be adding something to the pre-symbolic behaviors we found in our gorillas. Premack (1984) suggested that symbolic training may provide apes with ways of representing aspects of the environment in more abstract formats. This is reflected in their superior ability to solve certain kinds of problems and to engage in some form of referential communication with the symbols they have learned. It seems that, in a surprisingly Vygotskian mode – taking advantage of what we could call a "Zone of Proximal Evolution" – linguistic apes have learned to use symbols as referential tools. These newly acquired symbolic skills may be built upon pre-existing abilities for schematization and decontextualization of actions, like those exhibited by our symbolically naïve gorillas. One result would be the apparently more complex pretend examples reported in symbolic apes.

This interpretation would allow some integration of Piagetian theories of pretense with recent theory of mind views (e.g., Leslie, 1987; Perner, 1991). Piaget's account of the progressive construction of the symbolic function in human infants may actually describe some of the evolutionary steps (such as generating abstract, decontextualized representations of actions) that preceded the evolution of more specialized, hardwired symbolic capacities. Current human children would deploy such hardwired capacities alongside more general abilities for schema generalization and decontextualized action. Some apes (those subject to intensive human "education") may construct (or have constructed in them by their human caretakers) some sort of rudimentary semiotic function *à la* Piaget that allows them to show some surprisingly genuine-looking examples of pretense and other symbol-using behaviors. In these apes, human experimenters have not constructed the symbolic function(s) as known in modern humans, but rather have promoted some symbolic function(s) probably supported by partially different cognitive processes.

Apes that did not benefit from symbol training show a more limited range of pre-symbolic and pre-pretend behaviors. As illustrated by our observations, these may reflect the operation of basic mechanisms of schema generalization and functional object substitution ("acting instead"), which without being genuine pretense ("acting as if") can, however, give us a unique insight into how this human ability evolved.

Acknowledgments

Work funded by DGICYT Grant No. PB99–0377. Our gratitude to Madrid Zoo for giving permission to study their gorillas. Our thanks to Bob Mitchell for his patience and his, as usual, insightful and extremely helpful editorial comments.

WARREN P. ROBERTS AND MARK A. KRAUSE

19

Pretending culture: social and cognitive features of pretense in apes and humans

Understanding the origins of human symbolic behavior is an enduring quest. Little direct evidence addresses questions of how and why symbolic behavior originated, although lines of indirect evidence exist and numerous models and approaches have been offered (Lock & Peters, 1996). Among these has emerged a perspective that links our understanding of human origins with child cognitive development (Bruner, 1972; Gould, 1977; Parker & Gibson, 1979). This perspective, coupled with a focus on the evolution of cognition, has generated much recent research and scholarly attention (Bruner, Jolly & Sylva, 1976; Parker & Gibson, 1990; Mitchell, 1994a; Parker, Mitchell & Boccia, 1994; Parker & McKinney, 1999). This chapter addresses the evolution of child development with an emphasis on how its ecological contexts illuminate the evolution of a primary human symbolic domain – pretense. Further, we seek to map out connections between environmental influences on pretense and the role of pretense in the origin and persistent "reinvention" (Lock, 1980) of human symbolic culture.

Connections among play, children, the modern human mind, and culture are well appreciated. The primary adaptive aspects of our species are born within childhood, normally rich with opportunities for play and learning (Bruner, 1972), and symbolic play shares structural similarities with cultural phenomena such as myth and ritual (Bateson, 1955/1972; Turner, 1974; Geertz, 1976; Schwartzmann, 1978; Johnson, 1988). Often through pretend play (Goldman, 1998), children are transformative agents in symbolic culture (Valsiner, 1988), transforming even language (Bickerton, 1990, 1998). Unfortunately, many paleoanthropologists and psychologists studying the origin of symbolic culture focus largely on adult behaviors, treating symbolic culture as an outcome or invention of the modern human brain (Noble & Davidson, 1996; Deacon, 1997). The focus is often on symbols as adaptations (Barkow, Cosmides & Tooby, 1992) and links between symbols and tools (Gibson & Ingold, 1993). Among those utilizing developmental approaches, symbols are often viewed as emerging from structural processes canalized by epigenetic principles, especially as viewed through Piaget's theory (Parker & Gibson, 1990). Each view contributes to our understanding of the mosaic pattern of human evolution in relation to advanced cognitive abilities. Our approach is a systems perspective integrating paleoanthropology of childhood with the ecology of cognitive development in humans, our ancestors, and our closest living relatives, the great apes (see below).

The term "culture" has been appropriated variously by different disciplines, and its meanings have diversified to include social learning in seemingly noncultural animals such as budgerigars (Galef, Manzig, & Field, 1986). Culture was originally defined for anthropology by Tylor (1871, p. 1) as "That complex whole which includes knowledge, belief, art, law, morals, custom and any other capabilities and habits acquired by man as a member of society." Recent approaches to culture view it not as a monolithic phenomenon, but as composed of domains. (Similarly, the human brain, which at the individual level forms the substrate of cultural processes, is also not a monolithic general processor. Cognition appears as a mosaic of domains enhanced and facilitated by elaborate higher-order cognitive processes such as metacognition.) Some anthropologists refer to units of culture as "memes" (for example, Atran, 1998), a term derived from the biologist Dawkins (1976). Some core memes emerge in individuals from awareness of and interaction with the environment, little influenced by the social group. Core memes are socially established and maintained by basic acts of joint attention such as pointing, which are sufficient to elicit knowledge contained in other core memes (Atran, 1998). Memes for tool-use and tool-making can be acquired by simple social learning processes (Zentall, 1996) or through more complex socially guided processes (Boesch, 1991a; Boesch & Tomasello, 1998). More conventionalized symbolic memes (such as ethnobiological classifications; Berlin, 1992) require attention-directing activities to elicit shared knowledge; such memes are supported by innate cognitive and neurological domains (Atran, 1998), but the signs used rely on culturally attributed meaning. Other memes, especially those relating to religion, myth and language, are generated within and mediated by members of society. They are composed of and rely on symbolic processes.

Our systems approach to addressing the evolution of pretend play and symbolic culture synthesizes social science and biology, particularly ecology. Our approach is to situate child development within culture, viewing both as transforming and being transformed by each other (Bronfenbrenner, 1979). In ecology, "niche" refers to the role of an organism in its environment (Odum, 1993), including the organism's apprehension of the affordances (or possible utilizations) of its surroundings (Gibson, 1979). The systems approach, and the concept of a niche within which development occurs, together focus on changing transactions between individuals and their socio-cultural and physical environment (Super & Harkness, 1986; Gardner, Mutter & Kozmitzki, 1998).

Transaction occurs at the social interactional and neural levels. Fully three-quarters of the human brain develops after birth leading to what Shore (1996) describes as an ecological brain (for detailed information on many aspects of brain development see Gibson & Petersen, 1991). Due to a high degree of neural plasticity, the young brain goes through immense structural changes as it integrates, apprehends, and constructs its environment. Conversely, the young child comes into the world with a basic set of tools including temperament, reactivity, and bonding. With these tools the child immediately engages in feedback loops between itself and its social world with profound and lasting consequences at levels of analysis from molecular to behavioral (Scarr & McCartney, 1983; Gottlieb, 1992; 1998; Small, 1998). To examine niche structure, the development of pretense, and species adaptability, we will look at the ecological context within which symbolic behavior in great apes can emerge.

Pretend play

As this volume attests, human symbolic play exceeds that of apes in complexity and frequency (contra Fouts & Fouts, 1993), and the rarity of symbolic play among apes may be comparable to that of extremely impoverished humans (e.g., Wachs, 1993). Yet there are some examples of pretense in cross-fostered (human-reared) apes. The most often cited example is the chimpanzee Viki's play with an imaginary pull toy (Hayes, 1951). Other cross-fostered language-trained great apes show pretend play in treating teddy bears and blankets as if alive (*animation*), or treating one object as if it were another (*substitution*) (R. Gardner & Gardner, 1969; Patterson & Linden, 1981; B. Gardner & Gardner, 1985; Jensvold & Fouts, 1993; Miles, 1994; Miles, Mitchell & Harper, 1996; Matevia, Patterson &

Hillix, PIAC21). For example, the chimpanzee Washoe bathed her toy doll, signed to inanimate objects, and held a brush under her arm (something she normally does with a magazine) and signed BOOK IN. The chimpanzee Moja treated a purse as if it were a shoe, by removing an actual shoe from her foot and replacing it with the purse (Jensvold & Fouts, 1993).

The language-trained chimpanzees Moja and Tatu modified their appearances with clothing, masks, and make-up while looking at their mirror reflections (Roberts, Krause & Fouts, 1997). In one instance, Moja repeatedly placed a pair of sunglasses on her face, looked into a mirror, and signed GLASSES. Similarly, Tatu put on a hockey mask and stared at her image in a mirror. In other instances Moja looked at herself in a mirror while applying lip gloss, or used a small crayon in the same fashion. (Similar uses of a mirror for self-transformation are described by Hayes, 1951, for the home-reared language-comprehending chimpanzee Viki). The pretend play of home-reared, language-trained apes suggests a link between pretense and either language use, intensive human interaction, or both (Call & Tomasello, 1996; Miles et al., 1996; Matevia, Patterson & Hillix, PIAC21). The importance of language use and human interaction is evident when the behavior of cross-fostered apes is contrasted with the relatively low frequency of symbolic behavior found in laboratory and nursery-reared apes (Mignault, 1985; Call & Tomasello, 1996) and feral apes (Hayaki, 1985; Goodall, 1986; Parker & McKinney, 1999). Goodall (1986) reports one wild chimpanzee apparently using a twig to fish for ants where she likely understood there were none, and Hayaki (1985) described one wild chimpanzee instigating play with a clump of branches. While possibly suggestive of pretense by wild chimpanzees, these anecdotal accounts are far more ambiguous than the descriptions of pretense in human-reared chimpanzees.

The data from ape studies show that under special conditions apes can express some capacity for pretense and similar symbolic phenomena. Although feral apes show some symbolic behavior (Boesch, 1991b), the most advanced examples of pretense and symbolic behavior seem to occur in conditions where apes are treated "as if" they are human children, in human-structured environments emphasizing teaching, training, and great effort to teach skills (Hayes, 1951; Gardner & Gardner, 1969; Patterson & Linden, 1981; Miles, 1994). Human intervention may be essential – the chimpanzee Loulis, raised almost exclusively by the cross-fostered chimpanzee Washoe, showed less symbolic behavior than his adopted mother (Fouts, Hirsch & Fouts, 1982).

While some human groups employ elaborate teaching methods for advanced realms of knowledge, these are not universal and do not appear to be necessary for the overall acquisition of human symbolic culture. Nonliterate societies with simple technologies produce culturally competent individuals. Recent work indicates that pretense is a primary mechanism by which culture is reproduced or "reinvented" by younger generations (Fein, 1979; Galda, 1984; Goldman, 1998). In collaborative pretense, children create the transformational space in which they conventionalize, internalize, and enact the roles, codes, scripts, beliefs, models, and forms of human symbolic culture (Goldman, 1998). Goldman's (1998) work with the Huli shows symbolic culture and make-believe to be mutually referential, with culture instantiated in any act of pretense, and pretense transforming culture. Even in the West, with its high degree of institutional intervention in child learning, pretend play still forms the key arena for the development of skills supporting heavily instructed areas of learning such as literacy (Fein, 1979; Galda, 1984). These data support Vygotsky's (1935/1978) position that symbolic behavior can emerge in child–child interaction within the zone of proximal development (i.e., the "distance" between a child's problem-solving ability when alone and when collaborating with peers or guided by an adult). Given the essential nature of children's pretense in the reinvention of symbolic culture and behavior, we should ask how childhood evolved in our species.

The evolution of childhood

Given that chimpanzees and humans share about 98.4% of their DNA, biologists have wondered how to account for obvious species differences. The traditional explanation is that the differences lie in genes regulating development, suggesting that the course of human evolution has been altered by changes in the timing of development – i.e., heterochrony (Gould, 1977). Gould offered neoteny – the retention of juvenile features into adulthood – to account for the appearance and behavior of modern humans. However, neoteny is not sufficient because in many ways humans are overdeveloped not underdeveloped apes (McKinney & McNamara, 1991; Parker, 1996; Parker & McKinney, 1999). An alternative explanation is hypermorphosis – delay in the offset of developmental events, such that phases are longer (McKinney & McNamara, 1991). A better explanation, in our view, is that humans introduced an entirely new life stage into their development – that of childhood (Bogin, 1998).

Childhood is defined as the period following weaning prior to dietary maturity (eruption of molars) and the attainment of adult brain weight (around 7 years of age, when adult grammar emerges; Salzmann, 1998). During this time, children retain immature dentition, digestive tracts, and cognitive competency, and show rapid brain growth. Whereas apes go from infancy (or the period between birth and the cessation of maternal lactation) to juvenility (maternal independence prior to sexual maturity), childhood separates these periods in humans. Children under 5 years use 40–85% of resting metabolism to maintain brain activity, and thus require a diet rich in energy, lipids, and proteins; they also require extensive supervision for protection (Bogin, 1998). By the juvenile period, special foods are not required, brain and body growth are maintained by less than 50% of total energy needs, and physical strength and advanced cognitive processes are in place such that juveniles can protect themselves from predation and acquire much of their own food, as can juvenile apes (Bogin, 1998). Given that childhood is neither an extension of preweaning, nor an extension of independent and reproductively immature juvenility, it appears to be a unique life-stage. Is it adaptive?

The existence of childhood presents a demographic dilemma for humans. How can a species that produces few offspring and who require costly parental investment become more, rather than less, adaptable? After all, apes become sexually reproductive earlier than humans, yet chimpanzee populations barely maintain zero population growth and orangutans have negative population growth (Bogin, 1998).

The response to this dilemma is that childhood allows a decrease in infancy-related investments (Bogin, 1998). Similar to other mammalian species, humans introduce adult foods at about 2.1 times the birth weight, but we wean exceptionally early – at only 2.7 times birth weight. (By contrast, orangutans wean at 6.4 times birth weight, bonobos at 6.1, chimpanzees at 4.9, gorillas at 9.4, and the "average" primate at 4.6.) Freedom from nursing decreases birth intervals, and non-nursing children are relatively inexpensive to feed – a 5-year-old requires 22.7% less dietary energy per day than a 10-year-old. Children still need provisioning and protection, but babysitting is possible and can be accomplished by any competent human in a group, not just the mother, allowing her to increase resources for infants. Due to these and other factors, human children survive at a higher rate than young apes, allowing population growth.

The first major advance in population expansion may have arisen with *Homo erectus*. This was the first species of our genus to colonize successfully outside of Africa. Populations filled Africa, Europe and Asia, giving *H. erectus* a wider distribution than any preceding primate (Fleagle, 1998). Additionally, *H. erectus* achieved a larger-than-ape-sized brain, and shows evidence of a nonapelike (but still not fully modern) developmental pattern (Walker & Leakey, 1993). However, more light is shed on the importance of childhood and shifts in developmental timing by looking at more recently derived members of our genus – Neandertals and our own ancestors.

Among Neandertals, adult brain size was achieved by the age of 3–4 years (Stringer & Gamble, 1993). This suggests insufficient time between weaning and juvenility for the presence of childhood. Such patterns have long encouraged scholars to focus on childhood as a source for our advanced symbolic abilities. However, a clear problem emerges from this. Our anatomically modern ancestors appeared between 200 and 100 thousand years ago, yet we have no clear evidence of language, long-range planning, social organizational complexity, art, adornment or any other symbolic behavioral indicators until a mere 60–30 thousand years ago (Jolly & White, 1995). The best-known and well-documented evidence for the emergence of symbolic behavior comes from the European Upper Paleolithic Revolution, beginning about 40,000 years ago. If childhood existed for 100,000 years before the occurrence of symbolic culture, childhood is insufficient to explain human symbolic evolution, but it may well be a necessary ingredient.

Among Neandertals we see an apelike life history pattern and some symbolic behavior, the clearest example being burial of the dead which sometimes included tools (Stringer & Gamble, 1993; Trinkaus & Shipman, 1993; Mellars, 1996). Although once richly interpreted, these burials are now viewed as suggestive of only a small degree of symbolic processing (Gargett, 1989). If Neandertals had some symbolic processes, then social factors in Neandertal populations may have allowed juveniles to be treated "as if" they were children, thereby eliciting, from a more advanced brain, slightly more advanced but truly symbolic processes reminiscent of those seen in apes treated as if they were children. But these meager acts of symbolism were far outstripped by our ancestors shortly after they arrived in Upper Paleolithic (Ice Age) Europe.

Our ancestors arrived with the same tools and probable behaviors of Neandertals and (in some cases) late *H. erectus*. For possibly as long as 160 thousand years they lived in Africa with behavioral patterns similar to those of other archaic human groups, despite having the condition of childhood. Originally, childhood may have helped population growth, fueling expansion beyond Africa, but the cognitive promise of childhood remained in the realm of capacity. Among the most important factors promoting child symbolic development may have been another change in life history – the emergence of post-reproductive females. Among prehumans, individuals over the age of 35 comprise fewer than 10% of found remains. By contrast, in preindustrial, nonWestern groups, such as foragers, 50% of the population is over age 35 (Stringer & Gamble, 1993). When our ancestors arrived in Europe, they brought not only children, but something equally important – grandmothers.

The key adaptive advantage of childhood from a purely biological standpoint – increased survival and population growth – may have been magnified by the effects of women with no dependent offspring. The dependency of childhood is characterized by the need for protection and provisioning; among contemporary foraging groups, grandmothers provide both (Hawkes, O'Connell & Blurton-Jones, 1997). In safe, highly productive areas, children may forage for much of their own food with little risk from predation. Young physically able grandmothers, free of dependents, forage and support grandchildren, providing a selective advantage for menopause; older grandmothers, unable to lift heavy stones or dig for food, stay in camp, allowing indirect adult supervision and protection of children (Hawkes *et al.*, 1997). Unlike today, however, Upper Paleolithic Europe was not a safe environment and foraging was risky – predators were ubiquitous and climate highly unstable, including arctic periods. These conditions would tax mothers trying to both forage and protect young children, and would make food-gathering sometimes too strenuous and dangerous for children unable to extract resources from a treacherous environment. Thus the foraging and protection benefits of grandmothers would have allowed greater survival rates for humans (i.e., *Homo sapiens*) than for Neandertals, who went extinct shortly after the arrival of *H. sapiens*.

Systems approach to the changing ecology of childhood

The migration of our ancestors into Europe marked a major change in the ecology of childhood. Upper Paleolithic Europe was marked by frequent climatic changes including near arctic conditions and worse (Potts, 1996).

During the dramatic climate of the Middle to Upper Paleolithic transition, gathered foods often must have been very difficult to obtain. Foods gathered must have varied (as did animals hunted), and increased consumption of meat influenced social organization – reducing the age of menarche and decreasing interbirth intervals (Cachel, 1997). The consequent risk of sibling competition and drains on maternal resources may have amplified grandmother effects, and the need for protection from predation and the elements would have intensified the role of older people as protectors. Upper Paleolithic Europe is marked by increased sedentism and population size (Cachel, 1997) and is associated with regional stylism, diversification of technology, representational art, settlement planning, adornment and ornamentation, and other indicators of social division, organization, and demarcation of individual status and affiliation (Peters, 1996; Stringer & Gamble, 1993). This systems-level change accounts for and takes advantage of individual contributions. The effects of post-reproductive females and their kinship ties with grandchildren may have promoted ecological release, or a change in environment allowing the expression or elaboration of behavioral potential within the extreme conditions of the new continent. The factors influencing this release included a higher density of children, intense peer interaction, and a harsh environment where children could safely forage. (As stated above, survival rates among young apes are quite low, so that a large cohort of peers is unlikely.) Also, for chimpanzees, orangutans, and possibly our ancestors, a solution to periods of difficulty and deprivation is to fission into smaller groups, including at times solitary mothers with infants (Watts & Pusey, 1993; Goodall, 1986), thereby limiting the amount of peer interaction. With increased sedentism and group size, Upper Paleolithic children experienced unparalleled access to other children – a context ripe for development of joint pretense.

The argument for an impact of a changing developmental niche on the origin of symbolic culture has several strands (Roberts, 2001). One is that older women work while they stay in the camp, and it is slightly older children that actually care for younger children. The Gusii in Kenya are a prime example, where children as young as 5 years may be caregivers for younger children (Munroe & Munroe, 1975). In some societies, adults do not speak with children, but have slightly older children act as intermediaries in an age-dependent chain of command (Middleton, 1970). The absence of primary adult caregivers reduces adult anticipation of the needs of children and forces negotiation of internal states with slightly older children to clarify requirements. These negotiations occur within Vygotsky's zone of proximal development, the nexus by which symbolic behavior emerges in development. The need for such negotiations is also critical in children's pretend play (Galda, 1984). The developmental niche-changes in the Upper Paleolithic ecology of childhood included increased pressure for children to stay in protected areas, increased number of peers, increased care giving by peers, and decreased child foraging time, all resulting in increased opportunities for interaction involving play, collaboration, and conflict.

Pretense plays a strong role in the comprehension and production of semiotic functions and codes, including literacy (Piaget, 1945/1962; Fein, 1979; Galda, 1984; Garvey, 1990; Goldman, 1998). Pretend play interactions form the nexus for development of understanding symbols as symbols and for their construction, conventionalization, and deployment. By assigning real properties to the unreal (a rock is a bison, or a large bear is over there), children negotiate and align attributions. By assigning roles, they internalize and externalize social functions of others (you are the mother gathering mallows, and I am the old sick man), and engender conflicts which must be resolved (but I wanted to be the hunter, you can be the rabbit). These tasks require children to develop a rich notion of symbols as social objects, pivoting between real and unreal. In modern groups, children develop in a symbolic world, and such things as pretense and its cousin, prevarication, afford them a means of reinventing and transacting with their symbolic and social environments. Can this reinvention extend to the "invention" of symbolic culture?

There are reasons to believe that realization of a childhood niche and ecological release of cognitive potential is critical for understanding the origin of symbolic behavior (Roberts, 2001). We know from studies of "feral" children that language development has critical periods (e.g., Candland, 1993), and that adult brain weight and grammar are in place by the age of 7 years, the beginning of the juvenile period (Salzmann, 1998). We also know that environmental factors can release symbolic potential in relatively nonsymbolic species like apes. Further, the Middle to Upper Paleolithic transition may not have started at zero – Mousterian tools, burial of the dead, and other evidence suggest that while culture was static and limited, it was fundamentally symbolic. Arguably, limited protogrammatical language and protoculture (as in language-using apes) were in place (Parker & Gibson, 1979). Given a social symbolic platform of protogrammatical language, rich symbolic mimesis, or another simple symbolic system, it is possible that children, due to the unique but hitherto unutilized affordances of their developing brains, became central to the emergence of symbolic culture. Such a situation occurs even today, when children exposed to an impoverished, protogrammatical Pidgin linguistic system transform it into a rich, fully grammatical Creole language (Bickerton, 1990), an impressive act of cultural transformation.

Conclusion: the evolution of gullibility

Following Bruner (1972), we offer a caveat. Humans are a highly but not infinitely adaptable species. The evolution of symbolic behavior has allowed us to store and transmit knowledge, and afforded us the communicative flexibility of language, the social influence of adornment, and the power of art (Donald, 1991). Yet it is a double-edged sword. We are extremely plastic when it comes to acquiring knowledge. We are also extremely gullible. While many of our symbolic abilities are focused in benign or beneficial directions, we are also subject to epistemological pathologies. Examples of the destructive side of beliefs are present in the Ghost Dance revitalization movement which led to the first Wounded Knee Massacre, the collapse of the Aztecs in interpreting the Spaniards as fulfilled prophecy, the Comet Halle-Bopp tragedy, and Akhenaten's declaration that he was the living embodiment of the one true god which caused massive social upheaval in Egypt. Conflicting attribution and appropriation of meaning to the same props and locations caused countless deaths in the Crusades. Closer to home are beliefs in human manifest destiny and the right to ever-higher standards of living, which have resulted in massive population growth and environmental destruction. The chief difference between children's pretend play and adult symbolic culture seems to be that children frame pretense as not real, whereas adults only sometimes do (Walton, 1990) and often treat the unreal as real (de Rivera & Sarbin, 1998). We are adaptable, but if we are not wary of distinctions between real and unreal and responsible with the symbolic properties we attribute to ourselves and our world, we may find that the evolution of symbolic behavior was ultimately maladaptive. After merely 40,000 years, we may expire from a terminal case of gullibility.

20

Empathy in a bonobo

Pygmy chimpanzees or bonobos (*Pan paniscus*) have become best known for their varied and extensive repertoire of sexual behaviors and their use of sex in nonreproductive situations (Kano, 1986/1992; Furuichi, 1987; Hashimoto & Furuichi, 1994; de Waal, 1995; de Waal & Lanting, 1997). They are also quite adept at communicating their intentions and coordinating social activities, and many of these behaviors are closely linked to their highly cohesive social structure. In the wild, female genital–genital rubbing serves as a greeting among well-known individuals, as well as to form social bonds among unrelated individuals and ease tensions when entering a potentially competitive situation such as a fruit tree. Variations on a rocking gesture are used to request different kinds of close contact between individuals, ranging from close sitting to copulation (Kuroda, 1984; Ingmanson, 1992). Branch dragging is used to organize and direct group travel (Ingmanson, 1988, 1996), and play behavior is often signaled by carrying an object (Ingmanson, 1996). In captive studies, communication is also used to organize social activities – zoo bonobos use a clapping gesture to initiate grooming in a social group (Ingmanson, 1987, 1998) and have an extensive range of reconciliation behaviors (de Waal, 1989). Symbol-trained bonobos use lexicons on a keyboard to communicate both social and nonsocial desires, and respond appropriately to others' keyboard communications (Savage-Rumbaugh, 1984; Greenfield & Savage-Rumbaugh, 1993; Savage-Rumbaugh & Lewin, 1994). All of these behaviors coordinate social activities and reduce tension in the group by avoiding misunderstandings, thereby reducing aggression. With this species-typical emphasis on complex communication and social coordination among bonobos, it is logical to ask how far their social-cognitive abilities extend.

Like other apes raised among humans and taught symbol systems (Miles, 1986; 1990; Miles, Mitchell & Harper, 1996; Matevia, Patterson & Hillix, *PIAC21*), bonobos deceive both behaviorally and via the signs they learned, and engage in pretense and other imaginative activities (de Waal, 1986b; Savage-Rumbaugh & McDonald, 1988; Savage-Rumbaugh & Lewin, 1994). For example, Panbanisha enjoyed pretending monsters existed in another room (marking the sign "monster" on her keyboard) and encouraged others to participate in her game; at times she even put on a monster mask to chase her sister. Kanzi sometimes hid himself and objects from others, but also acted as if hiding "invisible" objects, then pretended to find them and sometimes offered them to others to watch their reactions. Whether any of these activities are simulations of others' actions is unclear, though I expect that, like human children, pretend actions are a mix of representations of self and other which allow the bonobo the chance of "trying out the world from another vantage point" (Bretherton, 1984, p. 10), and developing perspective-taking skills.

Because of its connection to perspective taking, pretense is likely to be related to empathy – "an understanding of (and sometimes sympathy with) the emotions of others" (Byrne, 1995, p. 111). But empathy is not easy to observe. Under experimental conditions in captivity, chimpanzees can judge *intentions* of experimenters and respond in seemingly appropriate ways (Tomasello & Call, 1997). Although bonobos have yet to be similarly tested, the intensity of their social interactions and their well-developed communication skills – potentially the result of strong selection pressures (Ingmanson, 1996) – lead one to expect capacities at least comparable to those of chimpanzees, with whom they share many characteristics (de Waal, 1988). Some anecdotes describe instances in which a bonobo apparently understood a potentially confusing social situation and tried to show that she understood another's perspective (Savage-Rumbaugh & Lewin, 1994), suggesting a basis for empathic skills.

Although I had not specifically set out to study this type of behavior, I also have observed instances where it seemed that the bonobos not only understood the motivations behind another individual's behavior, but also how another individual was feeling about a situation. It is especially this aspect of empathy, the recognition of another's feelings, for which Byrne (1995, p. 111) says evidence is lacking, at least in chimpanzees. In this chapter I describe an incident I observed among bonobos which I believe shows empathic understanding.

I observed bonobos at the Zoological Society of San Diego, USA, from the spring of 1986 through the summer of 1987, and again during the spring of 1988. During these periods I collected observational data on the two social groups there on a regular basis. The groups consisted of seven or eight individuals, with some variations in group composition over time. Each group was observed at least once a week, and often twice during the week. I collected data using both ad lib notes of general group behavior and focal animal sampling. With this technique, the behavior, social partners (if any), location in the enclosure, and distance from all other individuals in the group were recorded for the target individual at one-minute intervals for 30-minute sessions. I also recorded occurrences of some behavioral events that took place during the interval, including brief physical contacts, tool use, vocalizations, and sexual activity. This method allowed not only for easily quantifiable data, but also for a rich context for the behaviors.

One of the most impressive instances of apparent empathy and social understanding I have observed in the bonobos occurred in one of the captive groups at the San Diego Zoo in the spring of 1987. At the time, the group consisted of seven individuals who were part of an established social group and knew each other well. There were two young adult males, Kevin and Kalind, an adult female Loretta, her infant Lena, young adolescent males Kak Jr. and Akili, and juvenile female Lisa. Brothers Kevin (age 12 years) and Kalind (age 10 years) were only 18 months apart in age, and Kevin, as the elder, had always been the dominant of the two. However, as they reached their adult growth, Kalind was larger than Kevin, and had begun to try to assert his dominance over Kevin. For many months there had been sporadic tension between the two brothers, with Kalind periodically testing his ability to dominate Kevin. Kak was also a full brother to Kevin and Kalind, but at 7 years old was only just entering adolescence and not a threat to the dominance of either of his older brothers.

On this day Kevin and Kalind were again vying for position. I had been observing the group from mid-morning, and Kevin and Kalind had been engaging in frequent advance/retreat behavior, accompanied by many vocalizations, particularly high-pitched screams. The tension between them was obvious, affecting the entire group. All of the females and Akili kept to the edge of the large enclosure, well away from the two brothers. Kak, though, appeared to be hovering on the periphery of the activity between Kevin and Kalind. Kak remained 15–20 feet (*c.* 4–6 m) away from his brothers, staying oriented to them but not interacting with them directly. He appeared to be following his brothers' activities closely, without ever actually approaching them.

In the early afternoon, per routine, the keepers placed browse into the enclosure, consisting of an armful of leafy branches. The entire group rushed forward, with Loretta taking a large pile off to one side and Lena and Lisa accompanying her to share. Kevin took possession of most of the remaining browse, leaving only small pieces. As young adolescents, Akili and Kak were accustomed to this. But Kalind did not appear happy with the situation. He approached Kevin, who responded by gathering the browse and turning his back to Kalind, denying him access. When Kalind again approached, Kevin took his pile and moved several feet away. Kalind persisted in approaching Kevin, with Kevin retreating and avoiding him for almost 10 minutes. As a result, Kevin was unable to sit still and eat any of his browse. Kalind showed some piloerection and gave high-pitched vocalizations, clearly displaying frustration. As had been the pattern most of the day, Kak stayed away from the action while continuing to watch. The rest of the group remained on the opposite side of the enclosure.

After about 15 minutes from the time the browse had been provided, the maneuvering between Kevin and Kalind intensified, resulting in a physical scuffle. As is typical of many primate fights, the contact lasted only a few seconds, and was so rapid that I was unable to determine who had made the first move, or what precisely had occurred between them. The outcome, however, was clear. Kalind gathered up the browse and walked away from Kevin. He then sat down about 15 feet (c. 4.5 m) from where Kevin remained. With his back to Kevin, Kalind began eating the browse, making happy, chirping food vocalizations. All of this occurred in no more than 15–20 seconds. Kevin remained where he was sitting, with his shoulders and head slumped – the picture of dejection.

At this point Kak exhibited what I took to be empathy. Kak approached Kevin, placed his right hand on Kevin's shoulder, and looked him straight in the eyes for about 3 seconds. Kak then walked over to Kalind, sat by his side and groomed him.

For me, this was one of those moments when as a researcher you are astounded by what you are observing. It was not just that Kak recognized that something important had occurred and that a change in status needed to be acknowledged. That could have been accomplished simply by approaching and grooming Kalind. But first he approached Kevin and seemed to console him, using both physical contact and the direct eye gaze that is an important part of bonobo communication and social interaction. My impression was that Kak clearly understood not just the change in status, but how Kevin would feel about it. Kak was displaying his understanding of the situation, its outcome, and its effect on the principle participants.

Such perspective taking seems closely linked to empathy. Captive research with bonobos, with their unique features of their social behavior, certainly suggests that they have at least some psychological understanding of their social world. Knowledge of the context in which empathy occurs is vital to our understanding and even recognizing it. Such knowledge is much more difficult in the wild, where a more complete history of the interactions and relationships between individuals is less well-known. Without my knowledge of the situation between Kevin and Kalind over the previous weeks, as well as during the hours leading up to the incident described, I might have dismissed the fight and Kak's response to it as insignificant. But this was a major turning point in the long-term relationship between the two older brothers, which younger brother Kak understood.

A few hints of such complex social understanding in bonobos have come from field research at the site of Wamba, in what is now the Democratic Republic of the Congo. In particular, there have been situations where adult males have taken over much of the care of infants when orphaned or neglected by their mothers (Ingmanson, 1990; C. Hashimoto, pers. comm., August 1996). Such infant care may reflect complex social understanding and empathy on the part of the males toward the distressed infants, or a more basic understanding that the infant requires care and that they are able to provide it. From my own observations at Wamba, I lean toward the empathic interpretation. But only further long-term research with extensive contextual information will allow us to develop our understanding of bonobo social cognition and perspective-taking skills.

MARILYN L. MATEVIA, FRANCINE G. P. PATTERSON, AND
WILLIAM A. HILLIX

21

Pretend play in a signing gorilla

Consider the following examples of pretense:

Michael runs around his room dragging a small plastic wagon behind him, sometimes letting it "catch up," and then he screams in mock terror.

Chantek talks to his toy animals and offers them food and drink.

Viki walks about the house, "pulling" what seems an imaginary toy on a string. She stops occasionally and tugs at the "string," as if the "toy" is caught on something. Once, she "dipped" the item into the toilet bowl, raising and lowering the "string."

Koko lifts an empty toy teacup to her lips and makes loud "slurping" sounds, as if drinking.

Austin pretends to eat imaginary food with gusto, scooping out and swallowing large bites of nothing.

What could be more obvious as examples of pretense? In the first two examples, the pretender playfully attributes animate qualities to a blanket and toys and, in the second three, the pretender treats nothing as a toy, as tea, and as food. If Viki, Chantek, Michael, Koko, and Austen were human toddlers, most observers would not hesitate to call these episodes examples of "pretend play." But these are not human children, they are great apes: chimpanzees (Hayes, 1951; Savage-Rumbaugh, Shanker & Taylor, 1998); an orangutan (Miles, 1990); and lowland gorillas (Patterson & Kennedy, 1987). In this chapter, we present evidence of pretend play by great apes, focusing on one in particular – the sign-language-using gorilla Koko.

The problem of pretense

Pretense or make-believe begins in humans about when first words appear – at about 12 months of age (Flavell, 1985). Both pretense and language require an ability to use and manipulate symbols (Piaget, 1945/1962), and advances and delays in language development in human children appear to correlate positively with the frequency, sophistication, and complexity of pretend play (Lovell, Hoyle & Sidall, 1968; Bates et al., 1979). In their pretend enactments, children initially use familiar objects in familiar ways, and gradually develop by using novel or discrepant objects in familiar and unfamiliar scenarios, incorporating other children more directly into their pretense, planning more elaborate scenarios, and using more imagination and role playing, all of which are concomitant with their developing linguistic skills (Fein, 1975; Elder & Pederson, 1978; McCune-Nicolich, 1981; Pederson, Rook-Green & Elder, 1981; Cole & La Voie, 1985; Leslie, 1987; Garvey, 1990). Perhaps because of the intimate relationship between pretense and language in human development, theorists sometimes posit that pretense is not possible without language (Smith, 1982, 1996) and is, therefore, exclusively human (Flavell, 1985; Bickerton, 1990; Taylor & Carlson, PIAC12). Some researchers even argue that one can attribute pretense only if the pretender's utterances prior to the "pretense" reveal an intention to engage in pretense or symbolism (Huttenlocher & Higgins, 1978).

Yet most criteria for pretense do not require language. Such criteria usually include playfully re-enacting one's own or another's actions ("role playing"), treating an object as if it were a different object ("object substitution") or as if it can do things it cannot ("attribution of function"), treating an inanimate object (e.g., a doll) as if an animate being ("animation"), treating nothing as if something ("insubstantial material attribution"), and labeling oneself as other than one is ("character attribution") or a situation as other than it is ("insubstantial situation attribution") for playful purposes (Piaget, 1945/1962; Matthews, 1977; Leslie, 1987; McCune & Agayoff, PIAC3; Gómez & Martín-Andrade, PIAC18). Only the last of these criteria requires language, although certainly all are more easily recognized when children identify their ideas or intentions in relation to their actions and the objects they use in play. For this reason, language-trained great apes make promising subjects for the study of pretense in nonhuman primates (Jensvold & Fouts, 1993).

Koko

For nearly three decades, Koko, a 28-year-old female western lowland gorilla, has been learning and using sign language. She was born in the San Francisco Zoo, USA, in 1971, and at 12 months of age began her sign instruction while being reared much like a human child. She received intense social interaction and tutoring by one primary caregiver (Dr. Patterson, assisted by Dr. Ronald Cohn), with support by several secondary caregivers (long-term research assistants and volunteers) and a community of more transitional social partners (student interns, who worked with the project for one or more semesters at a time). She has been immersed in a richly linguistic environment, where structured vocabulary learning is supplemented with active narration, reading and storytelling. Her facility with sign language as a means of symbolic social communication, comparable to that of other great apes (Miles, 1999), has been demonstrated through her appropriate comprehension and use of word order, her novel combinations and invented signs to accommodate unfamiliar events and objects, her appropriate use of internal state terms, and her use of modulation and facial expression to convey changes in meaning or expectation of a response (Patterson, 1978a,b, 1979, 1980, 1984; Patterson & Linden, 1981; Patterson, Tanner & Mayer, 1988; Patterson & Cohn, 1990, 1994; Bonvillian & Patterson, 1993; 1999).

The examples that follow were selected from a small sample of our voluminous records; some have been previously reported. They are mostly drawn from our ad libitum daily diaries; two are from videotapes. Although we have numerous examples (some on film) of Koko making her monkey and ape dolls, plastic lizards, and assorted character dolls such as trolls "kiss" or "bite" one another unaccompanied by signing, in this presentation we focus on pretense in which signs were used. None of these episodes resulted from controlled test situations. Koko allows us to enter into her pretense, and will regularly make X do Y when asked, but asking her to pretend so that we can watch is almost certain to provoke "the silent treatment" – once she realizes she is being observed, she tends to stop what she is doing and look "embarrassed" or irritated. This apparent self-consciousness is a well-known characteristic of gorillas (Yerkes & Yerkes, 1929; Patterson & Cohn, 1994), and can be observed in some chimpanzees (Hayes, 1951) and human children. Surreptitious taping is sometimes possible (Jensvold & Fouts, 1993) from a single perspective in her indoor facilities (especially with some recent advances in affordable videotape technology), but historically the cameras could not be hidden and Koko simply moved away from the lens to regain privacy.

In the examples that follow, unless designated as "voice only," all observers communicated with Koko in a combination of sign and simultaneously spoken English; pauses in Koko's signing are represented by three periods between words, and the apparent end of an idea by a period at the end of utterances. Information about her actions or relevant to interpreting the signs is presented in parentheses. Except for Penny Patterson, all interactants named below are long-term research assistants.

Animation

Pretend play with dolls is frequent with Koko, and is present in other sign-using apes (Jensvold & Fouts, 1993; Miles, Mitchell & Harper, 1996). Koko pretends that dolls are companions, nurses them, signs to and on them, makes them interact (usually biting or kissing), supplies them with instructions, and takes part in the pretense herself, much as children do.

December 1977

With Cathy Ransom watching discreetly, 5-year-old Koko enacted what appeared to be an exchange between two toy gorillas, one pink and one blue.

> K: BAD BAD (looking at the pink gorilla). KISS (directed at the blue gorilla). CHASE TICKLE. (Koko hits the two gorilla dolls together, then she "joins in" and tussles with the two dolls. After awhile, she stops.) GOOD GORILLA, GOOD GOOD.

Koko now notices that Cathy is watching and abruptly leaves the dolls, a typical response.

January 1984

With Barbara Hiller, who had placed a "Smurf" doll into the gaping mouth of a dinosaur toy, 12-year-old Koko's laughter and actions in response to Barbara's comments again reveal her use of animation and her sense of humor.

> K: THAT TROUBLE THERE (pointing to the "Smurf" doll).
> B: Can you save the baby? (Koko takes the doll out of the dinosaur's mouth and kisses it.) You saved the baby.

Koko laughs and puts the doll back into the dinosaur's mouth. (Figure 21.1 presents a scene similar to the one just described, with Koko making an alligator bite another toy.)

February 1986

Koko (age 14 years) had been misbehaving with Joanne Tanner, and had been scolded by both Joanne and Penny Patterson. Both times, her afternoon snack was delayed. After the second delay, Koko went to the back of her room and retrieved three dolls – one animal and two human. She

Figure 21.1. Koko makes alligator toy bite another toy.

returned and sat with them next to Joanne, who recorded the following conversation:

> K: GOOD. (Koko kisses a doll and looks at Joanne.)
> J: What does a good baby do? (Voice only.)

Koko puts the smallest doll to her nipple, then has it kiss the largest doll on the mouth. She then kisses them all and makes them kiss each other. (Figure 21.2, a continuation of Figure 21.1, shows Koko in a playtime scene similar to the one just described, signing DRINK while holding an alligator to her nipple.)

April 1987

Penny Patterson recorded the following episode and conversation with 15-year-old Koko, who had settled in her room with a baby doll and a gorilla doll.

> K: DRINK NIPPLE BABY (cradling the baby doll). NIPPLES DRINK THERE (to the gorilla doll's mouth). YOU BOY (to baby doll in her lap; she might have signed, instead, YOU MIKE: the signs are similar and Penny couldn't distinguish which one Koko used. Koko pulls the neck of the baby doll).

Figure 21.2. Koko signs DRINK while holding alligator doll to nipple.

P: Careful with his head. Or her head. (Voice only.)
K: GORILLA HAVE (holding the doll on her chest). FOOT (signed on the baby's foot; then Koko mouths the foot. Koko uses the word "foot" to refer to human males; in this case, she might have been clarifying for Penny – who had been ambiguous about the human doll's gender). YOU (to the baby doll, which she then cradles at her stomach). THAT (on the gorilla doll's bellybutton; she then puts the baby doll on the gorilla doll's stomach). STOMACH (on the gorilla doll's stomach).

Koko then takes the baby doll, pulls its legs apart, shakes it, and cradles it. She seems frustrated. Then she strokes and kisses the baby doll, and cradles it. She retrieves a book and a small monkey clip toy, and mouths a label on the toy. Then she bites the monkey's face, kisses and cradles the monkey, and presses it face-to-face with the baby doll.

October 1988

Koko was lying down, holding her pink lizard toy, with Mitzi Phillips nearby.

K: TEETH.

Koko examines the toy lizard's mouth, kisses it, puts two lizards together as if kissing each other, then gives them a three-way kiss. She puts her hand into the toy's mouth and pulls it out and shakes her hand.

M: Oh, did it bite you?
K: BITE.
M: Oh no! Does it hurt? (Koko kisses her finger.) May I see that bad alligator? (Koko gives it to her.) Why did you bite Koko? (Mitzi puts the alligator to her ear, as if listening to its explanation. She then returns it to Koko, who kisses it repeatedly.)
K: ALLIGATOR. GORILLA. BITE. GORILLA NUT NUT NUT. STOMACH TOILET.

Later, Koko is eating her snack, with Mitzi taking notes. Koko again engages in animation, as well as what could be a labeling error, but is more likely (given her other, more certain uses) a labeling of another's object substitution.

K: NUT SANDWICH (for her peanut butter sandwich treat).
M: And what is on it?
K: APPLE.
M: Next?
K: BERRY BEAN (her compound sign for peas).
M: Does your alligator want some peas?
K: NO (emphatically).
Koko hands Mitzi a strawberry stem and Mitzi puts it on the alligator's head.
K: HAT

June 1991 (on videotape)

Penny Patterson brought an orangutan doll to 19-year-old Koko. As usual, Koko treated the doll as if it were alive. Even more intriguing, however, is that she molded the doll's hands into appropriate sign configurations to have it respond to a question from Penny (see Miles, 1990, for similar observations with the sign-using orangutan Chantek). This kind of sophisticated social pretending is exhibited by 4- and 5-year-old human children (see McCune & Agayoff, *PIAC3*). (Koko's name sign is a "K" signed

Figure 21.3. Koko molds chimpanzee doll to sign MOUTH.

at the shoulder; she compounded the name sign with the sign for "LOVE," so it is represented as KOKO-LOVE.)

> P: Look what I brought, Koko.
> K: NIPPLE, NIPPLE HAVE KOKO-LOVE. (Koko reaches for the doll and gets a few strands of Penny's hair, as well.)
> P: Ok, let's take just the baby and not my hair. (Koko hugs the doll to her chest.) Oh, you love.
> K: DRINK.
> P: Where does the baby drink? (voice and sign)
> K: DRINK (using the doll's thumb to form the sign on the doll's mouth). MOUTH (using the doll's index finger to form the sign on the doll's mouth). (See similar behavior in Figure 21.3.)
> P: (translating) "Drink"–the baby's signing, she's got the baby signing–"mouth" (Penny turns to camera). She's had the baby answer; the baby said "drink" (Penny uses Koko's doll to imitate Koko's actions with the doll) "mouth"! That's right!

March 1998 (on videotape)

DeeAnn Draper set up a video camera and left to retrieve a snack for 26-year-old Koko. Koko went inside and prepared a "nest." In another example of Koko's use of animation in her play, she picked up a small plastic gorilla and kissed it.

> K: KOKO-LOVE NIPPLE DRINK HURRY. (Koko then put the toy down.) THIS (pointing to the gorilla).

Ten minutes later, DeeAnn turned her back to Koko to adjust the camera. The camera caught Koko picking up the gorilla toy again, kissing it on the head and signing:

> K: ME.

Koko kisses it again, then drops it quickly when DeeAnn turns back to her and sees the last kiss.

> D: Is that a nice gorilla there?

Object substitution

The first instance is simple object substitution, using a bottle label as a hat. Note that Koko's caregiver suggests another object substitution, that of a plastic ring as jewelry. The second shows a sequence of one object substituting for a variety of objects. In addition, we provide a photographic sequence (from an unrecorded scene) showing Koko pointing to the mouth of a hippopotamus in a calendar photograph (Figure 21.4), then folding the calendar and placing it in her mouth, creating giant lips (Figure 21.5).

October 1995

Wendy Gordon brought 22-year-old Koko a small bottle of water. Koko drank it, and then bit the top plastic ring off the bottleneck and put it on her finger. Then she blew into the top of the bottle, as she had often seen her caregivers do.

> K: THAT (points repeatedly at the ring).
> W: You have jewelry!

Koko blows into the bottle again, then removes the label and puts it on her head.

> K: HAT.

Figure 21.4. Koko points to hippo's mouth.

Koko then holds the clear bottle up to her eye and peers into it; she looks through it sideways; then she bites it, biting down several times.

> W: That makes noise.
> K: THIRSTY.
> W: You just had all that water – you can eat in about 10 minutes, OK?

Koko tries on the ring again. At that point, another research associate JJ enters the hall at the other end of Koko's room.

> W: Show JJ your ring.
> K: (Doesn't move, but signs) THAT (repeatedly on her ring).

March 2000

Koko was on her chute entrance to the small yard, watching Penny take notes. At one point, she began tickling herself with the tail of a plastic alligator she held, then making the alligator bite the duck she also brought

Figure 21.5. Koko places calendar in mouth as giant lips.

outside. The alligator "bites" several times, the final bite an exaggerated flourish as Koko shakes each toy. Koko makes a kissing noise for Penny's attention, then signs:

K: LIPSTICK.

Penny places a dab of lipstick on the end of the alligator's tail. Koko applies the lipstick to her lips via the tail, purring and making loud lip-smacking noises and moving her lips together as a woman would to "blot" them after applying lipstick. Then Koko signs:

K: APPLY-LIPSTICK. (She smacks her lips loudly, twice.) APPLY-LIPSTICK LIP (to herself; then she smacks her lips again and purrs.)

Penny hands out sugar-free gum, saying to Koko as she does, "for being beautiful." Koko chews the gum with great exaggeration, pokes her finger into the gum and presses it into her upper molars (equivalent to her sign for "FAKETOOTH"). Then she presses the gum into a ball on her chin and rolls it around her chin with the palm of her hand. She puts two fingers over the gum to her upper molar, and signs:

> K: FROWN (with the gum held by her finger on her downturned lower lip).

Then Koko pokes the gum on her finger into a toy lobster's mouth, signing:

> K: LIP.

She shakes the gum out of the lobster's mouth, puts it back in her own and signs:

> K: CLAY.

She rubs her finger back and forth on the gum in her mouth and signs:

> K: LIPSTICK.

She has a loop of red yarn around her right arm; she fiddles with it, then signs:

> K: ARM VISIT.

She spits the gum out into her hand and blows forcefully onto it, then signs:

> K: TEETH.

Then she puts the gum up her nose and blows her nose, forcing the gum back out. She repeats this.

> P: Bad idea Koko! (A nearby research assistant adds, "That is what babies do!") You might have to go to the doctor!

Koko spits the gum back out of her mouth and signs:

> K: LIP. BITE (on her finger) BITE (on her finger, holding the sign while she waggles her head and makes a facial expression as if she is biting down hard). Koko then puts the gum back into her mouth. (This "bite" exchange was almost certainly her commentary on being scolded for putting the gum up her nose.)

In this one instance Koko used gum as lipstick, a filling, food for her toy, clay, and nasal mucus!

Object substitutions appear to be relatively infrequent among sign-using apes (Jensvold & Fouts, 1993). A few other instances are present in pretenses as described below.

Character attribution

In the next examples, Koko refers to herself as an elephant and a bird, and seems to draw an analogy between a rubber hose and her nose. She reveals the joke in labeling herself and her interactant as a bird when she declares that she is a "fake" and laughs about it. Note that she applies several internal state terms to herself.

February 1978

Working with Barbara Hiller, 6-year-old Koko had been nagging for juice drinks all afternoon, without much success. Hoping to entertain and distract Koko, Barbara finally provided her with a long rubber tube that could reach a pan of water outside the mesh dividing Koko's room; Barbara and the pan of water sat on the side opposite Koko. Barbara recorded the following exchange:

> K: SAD ELEPHANT.
> B: What do you mean?
> K: ELEPHANT.
> B: Are you a sad elephant?
> K: SAD . . . ELEPHANT ME. ELEPHANT LOVE THIRSTY.
> B: I thought you were a gorilla.
> K: ELEPHANT GORILLA THIRSTY.
> B: Are you a gorilla or an elephant?
> K: ELEPHANT ME ME. ELEPHANT STINK. TIME.
> B: Time for what?
> K: TIME KNOW COKE ELEPHANT GOOD ME.
> B: You want a drink, good elephant?
> K: DRINK FRUIT.

Barbara and Koko had not been talking about elephants that afternoon, but Barbara wondered if the fat rubber tube conjured the elephant imagery. She continued.

> B: What's this? (pointing to the tube).
> K: THAT ELEPHANT STINK.
> B: Is that why you're an elephant?
> K: THAT THERE (pointing first to the tube, then to her nose: THAT NOSE, then laughing). THAT THERE (pointing to a can of soda, then her glass).
> B: Who are you?

K: KOKO KNOW ELEPHANT DEVIL.
B: You're a devilish elephant?
K: GOOD ME THIRSTY.

October 1982

Barbara Hiller had just shown 10-year-old Koko a picture of a bird in a magazine.

K: THAT ME.
B: Is that really you?
K: KOKO GOOD BIRD.
B: I thought you were a gorilla.
K: KOKO BIRD.
B: You sure?
K: KOKO GOOD THAT (pointing to the bird).
B: OK, I must be a gorilla.
K: LIP BIRD YOU. ("LIP" is a sign Koko has consistently used when referring to or naming a human female.)
B: We're both birds?
K: GOOD.
B: Can you fly?
K: GOOD.
B: Show me.
K: FAKE BIRD CLOWN. (She laughs.)
B: You're teasing me. (Koko laughs again.) What are you really?
K: (laughs) GORILLA KOKO.

For Koko, simply naming oneself as an animal seems to be enough to enjoy the "fake" pretense; human children tend to try to act out the part as well (see Myers, *PIAC11*).

Animation and character attribution

In this instance, Koko acts as if she is frightened of a dinosaur when it appears that she is not. She also acknowledges pretense in calling the dinosaur "fake" and "faketooth" (the latter a nickname she customarily gives new visitors who display a filled or capped tooth during Koko's mandatory exploration of the visitor's dental features).

October 1983

Joanne Tanner brought a new toy dinosaur with her on this afternoon, and hid it behind her legs as she sat with 11-year-old Koko.

K: CANDY THERE.

Joanne shows that her hands are empty, then she pokes the gaping-mouthed dinosaur out from behind her legs. Koko jumps backward.

> J: I scared you! What's this?
> K: FAKETOOTH FAKE.
> J: Yes, it's a fake alligator. (Koko jumps a few more times as Joanne plays with the dinosaur. Plastic dinosaurs, lizards and alligators are all labeled "alligator" in Koko's sign vocabulary.) You like it? You want it?
> K: GOOD. (But she does not take it. Then Joanne makes it "bite" her own finger.)
> J: Ow!
> K: TOILET STINK.
> J: Give me your finger, Koko. (Koko instead extends a toy doll, letting it get bitten.) You funny, Koko, let monster bite doll instead of you. (Joanne continues pretending that the dinosaur is biting her.) Let's try being nice to it.
> K: YOU-BLEW-IT! (Koko occasionally blows aggressively in her companion's face when she is perturbed.)
> J: It's a nice animal. (Joanne kisses it.)
> K: FAKETOOTH.
> J: You want to kiss it, be nice? (Koko cautiously kisses the dinosaur, then quickly withdraws.) Time to go inside. Hurry or the monster will eat your lunch! (Joanne feeds Koko lettuce from the dinosaur's mouth.)
> K: EAT.

It would not be unreasonable to interpret Koko's behavior throughout this exchange as that of a witting accomplice to Joanne's play with the dinosaur, especially since she refers to the dinosaur as "faketooth" and "fake." On the other hand, she also exhibits some trepidation about the new toy. Perhaps similarly, children sometimes experience real fear after enacting pretend fear toward pretend monsters, and even for them the line between pretense and reality is not always easily drawn (DiLalla & Watson, 1988; Mitchell, 1993c; cf. Woolley, *PIAC8*). After a break during which Koko ate, napped, and worked on a counting and reading exercise with Joanne, she requested the dinosaur. Joanne recorded that Koko took the dinosaur outside, tossed it up and caught it repeatedly, twirled it by the tail, pounded her chest with it, then took it inside and made it "bite" a doll. Her playful attitude toward it suggests that it was never as threatening as she let on, or that she overcame her fears enough to play a different game with the dinosaur.

Insubstantial material attribution

This instance of treating nothing as something developed from a game of making funny faces. Her extravagant pantomime in this case, like Austin's in the introduction, reveals her intention. Note again the caregiver's suggesting an object substitution.

September 1987

Mitzi Phillips began this exchange when she saw a peanut shell on the floor.

> M: What's this? (She puts it on her head.) A silly hat!

Koko (age 15 years) responds by pulling down her lower eyelids and sticking out her tongue.

> M: That's a good silly face! Can you do this? (Mitzi puffs out her cheeks and then slaps them. Koko does the same. Koko then opens her mouth and waggles her tongue, while shaking her hands. Mitzi imitates.)
> M: Oh, Koko, I'm so glad no one can see us!
> K: (lies down and makes more faces) EAT.
> M: It's not time yet.
> K: TOILET (almost certainly an insult directed at Mitzi).

Koko next picks up a small bug, puts it between her lips, and blows it at Mitzi. She then gets her empty bowl and pretends to eat from it, takes an empty cup and holds it to her lips, making slurping sounds.

October 1988

In this example, Mitzi Phillips tells a story, and asks 16-year-old Koko to offer elements of the story. Koko participates in the pretense, supplying emotions and motives to the make-believe characters. Interestingly, the referents are not physically present.

> M: Do you want me to tell you a story?
> K: GOOD.
> M: What do you want it to be about?
> K: ALLIGATOR.
> M: Ok, you help me. Once an alligator was very hungry. "I want ice cream," he said. "I'll go get some." On his way he met – a what, Koko?
> K: BIRD.
> M: "Do you want to come with me to get something to eat?" "Yes," said the bird, and the bird rode along on the alligator. Where did he ride on, Koko?"

K: NOSE.
M: So off they went! Soon they met a – what?
K: CAT.
M: The cat said, "I am – " What was the cat? How did the cat feel, Koko?
K: HUNGRY.
M: When they got to the store, the man gave the bird bird-food, and the cat milk.
K: MILK KOKO-LOVE.
M: So did the cat! and the alligator got all the ice cream he could eat.
K: NO.
M: Yes! and the alligator was happy.
K: NO, SAD.
M: You want the alligator to be sad?
K: GOOD, SAD.
M: Ok, it's your story. So the ice cream man said, "I don't give food to alligators! Go away." The alligator was now hungry and sad.
K: SAD FROWN CRY BAD ALLIGATOR.
M: But since he had been nice and had given his friends a ride to town, the cat gave him milk, and the bird gave him bird food. And how did the alligator feel?
K: HAPPY.
M: Yes
K: RED EAT BAD.
M: Ok, more drama! So then the alligator went back to the store and said, "Now I'll have dessert." and he ate the ice cream man!
K: GOOD VISITOR BAD.
M: Did you like that story? (Mitzi said "story," but signed "book")
K: KOKO-LOVE GOOD BOOK. (She then got a scrap of browse and gave it to Mitzi.)
M: Because my story was so good?
K: EAT KOKO-LOVE. CAT SAD BIRD SAD EAT SAD ALLIGATOR BAD SAD KOKO-LOVE VISIT.
M: If you were an alligator, what would you eat?
K: KOKO GORILLA. VISITOR DO THERE (pointing to refrigerator). TIME EAT.

Animation, object substitution, and insubstantial material attribution

In this instance, 12-year-old Koko treats her toy cat as animate (and a named animate at that), wraps a sock on the cat's head and calls it a "hat," and acts as if a bowl contains a drinkable fluid.

February 1984

Barbara Hiller watched and took notes while Koko gave her toy cat a drink from an empty bowl.

> B: That's very nice.

Koko leans over and shares the imaginary drink with the cat, making loud slurping noises. She then picks up the toy cat.

> K: SOFT. (She then puts a sock on the toy cat's head.) HAT. (She then wraps a piece of cloth around the toy cat.) KOKO GOOD.
> B: You keep your cat nice and warm.
> K: (Kisses her toy cat.) CAT THAT.
> B: What do you call your cat?
> K: CAT TIGER CAT KNOW.
> B: That's a nice name, Tiger.
> K: CAT TIGER (emphatic).
> B: Sorry. Cat Tiger is a nice name.
> K: CAT TIGER KOKO GOOD.

Pretending not to know the correct sign

Although this type of pretense might also be called stubbornness, it is interesting nonetheless in relation to acting "as if." Sometimes when Koko is asked to make a sign, she not only refuses to produce it, but "almost" produces it. Here are some examples previously reported in Patterson & Linden (1981).

> Berry, for instance, is made by lightly grasping the thumb with the fingers of the other hand, and then pulling apart. Asked the sign for berry during one drill session, Koko grasped her index finger, middle finger, fourth finger, and little finger.
>
> (p. 75)

> Drink is one of the first signs Koko learned. It was made with the hand in a hitchhiking position but with the thumb touching the mouth... one day she persistently refused to make the sign when Barbara Hiller... requested that she do so. Barbara tried everything she could think of, but Koko would only sign sip, thirsty sip, and apple sip, anything but drink. Finally at the end of the day, Barbara said with some desperation, "Koko, please please sign drink for me." Koko leaned back against the counter and, grinning, executed the sign perfectly – to her ear.
>
> (p. 77)

[O]n January 1, 1978, I wanted Koko to sign shell. Out loud we asked Koko to sign shell, first showing her a shell. There was no response. "Forgot?" I asked. Still no response. Finally I sent Koko to her room and closed but did not lock the door. As I did so I said, "Well, I'll just take these goodies to Michael." At this point, Koko edged out of the door and, unprompted, signed shell.

(p. 80)

Similar examples are present in other apes (see Mitchell, 1999a,b).

Discussion

Koko's pretend play is characteristic of "enculturated apes" – apes who have been reared with extensive human interaction and tutelage (Tomasello, Savage-Rumbaugh & Kruger, 1993). These apes exhibit more compelling evidence of, and sophistication in, pretend play than those reared more exclusively with conspecifics. The few episodes we describe here do not offer enough information to provide a thorough developmental analysis of Koko's pretense, but they prompt us to look for more evidence of the ontogeny of her pretenses.

Koko plays with her toys much like human children do – sometimes acting maternally toward them, and sometimes making them interact with one another. Through her laughter, she shows comprehension of humorous or absurd situations such as the predicament of a doll in a dinosaur's mouth. Her frequent use of the word "fake" – both in examples above and in others we hope to compile and detail in the future – further supports our interpretation that Koko grasps the symbolism of pretend play (see Flavell, Flavell & Green, 1987).

Several researchers studying children point out that pretend play with a social partner requires an even higher-order level of mental representation than is required in solitary pretend play – in social play, the child must interpret his partner's actions in a nonliteral sense (Harris & Kavanaugh, 1993; Leslie, 1987; McCune-Nicolich, 1981; McCune & Agayoff, PIAC3) – an ability which seems required in perspective taking. Indeed, children who engage in social pretend play early in their development also perform better on tasks of false belief and affective understanding (Youngblade & Dunn, 1995). Leslie (1987) has even argued that the skills required in pretense are components of a theory of mind (though others disagree: Lillard, 1993b, PIAC7; Harris & Kavanaugh, 1993; Smith, PIAC9).

Koko regularly engages her caregivers in her pretense, although it is not clear whether she sees them as fulfilling a character role or believes that they know she is pretending. In Koko's pretend activities, she sometimes assigns internal state terms to herself and the human and animal toys with which she interacts. Although Koko has a vocabulary of internal state terms (Patterson & Cohn, 1990) and has appropriately used the sign for "pretend," she does not talk about pretending as a child might when saying "Pretend this alligator is biting you." In fact, like the 4.5- to 5-year-old children studied by Hall, Frank & Ellison (1995), Koko quite frequently uses the terms "know" and "think," but not "pretend." Whether or not this is the upper limit of Koko's pretense capabilities is at present unknown.

Ever since Premack & Woodruff (1978) asked if chimpanzees could have a theory of mind, investigators working with great apes have been hard-pressed to devise experimental methods adequate to meet a behaviorist challenge (Heyes, 1993, 1998). Meanwhile, the literature examining children's theory of mind is dominated by studies documenting children's verbal accounts of their observations, expectations, intentions, and perspectives. This is especially true of pretense studies. Nonetheless, there seems to be little question about the validity of these verbal accounts, despite indications that the children's reports of their own immediately past psychological states are often inaccurate (Gopnik, 1993). Some researchers have wryly stated that no experimental test of theory of mind in the great apes will be adequate unless they can be interviewed about their beliefs and perspectives (Smith, 1996). Koko and other language-using apes make that suggestion much less absurd, if we are willing to give as much credibility to their responses as we do to those of young children. The enculturated and language-using apes will continue to provide a fascinating link between species, one that will undoubtedly reveal more to us about the evolution of theory of mind.

Part IV

Prospects

22

Exploring pretense in animals and children

Pretenses appear in six forms of activity in this volume:
- self-pretense, as in reproducing the appearance of, feigning, or inhibiting one's own acts and emotions (for play, teasing, deception, communication, or evocation);
- object substitution (including replica toy play);
- animation of objects (including doll play);
- pretending about imaginary objects (supported by real objects or not);
- pretending to be (or act like) someone else;
- pretending to have (imaginary) companions.

These pretend types are intended to be neither exhaustive nor mutually exclusive, but rather to direct attention to the pretenses observable in animals (or potentially so). All of these activities are much more common in children than in animals, and how to interpret them in either is (as this volume attests) subject to disagreement, providing a fertile field for reflection and investigation. (Imaginative nonpretend actions are discussed in the first chapter.)

In this chapter, I summarize research on pretense and direct attention to future research on animal pretense. A pattern should become evident – scientists interested in discerning pretense in animals need to observe individual animals ontogenetically in order to understand and interpret their behaviors, and also need to employ knowledge of the types of behaviors animals of the observed species use in potentially pretend actions, to discern the precursors of these behaviors (Mitchell, 1986, 1987, 1990). Put simply, extensive longitudinal study incorporating contextual analysis, focal animal sampling, and an (at least partial) ethogram are essential for discerning pretense in animals. Exactly these concerns have been in place at the start of some longitudinal studies in the field (Breuggeman, 1973,

1978; Goodall, 1973, 1986; Smuts, 1985; Strum, 1987; see Quiatt, 1997), in captivity (Lorenz, 1932/1970; Parker, 1977, 1999; Gómez & Martín-Andrade, *PIAC18*), and at home (Guillaume, 1926/1971; Piaget, 1945/1962; Fein & Apfel, 1979a; Miles, 1986, 1991; Miles, Mitchell & Harper, 1996; Veneziano, *PIAC4*). Given that researchers studying young children find it difficult (and perhaps useless) to decide precisely when pretense begins (Fein & Apfel, 1979a; Fein & Moorin, 1985), researchers studying animals need not make this demarcation their focus.

Teasing and other self-simulations

Several authors included instances of replications of animals' own actions as pretense. In some cases these self-simulations were used deceptively, in others playfully. Gómez and Martín-Andrade longitudinally followed captive young gorillas as they developed new play patterns suggesting pretense, but they posit instead that the behaviors were likely reenactment of previous actions simply using different objects or in different contexts. By contrast, Russon and Zeller each argue from field observations that many self-simulations (often used deceitfully) show a well-developed understanding of how these simulations will be interpreted (or responded to), suggesting awareness of pretense. Evidence of such awareness includes laughing, obvious exaggeration, and deception itself (Piaget, 1945/1962; Valentine, 1942/1950; Nicolich, 1977; Bretherton, 1984; Jolly, 1988; Reddy, 1991; Mitchell, 1994a; McCune & Agayoff, *PIAC3*; Russon, *PIAC16*). McCune and Agayoff express dissatisfaction with interpreting deceptive encounters as pretense, suggesting that distinguishing among alternative interpretations is too difficult (see also Mitchell, 1993c; *PIAC1*). All of these views have points in their favor, but a program to examine whether self-simulations (deceptive or otherwise) are pretenses would seem to require longitudinal observations (like those of Gómez and Martín-Andrade) of similar behaviors in the same individuals over an extended period. Similarly, Guillaume's (1926/1971) and Piaget's (1945/1962) observations of their children's pretenses derived from knowledge of the children's repertoires of actions, so that the children's new actions could be compared to older actions. This approach has been employed in studying one child's numerous (usually linguistic) deceptions from 2.5 years of age, showing some to be creative pretenses typically not based on self-simulation, and highly reliant on language for their creativity (Newton, Reddy & Bull, 2000).

Teasing is a form of social exploration observed in both humans and animals (particularly dogs and apes), which sometimes involves self-simulation and apparent pretense (Groos, 1898, 1901; Hoyt, 1941; Valentine, 1942/1950; Adang, 1984; Byrne & Whiten, 1990; Reddy, 1991). One form of teasing common among animals is object-keepaway, where an animal acts "as if" to offer an object to another, but then doesn't allow the other to get it (Lindsay, 1880; Köhler, 1925/1976; Nissen & Crawford, 1936; Murphy, in Bateson, 1956; Mitchell & Thompson, 1991; Miles et al., 1996). Although differences exist between teasing and pretending, both suggest "an understanding at some level of the difference between serious and non-serious versions of the same interactions" which seems to "reveal the first signs of pretence" (Reddy, 1991, pp. 155–6). Another form of teasing, common in quite young children, is pretending to do something (e.g., cry, sleep) to see what happens, a form of contingency testing (Groos, 1901; Valentine, 1942/1950; Reddy, 1991). It would be interesting to know when this early pretending begins, and how it relates to other versions of the activity in the child's communicative repertoire and pretend play.

Object-keepaway may have an additional "as if" quality – as when bonobo Kanzi, unable to obtain a ball from his play partner, "feign[s] disinterest in the game by acting as though his attention has wandered to something else" (Savage-Rumbaugh & McDonald, 1988, p. 233). Dogs act similarly – one used such "subtle tricks as to drop [a ball] and lie down as though . . . not looking, and when a child approached it would jump up and seize the ball" (Murphy, in Bateson, 1956, p. 208). Upon inspection of videotaped interactions, I discovered that dogs' dropping a ball and looking away from it, or not attending to the ball when the other will not give it up, seem to express *real*, rather than feigned, inattention (Mitchell & Thompson, 1991). Distinguishing "being disinterested" and "feigning being disinterested" requires elaborate behavioral and contextual description.

People commonly tease children and dogs deceptively in play, in effect "teaching" the player that actions are sometimes only "as if" (Mitchell & Thompson, 1991; Nakano & Kanaya, 1993; Labrell, 1994). In my own research, one dog–human play pair engaged in a game we called "fakeout/avoid fakeout" (Mitchell & Thompson, 1991). In this game, the person acted as if to make a ball available to the dog (but only sometimes did) in various ways (e.g., acting as if to kick, throw, or allow him to get it), while the dog acted to avoid trying to act toward the ball until it was actually available (all the while maintaining a tense stance, ready to get the ball once available, but not before). The dog's actions to avoid responding to the fakeout suggest that it understood the pretense (Mitchell, 1991a). An interesting study would be to videotape the development of this pattern in detail with several dog–human pairs, to see how dogs respond when they "catch on." (Not all do.)

Object substitution and imaginary object use

Several researchers propose that children's object-related pretense develops from real objects, to substitute objects, to imaginary objects (Overton & Jackson, 1973; Elder & Pederson, 1978; Jackowitz & Watson, 1980; Rubin, Fein & Vandenberg, 1983; Bretherton, 1984; Copple, Cocking & Matthews, 1984; Fenson, 1984; Corrigan, 1987; Lyytinen, 1991; Boyatzis & Watson, 1993; Baudonnière *et al.*, PIAC5). This development indicates decreasing reliance on context, or "decontextualization." Substitute objects include (in order of increasing decontextualization) miniaturized or other versions of "real" objects used by adults (toys), objects prototypically similar to real objects, objects ambiguous in their similarity to real objects, and counterconventional objects: the real object "telephone" might be represented (respectively) by a toy phone; a banana; a block; and a toy car. Body parts can also substitute for objects (a pinkie finger and thumb sticking out of a fist to represent a phone). Object substitutions occur more easily with prototypical objects than with counterconventional ones, but are rare when realistic objects are available (Pederson, Rook-Green & Elder, 1981; Corrigan, 1987; Musatti & Mayer, 1987; Užgiris, 1999).

Imaginary objects – things represented via action or language, but not by something present – are distinguished by being object-supported (an empty cup containing imaginary water) or not (actions miming an object's presence, as if holding or using absent objects) (Fein, 1975; Ungerer *et al.*, 1981; Field, De Stefano & Koewler, 1982; Bretherton, 1984). Such miming might be called "body-supported imaginary object use," indicating a facility in using one's body as a prop unsupported by objects. Young children vary in their willingness to use imaginary objects – "patterners" (object-dependent pretenders) seem to need some object to represent another, whereas "dramatists" (object-independent pretenders) can more easily allow nothing to represent something (Matthews, 1977; Wolf & Grollman, 1982).

Object substitution and imaginary object use have diverse functions and variable forms (Musatti & Meyer, 1987). Although children seem to develop toward less need for prototypicality in objects used to substitute and greater ability to mime without objects, a developmental progression among types of substitute and imaginary object use is not consistently observed (Elder & Pederson, 1978; Fein & Apfel, 1979a; Wolf & Grollman, 1982; Crum *et al.*, 1983).

Whether imaginary object use is more cognitively complex than object substitution is questionable. The ages at which imaginary object use first occurs range from 7 months to 23 months and these overlap with ages of initial object substitution (Guillaume, 1926/1971; Valentine, 1942/1950; Piaget, 1945/1962; Fenson *et al.*, 1976; Jackowitz & Watson, 1980; Rubin *et al.*, 1983). In some cases imaginary object play seems to derive directly from familiar routines frequently enacted with real objects, suggesting self-simulation (Guillaume, 1926/1971; Valentine, 1942/1950). The contexts in which children are asked to pretend they are using imaginary objects influences whether they mime imaginary object use or use body parts to represent the object; how effective either method is in communicating the intended object influences both children's and adults' uses (Lyons, 1986). Indeed, miming imaginary object use can occur frequently in children's communicative actions, but infrequently in their play (Bretherton *et al.*, 1981a; Acredolo & Goodwyn, 1985). These findings make it difficult to interpret correlations between children's imaginary object pantomimes and their use of mental state terms (Nielsen & Dissanayake, 2000), as alternative methods might have resulted in more widespread use of miming. Indeed, autistic children, notable for their limited comprehension of mental states (see P. Mitchell, 1997), can competently mime imaginary object use (Curcio & Piserchia, 1978).

Examination of animals' early object substitutions, such as that of gorillas by Gómez and Martín-Andrade, suggest that apparent pretenses are simply well-known or innate schemas applied to novel objects. Further research is needed on these gorillas' continued development, to see if such play activities develop toward pretense, as observed in chimpanzees (Mignault, 1985) and children (Lézine, 1973; Fein & Apfel, 1979a). As Groos (1898) and Piaget (1945/1962) suggested, playful actions may serve as a springboard for pretense. Some object play may lead to blind alleys in relation to pretense (e.g., Huffman & Quiatt, 1986) while still providing us with fascinating details about animals. Other potential pretenses with objects, such as the pretend grooming and eating of plants (Goodall, 1973; Plooij, 1978; Russon, *PIAC16*), offer topics for further study. Field researchers such as Zeller and Russon, and laboratory researchers such as Matevia, Patterson and Hillix, have shown us the kinds of things animals might use objects to represent in pretense. With this information in mind, researchers interested in pretense need to watch particular animals longitudinally, noting when they use objects at all, so as to have evidence of precursors for potential instances of object substitution. Goodall's (1973; 1986) use of this technique led her to note that young Gombe chimpanzees learn elements of poking stems and sticks into insect holes gradually, initially playing randomly with these objects, and later playfully recreating adult preparation of these tools and poking them into various objects, activities suggestive of imitation and pretense.

Researchers might fruitfully create contexts with animals like those in which object substitution and imaginary object use occur in human children. For example, more elaborate instances of object substitution and imaginary object use may occur after experience with realistic objects, when realistic objects are unavailable, or when objects are removed. It may not be surprising that the only instance of object substitution in a macaque – use of a coconut to represent a baby (Breuggeman, 1973) – occurred when the macaque was apparently unable to interact with this baby.

Metarepresentation and secondary (second-order) representation

While many authors view some forms of pretense and self-recognition as indicating the presence of mental representations (Piaget, 1945/1962; see Mitchell, 1987; 1997b; PIAC2), some authors in this volume argue that these activities utilize metarepresentation or secondary representation (Baudonnière et al., PIAC5; Russon, PIAC16; Russon et al., PIAC17). This terminology can be confusing, subject to multiple divergent interpretations (Jarrold et al., 1994; Sperber, 2000). For example, second-order cognitions in Langer's (1993, 1996) terminology would appear to be second-order representations (Russon et al., PIAC17) but, contrary to this view, Langer himself regards them as simply representations that pave the way for metacognition. As Fein & Moorin (1985, p. 68) noted, "the concept of representation will require finer theoretical tuning if it is to continue to be useful," and the same seems true of secondary representation and metarepresentation.

Leslie (1987) introduced the idea of "metarepresentation" to refer to a decoupled (or secondary) representation of reality which he believed present in early pretense. In this view, representations (internal mental copies) of reality are "primary representations." In pretense, the primary representation is detached (decoupled) from its normal use, and becomes a "secondary representation" used to denote the primary representation (see Smith, PIAC9). Perner (1991), building on Leslie's view, accepted secondary representations in pretense, but disagreed that these are metarepresentations. Leslie (1988) himself later differentiated secondary representations from metarepresentations. (An imperfect translation into Piagetian terminology might be that primary representations are simply representations, secondary representations are accommodations or assimilations of representations, and metarepresentations are operations.) After initial enthusiasm that metarepresentation could be usefully applied to understand both animal deception and early human pretense (Whiten & Byrne, 1991), a more denuded notion of metarepresentation as "re-representation" has been suggested in relation to animals and young children (Whiten, 1996; 2000; Russon, PIAC16; Russon *et al.*, PIAC17). By contrast, the notion that metarepresentation is present in young children's pretense seems discredited among many developmental psychologists, who deny its relevance to early pretense and believe it present only in older children (Harris & Kavanaugh, 1993; Lillard, 1993a,b; Jarrold *et al.*, 1994; Smith, PIAC9). Rather than regarding eye-covering games, early pretenses and deceptions, and self-recognition as showing metarepresentation or secondary representation, I suggest that these activities (and similar ones by young children) indicate representational capacities (Piaget, 1945/1962), including elaborate and generalized abilities for scripts (event representations), self-imitation, and coordination among perceptual modalities (see next section; also Mitchell, 1994a, 1999a; Gómez & Martín-Andrade, PIAC18). Being representational, these activities by apes and children also exhibit a "Symbolic functioning [which] involves holding a representation in mind which differs from the percept held in view" (Russell, 1996, p. 166), much as Baudonnière *et al.* (PIAC5), Russon (PIAC16), and Russon *et al.* (PIAC17) contend. The eye-closing games of macaques suggest limited symbolic functioning in their seeming inability to coordinate among perceptual modalities (Russon *et al.*, PIAC17).

Pretending about others

Pretending to be another is a common activity of human children, as is animating objects such as dolls. Both are even present in some autistic children (Curcio & Piserchia, 1978; Riguet et al., 1981), but seem limited in animals. Sign-using and speech-comprehending great apes show the most extensive instances of such pretense, in animating dolls with stable personality traits, talking to or for dolls, acting out roles of people they interact with, and sometimes naming (or otherwise responding to) nothing as if an animate being (Hayes, 1951; Gardner & Gardner, 1969; Patterson & Linden, 1981; Miles, 1986, 1991; Savage-Rumbaugh & McDonald, 1988; Jensvold & Fouts, 1993; Miles et al., 1996; McCune & Agayoff, PIAC3; Matevia et al., PIAC21). Other potential but ambiguous instances of animating objects are two chimpanzees treating logs as dolls (Wrangham & Peterson, 1996; Matsuzawa, 1997), and a badger's and a chimpanzee's soliciting vegetation for play when each had no play partner (Eibl-Eibesfeldt, 1950/1978; Hayaki, 1985). The evidence for completely imaginary beings in unenculturated apes is suggestive but inconclusive (de Waal, 1982; Mitchell, 1993c, PIAC1), such that imaginary companions may be exclusively human (Taylor & Carlson, PIAC12) under normal circumstances.

In my own writings (Mitchell, 1990, 1993a,b, 1994, 1997a,b, 2000, 2002a), following Guillaume (1926/1971), I have argued that pretending to be another requires kinesthetic–visual matching, a skill which is present in self-recognition, generalized imitation of other's bodily actions, and recognition of imitation by others. Other researchers (Piaget, 1945/1962; Parker & Gibson, 1979; Užgiris, 1981; Gallup, 1970; 1985; Meltzoff, 1990; Parker, 1991; Whiten & Byrne, 1991; Hart & Fegley, 1994; Asendorpf, Warkentin & Baudonnière, 1996) have also called attention to part or all of this confluence of activities and psychological abilities, providing alternative (or similar) interpretations. Baudonnière et al. (PIAC5) provide evidence supporting the presence of kinesthetic–visual matching in the apparently simultaneous development of pretending to eat imaginary cake (after watching someone else model this activity) and self-recognition (see also Lewis & Ramsay, 1999).

Among nonhumans, the most convincing evidence indicates that great apes and bottlenose dolphins can pretend to be another; given that these organisms are also those capable of self-recognition, the common ability for kinesthetic–visual matching is implicated in both activities (Mitchell, 1993a, 1994a). (Many animals show proficiency in other perceptual matchings, such as visual–visual, tactile–visual, and kinesthetic–kinesthetic, which support some pretenses – see Mitchell, 1994a; Miles et al., 1996.) Like their human-raised conspecifics, wild and captive chimpanzees evince replications of others' actions suggestive of pretense; these include ant-fishing while hanging from a tree, walking oddly, and aggressively displaying (de Waal, 1982; Goodall, 1986; Custance, Whiten & Bard, 1995). Gorillas and orangutans provide similar evidence of pretending (Chevalier-Skolnikoff, 1977; Russon & Galdikas, 1993; 1995; Miles et al., 1996; Russon, 1996; Gómez & Martín-Andrade, PIAC18). At least some members of all great ape species (and even some lesser apes) show self-recognition, whereas monkeys do not (Parker, Mitchell & Boccia, 1994; Swartz, Sarauw & Evans, 1999; Ujhelyi et al., 2000; Roberts & Krause, PIAC19). The rare instances of pretense and body matching in macaques (Zeller, PIAC13) suggest that kinesthetic–visual matching occurs at exceedingly low levels among macaques, perhaps a variant available for selection. Like great apes, captive bottlenose dolphins (*Tursiops* species) offer evidence of self-recognition and pretense – in this case detailed imitations of humans and cohabiting pinnipeds (Tayler & Saayman, 1973; Marten & Psarakos, 1994; Reiss & Marino, 2001). Whether pretense is involved in intentional stranding play of another cetacean, killer whales, which is apparently derived through observation and teaching (Guinet & Bouvier, 1995), is uncertain, but these animals appear to show self-recognition (Delfour & Marten, 2001). African grey parrots provide interesting but ambiguous evidence of imitative action suggestive of pretending to be another (Moore, 1992; see Mitchell, 1997a; 2002b).

Based on the limited available evidence, it is likely that pretend abilities survived evolutionarily as a result of their utility in imaginative planning, apprenticeship, or both (Parker & Gibson, 1979; Parker, 1993, 2000; Mitchell, 1994a; Currie, 1995). Pretending animals might imagine themselves and others in visual images that can be translated into kinesthetically experienced actions, essentially putting plans into actions. Great apes use self-pretense and imitate complex actions of others in pretense to plan solutions to problems (including tool use for extractive foraging), and the development of pretense in children rapidly results in and utilizes abilities for planning and replicating others' actions (Musatti & Mayer, 1987; Mitchell, 1994a; McCune & Agayoff, PIAC3). Providing a complete account of the adaptive value of pretense requires knowledge of its extent across species and evolutionary contexts (Harvey & Pagel, 1991), a knowledge which we largely lack at present.

Conclusion

Much like children, animals appear to pretend about affiliative and aggressive activities, food-related behavior, tool and object use, odd or intriguing behaviors or events, and babies – the stuff of daily life. Like children's early pretend play about others (Bretherton, 1984), that of animals is reality based, in that they seem to reenact experienced or observed ("scripted") activities. Like children's pretenses (Fein & Moorin, 1985), those of animals may be attempts to master incompletely understood (rather than exceedingly well-known) material. To inculcate pretense, we need to recreate experimentally for animals those contexts in which their species members have shown a tendency to replicate actions suggestive of pretense, and see what they do. But we also need to gather ontogenetic information about all the forms of pretense described in these chapters in order to create an adequate comparative, developmental, and evolutionary theory about pretense (Parker & McKinney, 1999).

"I have said enough to justify my plea for new observations and for a reconsideration of hasty theories in the light of these." Sully (1896, pp. 27–8) wrote these words long ago concerning children's pretense. He elaborated:

> we are far from understanding the precise workings of imagination in children. We talk ... glibly about their play, their make-believe, their illusions; but how much do we really know of their state of mind when they act out a little scene of domestic life, or of the battle-field? ... I suspect that there must be a much wider and finer investigation of children's action and talk before we can feel quite sure that we have got at their mental whereabouts, and know how they feel when they pretend ...

Although we have learned a great deal about children's experience and understanding of pretense since Sully, his words seem appropriate to our current knowledge of animals' pretenses. The chapters in this volume should prompt new observations and studies of imaginative activities in children and diverse animal species, potentially resulting in reconsideration or elaboration of existing theories concerning their knowledge, abilities, and understanding. Ironically, the one thing which might limit such an endeavor is our own imagination.

References

Acredolo, L. P. & Goodwyn, S. W. (1985). Symbolic gesturing in language development. *Human Development*, **29**, 40–9.
Adang, O. (1984). Teasing in young chimpanzees. *Behaviour*, **88**, 98–122.
Aldis, O. (1975). *Play fighting*. New York: Academic Press.
Altmann, S. A. (Ed.) (1967). *Social communication among primates*. Chicago: University of Chicago Press.
Altmann, S. A. (1998). *Foraging for survival*. Chicago: University of Chicago Press.
Altmann, S. A. & Altmann, J. (1973). *Baboon ecology*. Chicago: University of Chicago Press.
Ames, L. B. (1966). Children's stories. *Genetic Psychology Monographs*, **73**, 337–96.
Amsterdam, B. (1972). Mirror self-image reactions before age two. *Developmental Psychobiology*, **5**, 297–305.
Anscombe, G. E. M. (1981). *Metaphysics and philosophy of mind*. Minneapolis: University of Minnesota.
Antinucci, F. & Visalberghi, E. (1986). Tool use in *Cebus apella*: a case study. *International Journal of Primatology*, **7**, 349–62.
Applebee, A. N. (1978). *The child's concept of story*. Chicago: University of Chicago Press.
Arlitt, A. H. (1928). *Psychology of infancy and early childhood*. New York: McGraw-Hill.
Aronson, J. N. & Golomb, C. (1999). Preschoolers' understanding of pretense and presumption of congruity between action and representation. *Developmental Psychology*, **35**, 1414–25.
Asendorpf, J. B. & Baudonnière, P. M. (1993). Self-awareness and other-awareness: mirror self-recognition and synchronic imitation among unfamiliar peers. *Developmental Psychology*, **29**, 88–95.
Asendorpf, J. B., Warkentin, V. & Baudonnière, P. M. (1996). Self-awareness and other-awareness. II: mirror self-recognition, social contingency awareness, and synchronic imitation. *Developmental Psychology*, **32**, 313–21.
Astington, J. W. & Jenkins, J. M. (1995). Theory of mind development and social understanding. *Cognition and Emotion*, **9**, 151–65.
Astington, J. W. & Jenkins, J. M. (1999). A longitudinal study of the relationship between language and theory of mind development. *Developmental Psychology*, **35**, 1311–20.

Atran, S. (1998). Folk biology and the anthropology of science: cognitive universals and cultural particulars. *Behavioral and Brain Sciences*, **21**, 547–609.

Austin, J. L. (1958/1979). Pretending. In *Philosophical papers*, ed. J. O. Urmson & G. J. Warnock, 3rd edn., pp. 253–271. Oxford: Oxford University Press.

Austin, J. L. (1962). *Sense and sensibilia*. Oxford: Oxford University Press.

Badrian, A. & Badrian, N. (1984). Social interaction of *Pan paniscus* in the Lomako Forest, Zaire. In *The pygmy chimpanzee*, ed. R. L. Susman, pp. 325–44. New York: Plenum Press.

Baldwin, J. M. (1894/1903). *Mental development in the child and the race*. New York: Macmillan.

Baldwin, J. M. (1898/1904). *The story of the mind*. New York: McClure, Philips & Co.

Baldwin, J. M. (1901/1960). *Dictionary of philosophy and psychology*, vol. II. Gloucester, MA: Peter Smith.

Baldwin, J. M. (1909). *Darwin and the humanities*. Baltimore: Review Publishing.

Barkow, J. H., Cosmides, L. & Tooby, J. (Eds) (1992). *The adapted mind*. Oxford: Oxford University Press.

Bartsch, K. (1996). Between desires and beliefs: young children's action predictions. *Child Development*, **67**, 1671–85.

Bartsch, K. & Wellman, H. M. (1995). *Children talk about the mind*. Oxford: Oxford University Press.

Bates, E., Benigni, L., Bretherton, I., Camaioni, L. & Volterra, V. (1979). *The emergence of symbols*. New York: Academic Press.

Bates, E., Bretherton, I., Shore, C. & McNew, S. (1983). Names, gestures, and objects: symbolization in infancy and aphasia. In *Children's language*, ed. K. E. Nelson, vol. 4, pp. 59–123. Hillsdale, NJ: Erlbaum.

Bateson, G. (1955/1972). A theory of play and fantasy. In *Steps to an ecology of mind*, ed. G. Bateson, pp. 177–93. New York: Ballantine Books.

Bateson, G. (1956). The message "This is play." In *Group processes: Transactions of the second conference*, ed. B. Schaffer, pp. 145–242. Madison, NJ: Josiah Macy Jr. Foundation.

Bateson, G. (1978). Theoretical contributions to the study of play. In *Play: anthropological perspectives*, ed. M. A. Salter, pp. 7–16. West Point, NY: Leisure Press.

Baudonnière, P. M. & Michel, J. (1988). L'imitation entre enfants au cours de la seconde année: changement de cible et / ou changement de fonction? [Imitation between toddlers during the second year of life: change of target and/or change in function?]. *Psychologie Française*, **33**, 29–35.

Beck, B. B. (1980). *Animal tool behavior*. New York: Garland, STM Press.

Bekoff, M. & Allen, C. (1998). Intentional communication and social play: how and why animals negotiate and agree to play. In *Animal play*, ed. M. Bekoff & J. A. Byers, pp. 45–96. Cambridge: Cambridge University Press.

Bekoff, M. & Byers, J. A. (Eds) (1998). *Animal play*. Cambridge: Cambridge University Press.

Berlin, B. (1992). *Ethnobiological classification*. Princeton: Princeton University Press.

Bertenthal, B. I. & Fischer, K. W. (1978). Development of self recognition in the infant. *Developmental Psychology*, **14**, 44–50.

Bertrand, M. (1976). Acquisition by a pigtail macaque of behavior patterns beyond the natural rerpertoire of species. *Zeitschrift für Tierpsychologie*, **42**, 139–69.

Béteille, A. (1994). Inequality and equality. In *Companion encyclopedia of anthropology*, ed. T. Ingold, pp. 1010–39. London: Routledge.

Biben, M. & Suomi, S. J. (1993). Lessons from primate play. In *Parent-child play*, ed. K. B. MacDonald, pp. 185–96. Albany: SUNY Press.

Bickerton, D. (1990). *Language and species*. Chicago: University of Chicago Press.

Bickerton, D. (1998). The creation and recreation of language. In *Handbook of evolutionary psychology*, ed. C. Crawford & D. L. Krebs, pp. 613–34. Hillsdale, NJ: Erlbaum.

Blades, M. & Spencer, C. (1994). The development of children's ability to use spatial representations. In *Advances in child development and behavior*, ed. H. W. Reese, pp. 157–99. San Diego: Academic Press.

Blake, J. (2000). *Routes to child language*. Cambridge: Cambridge University Press.

Boesch, C. (1991a). Teaching among wild chimpanzees. *Animal Behaviour*, 41, 530–2.

Boesch, C. (1991b). Symbolic communication in wild chimpanzees. *Human Evolution*, 6, 81–90.

Boesch, C. & Boesch, H. (1984). Mental maps in wild chimpanzees: an analysis of hammer transports for nutcracking. *Primates*, 25, 160–70.

Boesch, C. & Tomasello, M. (1998). Chimpanzee and human cultures. *Current Anthropology*, 39, 591–614.

Bogin, B. (1998). Evolutionary and biological aspects of childhood. In *Biosocial perspectives on children*, ed. C. Panter-Brick, pp. 10–44. Cambridge: Cambridge University Press.

Bonvillian, J. D. & Patterson, F. G. P. (1993). Early language acquisition in children and gorillas: vocabulary content and sign iconicity. *First Language*, 13, 315–38.

Bonvillian, J. D. & Patterson, F. G. P. (1999). Early sign-language acquisition: comparisons between children and gorillas. In *The mentalities of gorillas and orangutans*, ed. S. T. Parker, R. W. Mitchell & H. L. Miles, pp. 240–64. Cambridge: Cambridge University Press.

Bornstein, M. H., Haynes, O. M., O'Reilly, A. W. & Painter, K. M. (1996). Solitary and collaborative pretense play in early childhood: sources of individual variation in the development of representational competence. *Child Development*, 67, 2910–29.

Bornstein, M. H. & O'Reilly, A. W. (Eds) (1993). *The role of play in the development of thought*. San Francisco: Jossey-Bass.

Botvin, G. J. & Sutton-Smith, B. (1977). The development of structural complexity in children's fantasy narratives. *Developmental Psychology*, 13(4), 377–88.

Boucher, J. & Lewis, V. (1990). Guessing or creating? A reply to Baron-Cohen. *British Journal of Developmental Psychology*, 8, 205–6.

Boulanger-Balleyguier, G. (1967). Les étapes de la reconnaissance de soi devant le miroir [The stages of self-recognition in front of the mirror]. *Enfance*, 20, 91–116.

Bourchier, A. & Davis, A. (2000). Individual and developmental differences in children's understanding of the fantasy-reality distinction. *British Journal of Developmental Psychology*, 18, 353–68.

Bowlby, J. (2000). *Attachment*, 2nd edn. New York: Basic Books.

Boyatzis, C. J. & Watson, M. W. (1993). Preschool children's symbolic representation of objects through gestures. *Child Development*, 64, 729–35.

Boysen, S. T. & Berntson, G. G. (1989). Numerical competence in a chimpanzee (*Pan troglodytes*). *Journal of Comparative Psychology*, 103, 23–31.

Boysen, S. T., Berntson, G. G., Hannan, M. B. & Cacioppo, J. T. (1996). Quantity-based interference and symbolic representations in chimpanzees (*Pan troglodytes*). *Journal of Experimental Psychology: Animal Behavior Processes*, 22, 76–86.

Boysen, S. T., Mukobi, K. L. & Berntson, G. G. (1999). Overcoming response-bias using symbolic representations of number by chimpanzees (*Pan troglodytes*). *Animal Learning and Behavior*, **27**, 229–35.

Bråten, S. (Ed.) (1998). *Intersubjective communication and emotion in early ontogeny*. Cambridge: Cambridge University Press.

Braunwald, S. R. & Brislin, R. W. (1979). The diary method updated. In *Developmental pragmatics*, ed. E. Ochs & B. B. Schieffelin, pp. 21–42. New York: Academic Press.

Bretherton, I. (1984). Representing the social world in symbolic play: reality and fantasy. In *Symbolic play*, ed. I. Bretherton, pp. 3–41. New York: Academic Press.

Bretherton, I. (1989). Pretense: the form and function of make-believe play. *Developmental Review*, **9**, 383–401.

Bretherton, I., Bates, E., McNew, S., Shore, C., Williamson, C. & Beeghly-Smith, M. (1981a). Comprehension and production of symbols in infancy. *Developmental Psychology*, **17**, 728–36.

Bretherton, I. & Beeghly, M. (1982). Talking about internal states: the acquisition of an explicit theory of mind. *Developmental Psychology*, **18**, 906–921.

Bretherton, I., McNew, S. & Beeghly-Smith, M. (1981b). Early person knowledge as expressed in gestural and verbal communication: when do infants acquire a 'theory of mind'? In *Infant social cognition*, ed. M. E. Lamb & L. R. Sherrod, pp. 333–73. Hillsdale, NJ: Erlbaum.

Bretherton, I., O'Connell, B., Shore, C. & Bates, E. (1984). The effect of contextual variation in symbolic play: development from 20 to 28 months. In *Symbolic play*, ed. I. Bretherton, pp. 271–98. New York: Academic Press.

Breuer, J. & Freud, S. (1895/1955). *Studies on hysteria*. New York: Basic Books.

Breuggeman, J. A. (1973). Parental care in a group of free-ranging rhesus monkeys (*Macaca mulatta*). *Folia primatologica*, **20**, 178–210.

Breuggeman, J. A. (1978). The function of adult play in free-ranging *Macaca mulatta*. In *Social play in primates*, ed. E. O. Smith, pp. 169–91. New York: Academic Press.

Bronfenbrenner, U. (1979). *The ecology of human development*. Cambridge, MA: Harvard University Press.

Bronowski, J. & Bellugi, U. (1970). Language, name, and concept. *Science*, **168**, 669–73.

Brown, J. R., Donelan-McCall, N. & Dunn, J. (1996). Why talk about mental states? The significance of children's conversation with friends, siblings and mothers. *Child Development*, **67**, 836–49.

Bruell, M. J. & Woolley, J. (1998). Young children's understanding of diversity in pretense. *Cognitive Development*, **13**, 257–77.

Bruner, J. S. (1966). On cognitive growth. In *Studies in cognitive growth*, ed. J. Bruner, R. R. Olver, P. M. Greenfield, *et al.*, pp. 1–29. New York: Wiley.

Bruner, J. S. (1972). The nature and uses of immaturity. *American Psychologist*, **27**, 687–708.

Bruner, J. S. (1981). Intention in the structure of action and interaction. In *Advances in infancy research*, ed. L. P. Lipsitt & C. K. Rovee-Collier, vol. 1, pp. 41–56. Norwood, NJ: Ablex.

Bruner, J. S. (1990). *Acts of meaning*. Cambridge, MA: Harvard University Press.

Bruner, J. S., Jolly, A. & Sylva, K. (Eds) (1976). *Play: its role in development and evolution*. New York: Basic Books.

Budwig, N., Stein, S. & O'Brien, C. (2001). Non-agent subjects in early child language: a cross-linguistic comparison. In *Children's language*, ed. K. Nelson, A. Aksu-Koc & C. Johnson, vol. 10, pp. 49–67. Mahwah, NJ: Erlbaum.

Bühler, C. (1935). *From birth to maturity*. London: Kegan Paul.
Bühler, K. (1930). *The mental development of the child*. London: Kegan Paul, Trench, Trubner & Co.
Burton, F. D. (1992). The social group as information unit: cognitive behaviour, cultural processes. In *Social processes and mental abilities in non-human primates*, ed. F. D. Burton, pp. 31–60. Lewiston, NY: Edwin Mellen Press.
Byrne, R. W. (1995). *The thinking ape*. Oxford: Oxford University Press.
Byrne, R. W. & Russon, A. (1998). Learning by imitation: a hierarchical approach. *Behavioral and Brain Sciences*, 21, 667–721.
Byrne, R. W. & Whiten, A. (1990). Tactical deception in primates: the 1990 database. *Primate Report*, No. 27, 1–101.
Byrne, R. W. & Whiten, A. (1992). Cognition evolution in primates: evidence from tactical deception. *Man (N.S.)*, 27, 609–27.
Cachel, S. (1997). Dietary shifts and the European Upper Paleolithic Transition. *Current Anthropology*, 38, 579–604.
Call, J. & Tomasello, M. (1996). The effect of humans on the cognitive development of apes. In *Reaching into thought*, ed. A. E. Russon, K. A. Bard & S. T. Parker, pp. 371–403. Cambridge: Cambridge University Press.
Candland, D. K. (1993). *Feral children and clever animals*. Oxford: Oxford University Press.
Carey, S. (1985). *Conceptual change in childhood*. Cambridge, MA: MIT Press.
Carlson, S. M., Taylor, M. & Levin, G. R. (1998). The influence of culture on pretend play: the case of Mennonite children. *Merrill-Palmer Quarterly*, 44, 538–65.
Carpendale, J. I. & Chandler, M. J. (1996). On the distinction between false belief understanding and subscribing to an interpretive theory of mind. *Child Development*, 67, 1686–1706.
Carpenter, C. R. (1940). A field study in Siam of the behavior and social relations of the gibbon (*Hylobates lar*). *Comparative Psychology Monographs*, 16(84), 1–212.
Carroll, L. (1871/1946). *Through the looking glass*. New York: Grossett and Dunlap.
Casby, M. W. (1997). Symbolic play of children with language impairment: a critical review. *Journal of Speech, Language, and Hearing Research*, 40, 468–79.
Case, R. (1985). *Intellectual development*. New York: Academic Press.
Chance, M. R. A. & Jolly, C. J. (1970). *Social groups of monkeys, apes, and men*. New York: Dutton.
Chandler, M. (2001). Perspective taking in the aftermath of theory-theory and the collapse of the social role-taking literature. In *Working with Piaget*, ed. A. Tryphon & J. Vonèche, pp. 39–63. Hove, East Sussex: Psychology Press.
Chapman, M. (1987). A longitudinal study of cognitive representation in symbolic play, self recognition, and object permanence during the second year. *International Journal of Behavioral Development*, 10, 151–70.
Cheney, D. L. & Seyfarth, R. M. (1990). *How monkeys see the world*. Chicago: University of Chicago Press.
Chevalier-Skolnikoff, S. (1977). A Piagetian model for describing and comparing socialization in monkey, ape, and human infants. In *Primate biosocial development*, ed. S. Chevalier-Skolnikoff & F. E. Poirier, pp. 159–87. New York: Garland.
Chevalier-Skolnikoff, S. (1986). An exploration of the ontogeny of deception in human beings and nonhuman primates. In *Deception*, ed. R. W. Mitchell & N. S. Thompson, pp. 205–20. Albany: SUNY Press.
Chevalier-Skolnikoff, S. (1989). Spontaneous tool use and sensorimotor intelligence in

Cebus compared with other monkeys and apes. *Behavioral and Brain Sciences*, **12**, 561–627.

Christie, J. F. & Johnsen, E. P. (1985). Questioning the results of play training research. *Educational Psychologist*, **20**, 7–11.

Claparède, E. (1911/1975). *Experimental pedagogy*. New York: Arno Press.

Clark, E. & Hecht, B. F. (1983). Comprehension, production, and language acquisition. *Annual Review of Psychology*, **34**, 325–49.

Coates, S. W. & Moore, M. S. (1997). The complexity of early trauma: representation and transformation. *Psychoanalytic Inquiry*, **17**, 286–311.

Colburn, D. (1985, March 6). Inside Einstein's brain. *Washington Post, Health*, **1**(9), 17.

Cole, D. & La Voie, J. C. (1985). Fantasy play and related cognitive development in 2- to 6-year-olds. *Developmental Psychology*, **21**, 233–40.

Cole, M. (1996). *Cultural psychology*. Cambridge, MA: Harvard University Press.

Cooper, J. R., Roth, R. & Bloom, F. E. (1996). *The biochemical basis of neuropharmacology*. Oxford: Oxford University Press.

Copple, C. E., Cocking, R. R. & Matthews, W. S. (1984). Objects, symbols, and substitutes: the nature of cognitive activity during symbolic play. In *Child's play*, ed. T. D. Yawkey & A. D. Pellegrini, pp. 105–23. Hillsdale, NJ: Erlbaum.

Cords, M. & Killen, M. (1998). Conflict resolution in human and non-human primates. In *Piaget, evolution, and development*, ed. J. Langer & M. Killen, pp. 193–218. Mahwah, NJ: Erlbaum.

Corrigan, R. (1976). *Patterns of individual communication and cognitive development*. Unpublished doctoral dissertation, University of Denver.

Corrigan, R. (1987). A developmental sequence of actor-object pretend play in young children. *Merrill-Palmer Quarterly*, **33**, 87–106.

Corsaro, W. A. (1985). *Friendship and peer culture in the early years*. Norwood, NJ: Ablex Publishing.

Crawford, M. P. (1937). The cooperative solving of problems by young chimpanzees. *Comparative Psychology Monographs*, **14**(2), 1–88.

Crum, R. A., Thornburg, K., Benninga, J. & Bridge, C. (1983). Preschool children's object substitutions during symbolic play. *Perceptual and Motor Skills*, **56**, 947–55.

Csikszentmihalyi, M. (1991). *Flow*. New York: Harper Collins.

Csikszentmihalyi, M. (1997). *Creativity*. New York: Harper Collins.

Cunningham, A. (1921). A gorilla's life in civilization. *Zoological Society Bulletin*, **24**, 118–24.

Curcio, F. & Piserchia, E. A. (1978). Pantomimic representation in psychotic children. *Journal of Autism and Childhood Schizophrenia*, **8**, 181–9.

Currie, G. (1995). Imagination and simulation: aesthetics meets cognitive science. In *Mental simulation*, ed. M. Davies & T. Stone, pp. 151–69. Oxford: Blackwell.

Custance, D. M., Whiten, A. & Bard, K. A. (1995). Can young chimpanzees (*Pan troglodytes*) imitate arbitrary actions? Hayes & Hayes (1952) revisited. *Behaviour*, **132**, 837–59.

Custer, W. L. (1996). A comparison of young children's understanding of contradictory mental representations in pretense, memory, and belief. *Child Development*, **67**, 678–88.

Dansky, J. L. (1980). Make-believe: a mediator of the relationship between play and associative fluency. *Child Development*, **51**, 576–9.

Darwin, C. (1859/1902). *Origin of species*. New York: P. F. Collier & Sons.

Darwin, C. (1871/1896). *The descent of man*. New York: D. Appleton & Co.
Davidson, I. & Noble, W. (1989). The archaeology of perception: traces of depiction and language. *Current Anthropology*, 30, 125–51.
Davis, D. L., Woolley, J. D. & Bruell, M. J. (2002). Young children's understanding of pretense as a mental representation. *British Journal of Developmental Psychology*, 20. (In press.)
Davis, W. (1986). The origins of image making. *Current Anthropology*, 27, 193–215.
Dawkins, R. (1976). *The selfish gene*. Oxford: Oxford University Press.
Deacon, T. W. (1997). *The symbolic species*. New York: Norton.
de Gramont, P. (1990). *Language and the distortion of meaning*. New York: New York University Press.
Delfour, F. & Marten, K. (2001). Mirror image processing in three marine mammal species: killer whales (*Orcinus orca*), false killer whales (*Pseudorca crassidens*) and California sea lions (*Zalophus californianus*). *Behavioural Processes*, 53, 181–90.
DeLoache, J. S. (1987). Rapid change in the symbolic functioning of very young children. *Science*, 238, 1556–7.
DeLoache, J. S. (1990). Young children's understanding of models. In *Knowing and remembering in young children*, ed. R. Fivush & J. A. Hudson, pp. 94–126. New York: Cambridge University Press.
DeLoache, J. S. (1991). Symbolic functioning in very young children: understanding of pictures and models. *Child Development*, 62, 736–52.
DeLoache, J. S. (1995). Early symbol understanding and use. In *The psychology of learning and motivation*, ed. D. L. Medin, pp. 65–114. San Diego: Academic Press.
DeLoache, J. S., Miller, K. F. & Rosengren, K. S. (1997). The credible shrinking room: very young children's performance with symbolic and nonsymbolic relations. *Psychological Science*, 8, 308–13.
de Lorimier, S., Doyle, A-B. & Tessier, O. (1995). Social coordination during pretend play: comparisons with nonpretend play and effects on expressive content. *Merrill-Palmer Quarterly*, 41, 497–516.
de Rivera, J. & Sarbin, T. R. (Eds) (1998). *Believed-in imaginings*. Washington, DC: American Psychological Association.
Desmond, A. (1979). *The ape's reflexion*. New York: The Dial Press.
de Veer, M. W. & van den Bos, R. (1999). A critical review of methodology and interpretation of self-recognition research in nonhuman primates. *Animal Behaviour*, 58, 459–68.
DeVore, I. (Ed.) (1965). *Primate behavior*. New York: Holt, Rinehart and Winston.
de Waal, F. B. M. (1982). *Chimpanzee politics*. New York: Harper and Row.
de Waal, F. B. M. (1986a). Deception in the natural communication of chimpanzees. In *Deception*, ed. R. W. Mitchell & N. S. Thompson, pp. 221–44. Albany: SUNY Press.
de Waal, F. B. M. (1986b). Imaginative bonobo games. *Zoonooz*, 59, 6–10.
de Waal, F. B. M. (1988). The communication repertoire of captive bonobos (*Pan paniscus*), compared to that of chimpanzees. *Behaviour*, 106, 183–251.
de Waal, F. B. M. (1989). *Peacemaking among primates*. Cambridge, MA: Harvard University Press.
de Waal, F. B. M. (1995). Bonobo sex and society. *Scientific American*, 272(3), 82–8.
de Waal, F. B. M. & Lanting, F. (1997). *Bonobo*. Berkeley: University of California Press.
Dias, M. G. & Harris, P. L. (1988). The effect of make-believe play on deductive reasoning. *British Journal of Developmental Psychology*, 6, 207–21.

Dias, M. G. & Harris, P. L. (1990). The influence of the imagination on reasoning by young children. *British Journal of Developmental Psychology*, **8**, 305–18.

DiLalla, L. F. & Watson, M. W. (1988). Differentiation of fantasy and reality: preschoolers' reactions to interruptions in their play. *Developmental Psychology*, **24**, 286–91.

Dockett, S. (1998). Constructing understandings through play in the early years. *International Journal of Early Years Education*, **6**, 105–16.

Dolgin, K. G. & Behrend, D. A. (1984). Children's knowledge about animates and inanimates. *Child Development*, **55**, 1646–50.

Dolhinow, P. & Fuentes, A. (1999). *The nonhuman primates*. Mountain View, CA: Mayfield Publishing Co.

Donald, M. (1991). *Origins of the modern mind*. Cambridge, MA: Harvard University Press.

Dowson, T. A. (1998). Rock art: handmaiden to studies of cognitive evolution. In *Cognition and material culture*, ed. C. Renfrew & K. Scarre, pp. 67–76. Oxford: McDonald Institute for Archaeological Research.

Drucker, J. (1994). Constructing metaphors: the role of symbolization in the treatment of children. In *Children at play*, ed. A. Slade & D. P. Wolf, pp. 62–80. Oxford: Oxford University Press.

Dunn, J. (1988). *The beginnings of social understanding*. Cambridge, MA: Harvard University Press.

Dunn, J. (1991). Understanding others: evidence from naturalistic studies of children. In *Natural theories of mind*, ed. A. Whiten, pp. 51–61. Oxford: Blackwell.

Dunn, J., Bretherton, I. & Munn, P. (1987). Conversation about feeling states between mothers and their young children. *Developmental Psychology*, **23**, 132–9.

Dunn, J. & Dale, N. (1984). I a daddy: 2-year olds' collaboration in joint pretend play with sibling and with mother. In *Symbolic play*, ed. I. Bretherton, pp. 131–58. New York: Academic Press.

Dunn, J. & Wooding, C. (1977). Play in the home and its implications for learning. In *Biology of play*, ed. B. Tizard & D. Harvey, pp. 45–58. London: Heinemann.

Eibl-Eibesfeldt, I. (1950/1978). On the ontogeny of behavior of a male badger (*Meles meles* L.) with particular reference to play behavior. In *Evolution of play behavior*, ed. D. Müller-Schwarze, pp. 142–8. Stroudsburg, PA: Dowden, Hutchinson & Ross.

Eibl-Eibesfeldt, I. (1989). *Human ethology*. New York: Aldine.

El'konin, D. B. (1969). Some results of the study of the psychological development of preschool-age children. In *A handbook of contemporary Soviet psychology*, ed. M. Cole & I. Maltzman, pp. 163–208. New York: Basic Books.

Elder, J. L. & Pederson, D. R. (1978). Preschool children's use of objects in symbolic play. *Child Development*, **49**, 500–4.

Emory, G. R., Payne, R. G. & Chance, M. R. A. (1979). Observations on a newly described usage of the primate play face. *Behavioural Processes*, **4**, 67–71.

Estes, D., Wellman, H. M. & Woolley, J. D. (1989). Children's understanding of mental phenomena. In *Advances in child development and behavior*, ed. H. Reese, pp. 41–86. New York: Academic Press.

Ewer, R. F. (1968). *Ethology of mammals*. London: Elek Science.

Fagen, R. (1981). *Animal play behavior*. Oxford: Oxford University Press.

Fagot, J. (Ed.) (1999). *Picture perception in animals*. *Cahiers de Psychologie Cognitive/Current Psychology of Cognition*, **18**(5–6).

Farver, J. A. M. (1993). Cultural differences in scaffolding pretend play: a comparison of American and Mexican-American mother-child and sibling-child dyads. In *Parent–child play*, ed. K. MacDonald, pp. 349–66. Albany: SUNY Press.

Farver, J. A. M. & Howes, C. (1993). Cultural differences in American and Mexican mother–child pretend play. *Merrill-Palmer Quarterly*, **39**, 344–58.

Farver, J. A. M. & Shin, Y. L. (1997). Social pretend play in Korean- and Anglo-American preschoolers. *Child Development*, **68**, 544–56.

Fedigan, L. M. (1982). *Primate paradigms*. Montreal: Eden Press.

Fein, G. G. (1975). A transformational analysis of pretending. *Developmental Psychology*, **11**, 291–6.

Fein, G. G. (1979). Play and the acquisition of symbols. In *Current topics in early childhood education*, ed. L. Katz, pp. 195–225. Norwood, NJ: Ablex.

Fein, G. G. (1981). Pretend play in childhood: an integrative review. *Child Development*, **52**, 1095–118.

Fein, G. G. (1989). Mind, meaning, and affect: proposals for a theory of pretense. *Developmental Review*, **9**, 345–63.

Fein, G. G. (1995). Toys and stories. In *The future of play theory*, ed. A. D. Pellegrini, pp. 151–64. NY: SUNY Press.

Fein, G. G. & Apfel, N. (1979a). Some preliminary observations on knowing and pretending. In *Symbolic functioning in childhood*, ed. M. Smith & M. B. Franklin, pp. 87–100. Hillsdale, NJ: Erlbaum.

Fein, G. G. & Apfel, N. (1979b). The development of play: style, structure, and situation. *Genetic Psychology Monographs*, **99**, 231–50.

Fein, G. G. & Glaubman, R. (1992). Commentary. *Human Development*, **36**, 247–52.

Fein, G. G. & Moorin, E. R. (1985). Confusion, substitution, and mastery: pretend play during the second year of life. In *Children's language*, ed. K. E. Nelson, vol. 5, pp. 61–76. Hillsdale, NJ: Erlbaum.

Feitelson, D. (1977). Cross cultural studies of representational play. In *Biology of play*, ed. B. Tizard & D. Harvey, pp. 6–14. London: Heinemann.

Feldman, C. F., Bruner, J., Renderer, B. & Spitzer, S. (1990). Narrative comprehension. In *Narrative thought and narrative language*, ed. B. K. Britton & A. D. Pellegrini, pp. 1–78. Hillsdale, NJ: Erlbaum.

Fellner, W. & Bauer, G. B. (1999, December). Synchrony between a mother-calf pair of bottlenose dolphins (*Tursiops truncatus*). Poster presented at the 13th Biennial Conference of the Biology of Marine Mammals, Wailea, HI.

Fenson, L. (1984). Developmental trends for action and speech in pretend play. In *Symbolic play*, ed. I. Bretherton, pp. 249–70. London: Academic Press.

Fenson, L., Kagan, J., Kearsley, R. B. & Zelazo, P. R. (1976). The developmental progression of manipulative play in the first two years. *Child Development*, **47**, 232–6.

Fenson, L. & Ramsay, D. (1981). Effects of modeling actions on the play of twelve-, fifteen-, and nineteen-month-old children. *Child Development*, **52**, 1028–36.

Ferenczi, S. (1913/1916). A little chanticleer. In *Contributions to psychoanalysis*, ed. E. Jones (trans.), pp. 204–13. Boston: R. G. Badger.

Fernandez, J. W. (1986). *Persuasions and performances*. Bloomington: Indiana University Press.

Field, T., De Stefano, L., Koewler, J. H., III. (1982). Fantasy play of toddlers and preschoolers. *Developmental Psychology*, **18**, 503–8.

Fiese, B. H. (1990). Playful relationships: a contextual analysis of mother–child interaction and symbolic play. *Child Development*, **61**, 1648–56.

First, E. (1994). The leaving game, or I'll play you and you play me: the emergence of dramatic role play in 2-year-olds. In *Children at play*, ed. A. Slade & D. P. Wolf, pp. 111–32. Oxford: Oxford University Press.

Flaum, M., Goodall, J. & Wells, O. (1965). *Miss Goodall and the wild chimpanzees* (videotape). Washington: National Geographic Society.

Flavell, J. H. (1985). *Cognitive development* (rev. edn.). Englewood Cliffs, NJ: Prentice-Hall.

Flavell, J. H. (1988). The development of children's knowledge about the mind: from cognitive connections to mental representations. In *Developing theories of mind*, ed. J. W. Astington, P. L. Harris & D. R. Olson, pp. 244–71. Cambridge: Cambridge University Press.

Flavell, J. H., Flavell, E. R. & Green, F. L. (1987). Young children's knowledge about the apparent-real and pretend-real distinctions. *Developmental Psychology*, **23**, 816–22.

Flavell, J. H. & Miller, P. H. (1998). Social cognition. In *Handbook of child psychology*, vol. 2. *Cognition, perception, and language development*, ed. D. Kuhn & R. S. Siegler, pp. 851–98. New York: Wiley.

Fleagle, J. G. (1998). *Primate adaptation and evolution*. San Diego: Academic Press.

Folven, R. J., Bonvillian, J. D. & Orlansky, M. D. (1984/1985). Communicative gestures and early sign language acquisition. *First Language*, **5**, 129–44.

Fonagy, P., Redfern, S. & Charman, T. (1997). The relationship between belief-desire reasoning and a projective measure of attachment security (SAT). *British Journal of Developmental Psychology*, **15**, 51–61.

Forguson, L. & Gopnik, A. (1988). The ontogeny of common sense. In *Developing theories of mind*, ed. J. W. Astington, P. L. Harris & D. R. Olson, pp. 226–43. New York: Cambridge University Press.

Formanek-Brunell, M. (1993). *Made to play house*. New Haven: Yale University Press.

Forys, S. & McCune-Nicolich, L. (1984). Shared pretend: sociodramatic play at three years. In *Symbolic play*, ed. I. Bretherton, pp. 159–94. New York: Academic Press.

Fouts, R. S. & Fouts, D. H. (1993). Chimpanzees' use of sign language. In *The great ape project*, ed. P. Cavalieri & P. Singer, pp. 28–42. New York: St. Martin's Press.

Fouts, R. S., Hirsch, A. D. & Fouts, D. H. (1982). Cultural transmission of a human language in a chimpanzee mother-infant relationship. In *Psychobiological perspectives*, ed. H. E. Fitzgerald, J. A. Mullins & P. Page, vol. 3, pp. 159–93. New York: Plenum Press.

Freud, S. (1920/1972). *Beyond the pleasure principle*. New York: Bantam Books.

Fragaszy, D. M. & Visalberghi, E. (1990). Social processes affecting the appearance of innovative behaviors in capuchin monkeys. *Folia Primatologica*, **54**, 155–65.

Fromberg, D. P. & Bergen, D. (Eds) (1998). *Play from birth to twelve and beyond*. New York: Garland.

Furuichi, T. (1987). Sexual swelling, receptivity, and grouping of wild pygmy chimpanzee females at Wamba, Zaire. *Primates*, **28**, 309–18.

Galda, L. (1984). Narrative competence: play, storytelling, and story comprehension. In *The development of oral and written language in social contexts*, ed. A. D. Pellegrini & T. Yawkey, pp. 105–16. Norwood, NJ: Ablex.

Galdikas, B. M. F. & Vasey, P. (1992). Why are orangutans so smart? Ecological and social hypotheses. In *Social processes and mental abilities in non-human primates*, ed. F. D. Burton, pp. 183–224. Lewiston, NY: Edwin Mellen Press.

Galef, B. G., Manzig, L. A. & Field, R. M. (1986). Imitation learning in budgerigars: Dawson and Foss (1965) revisited. *Behavioural Processes*, 13, 191–202.

Gallup, G. G., Jr. (1970). Chimpanzees: self-recognition. *Science*, 167, 86–7.

Gallup, G. G., Jr. (1985). Do minds exist in species other than our own? *Neurosciences and Biobehavioral Review*, 9, 631–41.

Gallup, G. G., Jr., Povinelli, D. J., Suarez, S. D., Anderson, J. R., Lethmate, J. & Menzel, E. W., Jr. (1995). Further reflections on self-recognition in primates. *Animal Behaviour*, 50, 1525–32.

Gardner, B. T. & Gardner, R. A. (1985). Signs of intelligence in cross-fostered chimpanzees. *Philosophical Transcripts of the Royal Society, London*, 308, 159–76.

Gardner, H. W., Mutter, J. D. & Kosmitzki, C. (1998). *Lives across cultures*. Boston: Allyn and Bacon.

Gardner, R. A. & Gardner, B. T. (1969). Teaching sign language to a chimpanzee. *Science*, 165, 664–72.

Gardner, R. A. & Gardner, B. T. (1978). Comparative physiology and language acquisition. *Annals of the New York Academy of Science*, 309, 37–76.

Gargett, R. H. (1989). Grave shortcomings: the evidence for Neanderthal burial. *Current Anthropology*, 30, 157–90.

Garvey, C. (1990). *Play*. Cambridge, MA: Harvard University Press.

Garvey, C. & Kramer, T. (1989). The language of social pretend play. *Developmental Review*, 9, 364–82.

Gaskins, S. (1996). How Mayan parental theories come into play. In *Parents' cultural belief systems*, ed. S. Harkness & C. Super, pp. 345–63. New York: Guilford Press.

Gautier-Hion, A. C. (1971a). L'ecologie du talapoin du Gabon [Ecology of the talapoins of Gabon]. *Tene Vie*, 25, 427–90.

Gautier-Hion, A. (1971b). Répertoire comportemental du Talapoin (*Miopithecus talapoin*) [Behavioral repertoire of the talapoins]. *Biologia Gabonica*, 7, 295–391.

Geertz, C. (1976). Deep play: a description of the Balinese cockfight. In *Play*, ed. J. S. Bruner, A. Jolly & K. Sylva, pp. 656–74. New York: Basic Books.

Gelman, R., Spelke, E. S. & Meck, E. (1983). What preschoolers know about animate and inanimate objects. In *The acquisition of symbolic skills*, ed. D. Rogers & J. Sloboda, pp. 297–326. New York: Plenum Press.

Gentner, D. (1988). Metaphor as structure mapping: the relational shift. *Child Development*, 59, 47–59.

Gentner, D., Rattermann, M. J., Markman, A. & Kotosky, L. (1995). Two forces in the development of relational similarity. In *Developing cognitive competence*, ed. T. J. Simon & G. S. Halford, pp. 263–313. Hillsdale, NJ: Erlbaum.

Gerow, L. E., Taylor, M. & Moses, L. J. (1999). *Children's understanding that pretense involves mental representations*. Unpublished manuscript, University of Oregon.

Gesell, A. & Thompson, N. (1934). *Infant behavior*. New York: McGraw-Hill.

Gibson, J. J. (1979). *The ecological approach to visual perception*. Boston: Houghton-Mifflin.

Gibson, K. R. (1993). Generative interplay between technical capacities, social relations, imitation and cognition. In *Tools, language, and cognition in human evolution*, ed. K. R. Gibson & T. Ingold, pp. 131–7. Cambridge: Cambridge University Press.

Gibson, K. R. & Ingold, T. (Eds) (1993). *Tools, language, and cognition in human evolution*. Cambridge: Cambridge University Press.

Gibson, K. R. & Petersen, A. C. (1991). *Brain maturation and cognitive development*. New York: Aldine De Gruyter.

Giffin, H. (1984). The coordination of meaning in the creation of shared make-believe play. In *Symbolic play*, ed. I. Bretherton, pp. 73–100. London: Academic Press.
Goldman, L. R. (1998). *Child's play*. Oxford: Berg.
Golinkoff, R. (1993). When is communication 'a meeting of minds'? *Journal of Child Language*, 20, 199–207.
Golomb, C. & Bonen, S. (1981). Playing games of make-believe: the effectiveness of symbolic play training with children who failed to benefit from early conservation training. *Genetic Psychology Monographs*, 104, 137–59.
Golomb, C. & Cornelius, C. B. (1977). Symbolic play and its cognitive significance. *Developmental Psychology*, 13, 246–52.
Golomb, C., Gowing, E. D. & Friedman, L. (1982). Play and cognition: studies of pretense play and conservation of quantity. *Journal of Experimental Child Psychology*, 33, 257–79.
Golomb, C. & Kuersten, R. (1996). On the transition from pretense play to reality: what are the rules of the game? *British Journal of Developmental Psychology*, 31, 800–10.
Gómez, J. C. (1992). *El desarrollo de la comunicación intencional en el gorila* [The development of intentional communication in the gorilla]. Unpublished Ph.D. Dissertation, Universidad Autónoma de Madrid.
Gómez, J. C. (1999). Development of sensorimotor intelligence in infant gorillas. In *The mentalities of gorillas and orangutans*, ed. S. T. Parker, R. W. Mitchell & H. L. Miles, pp. 160–78. Cambridge: Cambridge University Press.
Göncü, A. (Ed.) (1999). *Children's engagement in the world*. New York: Cambridge University Press.
Göncü, A., Tuermer, U., Jain, J. & Johnson, D. (1999). Children's play as cultural activity. In *Children's engagement in the world*, ed. A. Göncü, pp. 173–202. Cambridge: Cambridge University Press.
Goodall, J. van Lawick- (1971). *In the shadow of man*. Boston: Houghton-Mifflin.
Goodall, J. van Lawick- (1973). Cultural elements in a chimpanzee community. In *Precultural primate behavior*, ed. E. W. Menzel, Jr., pp. 144–84. Basel: Karger.
Goodall, J. (1986). *The chimpanzees of Gombe*. Cambridge, MA: Harvard University Press.
Gopnik, A. (1993). How we know our minds: the illusion of first-person knowledge of intentionality. *Behavioral and Brain Sciences*, 16, 1–14.
Gopnik, A. & Astington, J. W. (1988). Children's understanding of representational change and its relation to the understanding of false belief and the appearance-reality distinction. *Child Development*, 59, 26–37.
Gordon, D. E. (1993). The inhibition of pretend play and its implications for development. *Human Development*, 36, 215–34.
Gottlieb, G. (1992). *Individual development and evolution*. Oxford: Oxford University Press.
Gottlieb, G. (1998). Normally occurring environmental and behavioral influences on gene activity: from central dogma to probabilistic epigenesis. *Psychological Review*, 105, 792–802.
Gouin-Décarie, T., Pouliot, T. & Poulin-Dubois, D. (1983). Image spéculaire et genèse de la reconnaissance de soi: une analyse hiérarchique [Specular image and the genesis of self-recognition: a hierarchical analysis]. *Enfance*, 1–2, 99–115.
Gould, S. J. (1977). *Ontogeny and phylogeny*. Cambridge, MA: Harvard University Press.
Gould, J. L. & Gould, C. G. (1988). *The honey bee*. New York: Scientific American Library.
Gouzoules, S., Gouzoules, H. & Marler, P. (1984). Rhesus monkey (*Macaca mulatta*)

scream vocalizations: representational signalling in the recruitment of agnostic aid. *Animal Behaviour*, **32**, 182–93.

Greenfield, P. M. & Savage-Rumbaugh, E. S. (1993). Comparing communicative competence in child and chimp: the pragmatics of repetition. *Journal of Child Language*, **20**, 1–26.

Gregory, R. L. (1987). Mirror reversal. In *The Oxford companion to the mind*, ed. R. L. Gregory & O. L. Zangwill, pp. 491–93. Oxford: Oxford University Press.

Grice, H. P. (1982). Meaning revisited. In *Mutual knowledge*, ed. N. V. Smith, pp. 223–43. New York: Academic Press.

Grize, J.-B. (1996). *Logique naturelle et communications* [Natural logic and communications]. Paris: Presses Universitaires de France.

Groos, K. (1898). *The play of animals*. New York: D. Appleton & Co.

Groos, K. (1901). *The play of man*. New York: D. Appleton & Co.

Guillaume, P. (1925). *L'imitation chez l'enfant* [Imitation in children]. Paris: Alcan.

Guillaume, P. (1926/1971). *Imitation in children* [Trans. E. P. Halperin]. Chicago: University of Chicago. (Original work: P. Guillaume (1925). *L'imitation chez l'enfant*.)

Guinet, C. & Bouvier, J. (1995). Development of intentional stranding hunting techniques in killer whale (*Orcinus orca*) calves at Crozet Archipelago. *Canadian Journal of Zoology*, **73**, 27–33.

Guthrie, S. E. (1993). *Faces in the clouds*. Oxford: Oxford University Press.

Hahn, E. (1982). Annals of zoology; Gorillas – Part I. *New Yorker*, **58**(5), 39–62.

Haight, W. L., Masiello, T., Dickson, L., Heckeby, E. & Black, J. (1994). The everyday contexts and social functions of spontaneous mother-child pretend play in the home. *Merrill-Palmer Quarterly*, **40**, 509–22.

Haight, W. L. & Miller, P. J. (1992). The development of everyday pretend play: a longitudinal study of mothers' participation. *Merrill-Palmer Quarterly*, **38**, 331–49.

Haight, W. L. & Miller, P. J. (1993). *Pretending at home*. Albany: SUNY Press.

Haight, W. L., Parke, R. & Black, J. (1997). Mothers' and fathers' beliefs about and spontaneous participation in their toddlers' pretend play. *Merrill-Palmer Quarterly*, **43**, 271–90.

Haight, W. L., Wang, X., Han-tih, H., Williams, K. & Mintz, J. (1999). Universal, developmental, and variable aspects of young children's play: a cross-cultural comparison of pretending at home. *Child Development*, **70**, 1477–88.

Hall, G. S. (1914). *Aspects of child life and education*. Boston: Ginn & Co.

Hall, G. S. (1928). *Youth*. New York: D. Appleton & Co.

Hall, W. S., Frank, R. & Ellison, C. (1995). The development of pretend language: toward an understanding of the child's theory of mind. *Journal of Psycholinguistic Research*, **24**, 231–54.

Hanson, K. (1986). *The self imagined*. New York: Routledge & Kegan Paul.

Harlow, H. F. (1971). *Learning to love*. San Francisco: Albion Publishing Company.

Harris, P. L. (1991). The work of the imagination. In *Natural theories of mind*, ed. A. Whiten, pp. 283–304. Oxford: Blackwell.

Harris, P. L. (1992). From simulation to folk psychology. *Mind and Language*, **7**, 120–44.

Harris, P. L. (1994). Understanding pretence. In *Children's early understanding of mind*, ed. C. Lewis & P. Mitchell, pp. 235–59. Hove: Erlbaum.

Harris, P. L. (1998). Fictional absorption: emotional responses to make-believe. In *Intersubjective communication and emotion in early ontogeny*, ed. S. Bråten, pp. 336–53. Cambridge: Cambridge University Press.

Harris, P. L. (2000). *The work of the imagination*. Oxford: Blackwell.
Harris, P. L., Brown, E., Marriott, C., Whittall, S. & Harmer, S. (1991). Monsters, ghosts, and witches: testing the limits of the fantasy-reality distinction in young children. *British Journal of Developmental Psychology*, 9, 105–24.
Harris, P. L. & Kavanaugh, R. D. (1993). Young children's understanding of pretense. *Monographs of the Society for Research in Child Development*, 58 (1, serial no. 231), 1–108.
Harrisson, B. (1961). A study of orang-utan behavior in the semi-wild state. *International Zoo Yearbook*, 3, 57–68.
Harrisson, B. (1962). *Orangutan*. Singapore: Oxford University Press.
Hart, D. & Fegley, S. (1994). Social imitation and the emergence of a mental model of self. In *Self-awareness in animals and humans*, ed. S. T. Parker, R. W. Mitchell & M. L. Boccia, pp. 149–65. New York: Cambridge University Press.
Harvey, P. H. & Pagel, M. D. (1991). *The comparative method in evolutionary biology*. Oxford: Oxford University Press.
Hashimoto, C. & Furuichi, T. (1994). Social role and development of noncopulatory sexual behavior of wild bonobos. In *Chimpanzee cultures*, ed. R. W. Wrangham, W. C. McGrew, F. B. M. de Waal & P. G. Heltne, pp. 155–68. Cambridge, MA: Harvard University Press.
Hatfield, E., Cacioppo, J. T. & Rapson, R. L. (1994). *Emotional contagion*. New York: Cambridge University Press.
Hawkes, K., O'Connell, J. F. & Blurton-Jones, N. G. (1997). Hadza women's time allocation, offspring provisioning, and the evolution of long postmenopausal life spans. *Current Anthropology*, 38, 551–78.
Hayaki, H. (1985). Social play of juvenile and adolescent chimpanzees in the Mahale Mountain National Park, Tanzania. *Primates*, 26, 343–60.
Hayes, C. (1951). *The ape in our house*. New York: Harper and Brothers.
Hayes, K. J. & Hayes, C. (1955). The cultural capacity of chimpanzee. In *The non-human primates and human evolution*, ed. J. A. Gavan, pp. 110–25. Detroit: Wayne University Press.
Hayes, K. J. & Nissen, C. (1971). Higher mental functions of a home-raised chimpanzee. In *Behavior of nonhuman primates*, ed. A. M. Schrier & F. Stollnitz, vol. 4, pp. 60–115. New York: Academic Press.
Heyes, C. M. (1993). Anecdotes, trapping and triangulation: do animals attribute mental states? *Animal Behaviour*, 46, 177–88.
Heyes, C. M. (1994). Reflections on self-recognition in primates. *Animal Behaviour*, 47, 909–19.
Heyes, C. M. (1998). Theory of mind in nonhuman primates. *Behavioral and Brain Sciences*, 21, 101–48.
Hickerson, P. (1996). *Do orang-utans engage in pretence? A preliminary observational study of eye-covering behavior in captive orang-utans* (Pongo pygmaeus). Unpublished BA thesis, Glendon College-York University, Toronto, Canada.
Hickling, A. K., Wellman, H., M. & Gottfried, G. M. (1997). Preschoolers' understanding of others' mental attitudes towards pretend happenings. *British Journal of Developmental Psychology*, 15, 339–54.
Hinde, R. A. (1958). The nest-building behaviour of domesticated canaries. *Proceedings of the Zoological Society of London*, 131, 1–48.
Hirschfeld, L. & Gelman, S. (Eds) (1994). *Mapping the mind*. Cambridge: Cambridge University Press.

Hoppe-Graff, S. (1993). Individual differences in the emergence of pretend play. *Contributions to Human Development*, 23, 57–70.

Hornaday, W. T. (1879). On the species of Bornean orangs, with notes on their habits. *Proceedings of the American Association for the Advancement of Science*, 28, 438–55.

Howe, N., Petrakos, H. & Rinaldi, C. M. (1998). "All the sheeps are dead. He murdered them.": sibling pretense, negotiation, internal state language, and relationship quality. *Child Development*, 69, 182–91.

Howes, C. (Ed.) (1992). *The collaborative construction of pretend.* Albany: SUNY Press.

Howes, C. & Matheson, C. C. (1992). Sequences in the development of competent play with peers: social and pretend play. *Developmental Psychology*, 28, 961–74.

Hoyt, A. M. (1941). *Toto and I.* Philadelphia: J. B. Lippincott.

Huffman, M. A. & Quiatt, D. (1986). Stone handling by Japanese macaques (*Macaca fuscata*): implications for tool use of stone. *Primates*, 27, 427–37.

Humphrey, N. K. (1984). *Consciousness regained.* Oxford: Oxford University.

Hutt, S. J., Tyler, S., Hutt, C. & Christopherson, H. (1989). *Play, exploration and learning.* London: Routledge.

Huttenlocher, J. & Higgins, E. T. (1978). Issues in the study of symbolic play development. *Minnesota Symposia on Child Psychology*, 11, 98–140.

Huxley, T. H. (1863). *Man's place in nature and other anthropological essays.* Akron, OH: Werner Company.

Ingmanson, E. J. (1987). Clapping behavior: nonverbal communication during grooming in a gorup of captive pygmy chimpanzees (*Pan paniscus*). *American Journal of Physical Anthropology*, 72, 173–4.

Ingmanson, E. J. (1988). The context of object manipulation by captive pygmy chimpanzees (*Pan paniscus*). *American Journal of Physical Anthropology*, 76, 224.

Ingmanson, E. J. (1990). Male care of infants in *Pan paniscus* at Wamba, Zaire. *American Journal of Primatology*, 20, 200–1.

Ingmanson, E. J. (1992). *Pan paniscus* and the social context of complex communication. Paper presented at the VIIIth Annual Meeting of the Language Origins Society, Cambridge, UK.

Ingmanson, E. J. (1996). Tool-using behavior in wild *Pan paniscus*: social and ecological considerations. In *Reaching into thought*, ed. A. E. Russon, K. A. Bard & S. T. Parker, pp. 190–210. Cambridge: Cambridge University Press.

Ingmanson, E. J. (1998). Cultural transmission of a communicative gesture in a group of *Pan paniscus* (bonobos). Paper presented at the Napoli Social Learning Conference, Naples, Italy.

Inhelder, B., Lezine, I., Sinclair, H. & Stambak, M. (1972). Les débuts de la fonction symbolique [The beginnings of the symbolic function]. *Archives de Psychologie*, 41, 187–243.

Jackowitz, E. R. & Watson, M. W. (1980). Development of object transformations in early pretend play. *Developmental Psychology*, 16, 543–9.

James, W. (1890). *The principles of psychology.* New York: Henry Holt.

Jarrold, C., Boucher, J., Smith, P. K. & Harris, P. (1993). Symbolic play in autism: a review. *Journal of Austism and Developmental Disorders*, 23, 281–307.

Jarrold, C., Carruthers, P., Smith, P. K. & Boucher, J. (1994). Pretend play: is it metarepresentational? *Mind and Language*, 9, 445–68.

Jensvold, M. L. A. & Fouts, R. S. (1993). Imaginary play in chimpanzees (*Pan troglodytes*). *Human Evolution*, 8, 217–27.

Jézéquel, J. L. & Baudonnière, P. M. (1987). Self-recognition in early childhood: a concept revisited through knowledge of reflecting properties of the mirror. Poster presented at the British Psychological Society, Developmental Section, University of York, England.

Johnson, C. N. & Harris, P. L. (1994). Magic: special but not excluded. *British Journal of Developmental Psychology*, **12**, 35–51.

Johnson, C. N. & Wellman, H. M. (1982). Children's developing conceptions of the mind and the brain. *Child Development*, **52**, 222–34.

Johnson, N. B. (1988). Prehistoric European decorated caves: structured earth environments, initiation, and rites of passage. In *Child development within culturally structured environment*, ed. J. Valsiner, vol. 2, pp. 227–67. Norwood, NJ: Ablex.

Jolly, A. (1972). *The evolution of primate behavior*. New York: Macmillan.

Jolly, A. (1988). The evolution of purpose. In *Machiavellian intelligence*, ed. R. W. Byrne & A. Whiten, pp. 363–78. Oxford: Clarendon Press.

Jolly, C. J. & White, R. (1995). *Physical anthropology and archeology*, 5th edn. New York: McGraw Hill.

Jones, W. (1889). *Glimpses of animal life*. London: Elliot Stock.

Joseph, R. M. (1998). Intention and knowledge in preschoolers' conception of pretend. *Child Development*, **69**, 966–80.

Kagan, J. (1981). *The second year*. Cambridge: Cambridge University Press.

Kahn, P. H. (1999). *The human relationship with nature*. Cambridge, MA: MIT Press.

Kano, T. (1986/1992). *The last ape*. Stanford: Stanford University Press.

Katcher, A. & Wilkins, G. (1993). Dialogue with animals: its nature and culture. In *The biophilia hypothesis*, ed. S. R. Kellert & E. O. Wilson, pp. 173–97. Washington, DC: Island Press.

Kavanaugh, R. D. & Harris, P. L. (1991, September). *Comprehension and production of pretend language by two-year-olds*. Paper presented at the annual meeting of the Developmental Section, British Psychological Society, Cambridge, UK.

Kavanaugh, R. D. & Taylor, M. (1999). Adult memories of childhood imaginary companions. (Unpublished data.)

Kavanaugh, R. D., Whittington, S. & Cerbone, M. J. (1983). Mothers' use of fantasy in speech to young children. *Journal of Child Language*, **10**, 45–55.

Kaye, K. (1980). Why we don't talk "baby talk" to babies. *Journal of Child Language*, **7**, 489–507.

Keller, H. (1902/1965). *The story of my life*. Clinton, MA: Airmont Books.

Kellogg, W. N. & Kellogg, L. A. (1933/1967). *The ape and the child*. New York: Hafner Publishing Co.

Kendon, A. (1991). Some considerations for a theory of language origins. *Man*, **26**, 199–221.

King, B. J. (1994). *The information continuum*. Sante Fe: School of American Research Press.

King, C. E. (1979). *Antique toys and dolls*. New York: Rizzoli.

Köhler, W. (1925/1976). *The mentality of apes*, 2nd edn. New York: Liveright.

Konner, M. (1976). Relationships among infants and juveniles in comparative perspective. *Social Sciences Information*, **13**, 371–402.

Kuczaj, S. A. (1981). Factors influencing children's hypothetical reference. *Journal of Child Language*, **8**, 131–7.

Kuhlmeier, V. A. & Boysen, S. T. (2002). Contributions of spatial and object features to scale model comprehension by chimpanzees (*Pan troglodytes*). *Psychological Science*, **13**, 60–3.

Kuhlmeier, V. A., Boysen, S. T. & Mukobi, K. L. (1999). Scale model comprehension by chimpanzees (*Pan troglodytes*). *Journal of Comparative Psychology*, **113**, 396–402.

Kummer, H. (1968). *Social organization of hamadryas baboons*. Chicago: University of Chicago Press.

Kupfermann, K. (1977). A latency boy's identity as a cat. *Psychoanalytic Study of the Child*, **32**, 363–85.

Kuroda, S. (1984). Interaction over food among pygmy chimpanzees. In *The pygmy chimpanzee*, ed. R. L. Susman, pp. 301–24. New York: Plenum Press.

Kuroda, S. (1984). Rocking gesture as communicative behavior in the wild pygmy chimpanzees in Wamba, central Zaire. *Journal of Ethology*, **2**, 127–37.

Labrell, F. (1994). A typical interaction behaviour between fathers and toddlers: teasing. *Early Development and Parenting*, **3**, 125–30.

Ladygina-Kots, N. N. (1935/1982). Infant ape and human child: their instincts, emotions, games, and expressive movements. *Storia e Critica della Psicologia*, **3**, 113–89.

Lang, E. M. (1963). *Goma, the baby gorilla*. Garden City, NY: Doubleday Books.

Langer, J. (1993). Comparative cognitive development. In *Tools, language, and cognition in human evolution*, ed. K. R. Gibson & T. Ingold, pp. 300–13. Cambridge: Cambridge University Press.

Langer, J. (1996). Heterochrony and the evolution of primate cognitive development. In *Reaching into thought*, ed. A. E. Russon, K. A. Bard & S. T. Parker, pp. 257–77. Cambridge: Cambridge University Press.

Leevers, H. J. & Harris, P. L. (1999). Persisting effects of instruction on young children's syllogistic reasoning with incongruent and abstract premises. *Thinking and Reasoning*, **5**, 145–73.

Leslie, A. M. (1987). Pretense and representation: the origins of "theory of mind." *Psychological Review*, **94**, 412–26.

Leslie, A. M. (1988). Some implications of pretense for mechanisms underlying the child's theory of mind. In *Developing theories of mind*, ed. J. W. Astington, P. L. Harris & D. R. Olson, pp. 19–46. New York: Cambridge University Press.

Lewis, C., Freeman, N. H., Kyriakidou, C., Maridaki-Kassotaki, K. & Berridge, D. M. (1996). Social influence on false belief access: specific sibling influences or general apprenticeship? *Child Development*, **67**, 2930–47.

Lewis, M. & Brooks-Gunn, J. (1979). *Social cognition and the acquisition of self*. New York: Plenum Press.

Lewis, M. & Ramsay, D. (1999). Intentions, consciousness and pretend play. In *Developing theories of intention*, ed. P. Zelazo, J. Astington & D. R. Olson, pp. 77–94. Mahwah, NJ: Erlbaum.

Lewis, V. & Boucher, J. (1988). Spontaneous, instructed and elicited play in relatively able autistic children. *British Journal of Developmental Psychology*, **6**, 325–39.

Leyhausen, P. (1952/1973). Theoretical considerations in criticism of the concept of the "displacement movement." In *Motivation of human and animal behavior*, K. Lorenz & P. Leyhausen, pp. 59–69. New York: Van Nostrand.

Leyhausen, P. (1965/1973). On the function of the relative hierarchy of moods. In *Motivation of human and animal behavior*, K. Lorenz & P. Leyhausen, pp. 144–247. New York: Van Nostrand.

Lézine, I. (1973). The transition from sensorimotor to earliest symbolic function in early development. *Research Publication of the Association for Research in Nervous and Mental Diseases*, **51**, 221–32.

Lillard, A. S. (1993a). Pretend play skills and the child's theory of mind. *Child Development*, **64**, 348–71.

Lillard, A. S. (1993b). Young children's conceptualization of pretense: action or mental representational state. *Child Development*, **64**, 372–86.

Lillard, A. S. (1994). Making sense of pretence. In *Children's early understanding of mind*, ed. C. Lewis & P. Mitchell, pp. 211–34. Hove: Erlbaum.

Lillard, A. S. (1996). Body or mind: children's categorizing of pretense. *Child Development*, **67**, 1717–34.

Lillard, A. S. (1998a). Playing with a theory of mind. In *Multiple perspectives on play in early childhood education*, ed. O. Saracho & B. Spodek, pp. 11–33. Albany: SUNY Press.

Lillard, A. S. (1998b). Wanting to be it: children's understanding of intentions underlying pretense. *Child Development*, **69**, 981–93.

Lillard, A. S. (2001a). Pretend play as Twin Earth. *Developmental Review*. (In press.)

Lillard, A. S. (2001b). Pretending, understanding pretense, and understanding minds. In *Play and culture studies*, ed. S. Reifel, vol. 3, pp. 233–54. Norwood, NJ: Ablex.

Lillard, A. S. & Flavell, J. H. (1992). Young children's understanding of different mental states. *Developmental Psychology*, **28**, 626–34.

Lillard, A. S. & Joffre, K. (1999). *The robustness of children's conceptualization of pretense as action*. Unpublished manuscript, University of Virginia.

Lillard, A. S. & Sobel, D. (1999). Lion Kings or puppies: children's understanding of pretense. *Developmental Science*, **2**, 75–80.

Lillard, A. S., Zeljo, A., Curenton, S. & Kaugars, A. S. (2000). Children's understanding of the animacy constraint on pretense. *Merrill-Palmer Quarterly*, **46**, 21–44.

Limongelli, L., Boysen, S. T. & Visalberghi, E. (1995). Comprehension of cause-effect relations by tool-using chimpanzees (*Pan troglodytes*). *Journal of Comparative Psychology*, **109**, 18–26.

Lindburg, D. G. (Ed.) (1980). *The macaques*. New York: Van Nostrand Reinhold.

Lindsay, W. L. (1880). *Mind in the lower animals in health and disease*, vol. 1. New York: D. Appleton & Co.

Liss, M. B. (Ed.) (1983). *Social and cognitive skills*. New York: Academic Press.

Lock, A. (Ed.) (1978). *Action, gesture and symbol*. London: Academic Press.

Lock, A. (1980). *The guided reinvention of language*. London: Harcourt Brace and Co.

Lock, A. & Peters, C. R. (Eds) (1996). *The handbook of human symbolic evolution*. Oxford: Clarendon Press.

Lorenz, K. (1932/1970). A consideration of methods of identification of species-specific instinctive behaviour patterns in birds. In *Studies in animal and human behavior*, K. Lorenz, vol. 1, pp. 57–100. Cambridge, MA: Harvard University Press.

Lorenz, K. (1935/1970). Companions as factors in the bird's environment. In *Studies in animal and human behavior*, K. Lorenz, vol. 1, pp. 101–258. Cambridge, MA: Harvard University Press.

Lorenz, K. (1937/1970). The establishment of the instinct concept. In *Studies in animal and human behavior*, K. Lorenz, vol. 1, pp. 259–315. Cambridge, MA: Harvard University Press.

Lorenz, K. (1942/1970). Inductive and teleological psychology. In *Studies in animal and human behavior*, K. Lorenz, vol. 1, pp. 351–70. Cambridge, MA: Harvard University Press.

Lorenz, K. (1950/1971). Part and parcel in animal and human societies. In *Studies in animal and human behavior*, K. Lorenz, vol. 2, pp. 115–95. Cambridge, MA: Harvard University Press.

Lorenz, K. (1953/1977). *Man meets dog*. Harmondsworth: Penguin Books.
Lorenz, K. (1956). Plays and vacuum activities. In *L'instinct dans le comportement des animaux et de l'homme* [Instincts in the behavior of animals and man], ed. Foundation Singer-Polignac, pp. 633–7. Paris: Masson et Cie.
Lovell, L., Hoyle, H. W. & Sidall, M. Q. (1968). A study of some aspects of the play and language of young children with delayed speech. *Journal of Child Psychiatry*, 9, 41–50.
Lowe, M. (1975). Trends in the development of representational play in infants from one to three years: an observational study. *Journal of Child Psychology and Psychiatry*, 16, 33–47.
Lucariello, J. (1990). Canonicality and consciousness in child narrative. In *Narrative thought and narrative language*, ed. B. K. Britton & A. D. Pellegrini, pp. 131–40. Hillsdale, NJ: Erlbaum.
Luquet, G. H. (1927). *Le dessin enfantin* [Childish drawing]. Paris: Alcan.
Lyons, B. G. (1986). Zone of potential development for 4-year-olds attempting to simulate the use of absent objects. *Occupational Therapy Journal of Research*, 6, 33–46.
Lyons, J. (1977). *Semantics*, vol. 1. Cambridge: Cambridge University Press.
Lyytinen, P. (1991). Developmental trends in children's pretend play. *Child: Care, Health and Development*, 17, 9–25.
MacKinnon, J. (1974). The behavior and ecology of wild orangutans. *Animal Behaviour*, 22, 3–74.
MacWhinney, B. & Snow, C. (1985). The child language data exchange system. *Journal of Child Language*, 12, 271–96.
Main, M. (1983). Exploration, play and cognitive functioning related to infant-mother attachment. *Infant Behavior and Development*, 6, 167–74.
Mandler, J. (1992). How to build a baby. II. Conceptual primitives. *Psychological Review*, 99, 587–604.
Mandler, J. (1998). Representation. In *Handbook of child psychology*, vol. 2. *Cognition, perception, and language development*, ed. D. Kuhn & R. S. Siegler, pp. 255–308. New York: John Wiley & Sons, Inc.
Mandler, J. M. (1984). *Stories, scripts and scenes*. Hillsdale, NJ: Erlbaum.
Maple, T. L. & Hoff, M. P. (1982). *Gorilla behavior*. New York: Van Nostrand Reinhold.
Markovits, H., Venet, M., Janveau-Brennan, G., Malfait, N., Pion, N. & Vadeboncoeur, I. (1996). Reasoning in young children: fantasy and information retrieval. *Child Development*, 67, 2857–72.
Marten, K. & Psarakos, S. (1994). Evidence of self-awareness in the bottlenose dolphin (*Tursiops truncatus*). In *Self-awareness in animals and humans*, ed. S. T. Parker, R. W. Mitchell & M. L. Boccia, pp. 361–79. Cambridge: Cambridge University Press.
Martini, M. (1994). Peer interactions in Polynesia: a view from the Marquesas. In *Children's play in diverse cultures*, ed. J. L. Roopnarine, J. E. Johnson & F. H. Hooper, pp. 73–103. Albany: SUNY Press.
Matsuzawa, T. (1991). Nesting cups and metatools in chimpanzees. *Behavioral and Brain Sciences*, 14, 570–1.
Matsuzawa, T. (1997). The death of an infant chimpanzee at Bossou, Guinea. *Pan African News*, 4(1), 4–6.
Matthews, W. S. (1977). Modes of transformation in the initiation of fantasy play. *Developmental Psychology*, 13, 212–16.

Mauro, J. (1991). *The friend that only I can see: a longitudinal investigation of children's imaginary companions*. Unpublished doctoral dissertation, University of Oregon.

McCune, L. (1993). The development of play as the development of consciousness. In *The role of play in the development of thought*, ed. M. Bornstein & A. W. O'Reilly, pp. 67–80. San Francisco: Jossey-Bass.

McCune, L. (1995). A normative study of representational play at the transition to language. *Developmental Psychology*, 31, 198–206.

McCune, L. (1999). The transition to language in human infants: a human model for the development of vocal communication in other primate species. In *The origins of language*, ed. B. King, pp. 269–306. Santa Fe: School of American Research Press.

McCune, L. & Vihman, M. M. (2001). Early phonetic and lexical development: a productivity approach. *Journal of Speech, Language, and Hearing Research*, 44, 670–84.

McCune, L., Vihman, M. M., Roug-Hellichius, L., Delery, D. B. & Gogate, L. (1996). Grunt communication in human infants. *Journal of Comparative Psychology*, 110, 27–37.

McCune-Nicolich, L. (1981). Toward symbolic functioning: structure and early pretend games and potential parallels with language. *Child Development*, 52, 785–97.

McCune-Nicolich, L. & Bruskin, C. (1982). Combinatiorial competency in symbolic play and language. In *The play of children*, ed. K. Rubin & D. Pepler, pp. 5–22. New York: Karger.

McKinney, M. & McNamara, K. (1991). *Heterochrony*. New York: Plenum Press.

Mead, G. H. (1934/1974). *Mind, self and society*. Chicago: University of Chicago Press.

Meins, E. (1997). *Security of attachment and the social development of cognition*. Hove: Psychology Press.

Mellars, P. (1996). *The Neanderthal legacy*. Princeton: Princeton University Press.

Meltzoff, A. N. (1990). Foundations for developing a concept of self: the role of imitation in relating self to other and the value of social mirroring, social modeling and self practice in infancy. In *The self in transition*, ed. D. Cicchetti & M. Beeghly, pp. 139–64. Chicago: University of Chicago Press.

Meltzoff, A. N. & Moore, M. K. (1998). Infant intersubjectivity: broadening the dialogue to include imitation, identity and intention. In *Intersubjective communication and emotion in early ontogeny*, ed. S. Bråten, pp. 47–62. Cambridge: Cambridge University Press.

Menzel, E. W., Jr. (1973). Leadership and communication in young chimpanzees. In *Precultural primate behavior*, ed. E. W. Menzel, Jr., pp. 192–225. Basel: S. Karger.

Menzel, E. W., Jr., Premack, D. & Woodruff, G. (1978). Map reading by chimpanzees. *Folia Primatologica*, 29, 241–9.

Menzel, E. W., Jr., Savage-Rumbaugh, E. S. & Lawson, J. (1985). Chimpanzee (*Pan troglodytes*) problem solving with the use of mirrors and televised equivalents of mirrors. *Journal of Comparative Psychology*, 99, 211–17.

Meyer-Holzapfel, M. (1956/1978). On the readiness for play and the instinctive activities. In *Evolution of play behavior*, ed. D. Müller-Schwarze, pp. 252–68. Stroudsburg, PA: Dowden, Hutchinson & Ross.

Middleton, J. (Ed.) (1970). *From child to adult*. Garden City, NY: Natural History Press.

Mignault, C. (1985). Transition between sensorimotor and symbolic activities in nursery-reared chimpanzees (*Pan troglodytes*). *Journal of Human Evolution*, 14, 747–58.

Miles, H. L. (1986). How can I tell a lie? Apes, language, and the problem of deception. In *Deception*, ed. R. W. Mitchell & N. S. Thompson, pp. 245–66. Albany: SUNY Press.

Miles, H. L. (1990). The cognitive foundations for reference in a signing orangutan. In *"Language" and intelligence in monkeys and apes*, ed. S. T. Parker & K. R. Gibson, pp. 511–39. New York: Cambridge University Press.

Miles, H. L. (1991). The development of symbolic communication in apes and early hominids. In *Studies in language origins*, ed. W. von Raffler-Engel, J. Wind & A. Jonker, vol. 2, pp. 9–20. Amsterdam: John Benjamins.

Miles, H. L. (1994). ME CHANTEK: the development of self-awareness in a signing orangutan. In *Self-awareness in animals and humans*, ed. S. T. Parker, R. W. Mitchell & M. L. Boccia, pp. 254–72. New York: Cambridge University Press.

Miles, H. L. (1999). Symbolic communication with and by great apes. In *The mentalities of gorillas and orangutans*, ed. S. T. Parker, R. W. Mitchell & H. L. Miles, pp. 197–210. New York: Cambridge University Press.

Miles, H. L., Mitchell, R. W. & Harper, S. E. (1996). Simon says: the development of imitation in an enculturated orangutan. In *Reaching into thought*, ed. A. E. Russon, K. A. Bard & S. T. Parker, pp. 278–99. New York: Cambridge University Press.

Millar, S. (1968). *The psychology of play*. Middlesex, UK: Penguin Books.

Miller, P., Fung, H. & Mintz, J. (1996). Self construction through narrative practices: a Chinese and American comparison. *Ethos*, 24, 1–44.

Miller, P. & Garvey, C. (1984). Mother-baby role play: its origins in social support. In *Symbolic play*, ed. I. Bretherton, pp. 101–30. New York: Academic Press.

Mills, E. (1919/1976). *The grizzly*. Sausalito: Comstock.

Mills, E. (1921). The beaver claims his birthright. *Outlook*, 128, 479–81.

Mitchell, P. (1996). *Acquiring a conception of mind*. Hove, East Sussex: Psychology Press.

Mitchell, P. (1997). *Introduction to theory of mind*. London: Edward Arnold.

Mitchell, R. W. (1986). A framework for discussing deception. In *Deception*, ed. R. W. Mitchell & N. S. Thompson, pp. 3–40. Albany: SUNY Press.

Mitchell, R. W. (1987). A comparative-developmental approach to understanding imitation. In *Perspectives in ethology*, ed. P. P. G. Bateson & P. H. Klopfer, vol. 7, pp. 183–215. New York: Plenum Press.

Mitchell, R. W. (1988). Ontogeny, biography, and evidence for tactical deception. *Behavioral and Brain Sciences*, 11, 259–60.

Mitchell, R. W. (1990). A theory of play. In *Interpretation and explanation in the study of animal behavior*, ed. M. Bekoff & D. Jamieson, vol. 1, pp. 197–227. Boulder, CO: Westview Press.

Mitchell, R. W. (1991a). Bateson's concept of "metacommunication" in play. *New Ideas in Psychology*, 9, 73–87.

Mitchell, R. W. (1991b). Deception and hiding in captive lowland gorillas (*Gorilla gorilla gorilla*). *Primates*, 32, 523–27.

Mitchell, R. W. (1993a). Mental models of mirror-self-recognition: two theories. *New Idea in Psychology*, 11, 295–325.

Mitchell, R. W. (1993b). Recognizing one's self in a mirror? A reply to Gallup and Povinelli, De Lannoy, Anderson, and Byrne. *New Ideas in Psychology*, 11, 351–77.

Mitchell, R. W. (1993c). Animal as liars: the human face of nonhuman duplicity. In *Lying and deception in everyday life*, ed. M. Lewis & C. Saarni, pp. 59–89. New York: Guilford Press.

Mitchell, R. W. (1994a). The evolution of primate cognition: simulation, self-knowledge and knowledge of other minds. In *Hominid culture in primate perspective*, ed. D. Quiatt & J. Itani, pp. 177–232. Boulder: University of Colorado Press.

Mitchell, R. W. (1994b). Review of *The biophilia hypothesis*. *Anthrozoös*, 7, 212–14.

Mitchell, R. W. (1995). Self-recognition, methodology and explanation: a comment on Heyes (1994). *Animal Behaviour*, 51, 467–9.

Mitchell, R. W. (1996). The psychology of human deception. *Social Research*, 63, 819–61.

Mitchell, R. W. (1997a). A comparison of the self-awareness and kinesthetic-visual matching theories of self-recognition: autistic children and others. *Annals of the New York Academy of Sciences*, 818, 39–62.

Mitchell, R. W. (1997b). Kinesthetic-visual matching and the self-concept as explanations of mirror-self-recognition. *Journal for the Theory of Social Behavior*, 27, 101–23.

Mitchell, R. W. (1997c). Anthropomorphic anecdotalism as method. In *Anthropomorphism, anecdotes, and animals*, ed. R. W. Mitchell, N. S. Thompson & H. L. Miles, pp. 151–69. Albany: SUNY Press.

Mitchell, R. W. (1997d). Anthropomorphism and anecdotes: a guide for the perplexed. In *Anthropomorphism, anecdotes, and animals*, ed. R. W. Mitchell, N. S. Thompson & H. L. Miles, pp. 407–27. Albany: SUNY Press.

Mitchell, R. W. (1999a). Deception and concealment as strategic script violation in great apes and humans. In *The mentalities of gorillas and orangutans*, ed. S. T. Parker, R. W. Mitchell & H. L. Miles, pp. 295–315. Cambridge: Cambridge University Press.

Mitchell, R. W. (1999b). Scientific and popular conceptions of the psychology of great apes from the 1790s to the 1970s: déjà vu all over again. *Primate Report*, 53, 1–118.

Mitchell, R. W. (2000). A proposal for the development of a mental vocabulary, with special reference to pretense and false belief. In *Children's reasoning and the mind*, ed. K. Riggs & P. Mitchell, pp. 37–65. Hove: Psychology Press.

Mitchell, R. W. (2001). Americans' talk to dogs during play: similarities and differences with talk to infants. *Research on Language and Social Interaction*, 34, 182–210.

Mitchell, R. W. (2002a). Kinesthetic-visual matching, imitation, and self-recognition. In *The cognitive animal*, ed. M. Bekoff, C. Allen & G. Burghardt, Cambridge, MA: MIT Press. (In press.)

Mitchell, R. W. (2002b). Imitation as a perceptual process. In *Imitation in animals and artifacts*, ed. C. L. Nehaniv & K. Dautenhahn, Cambridge, MA: MIT Press. (In press.)

Mitchell, R. W. & Anderson, J. (1993). Discrimination learning of scratching, but failure to obtain imitation and self-recognition in a long-tailed macaque. *Primates*, 34, 301–9.

Mitchell, R. W. & Hamm, M. (1997). The interpretation of animal psychology: anthropomorphism or behavior reading? *Behaviour*, 134, 173–204.

Mitchell, R. W. & Neal, M. (1999). *Children's understanding of their own and others' pretense and false belief*. Unpublished manuscript, Eastern Kentucky University.

Mitchell, R. W. & Thompson, N. S. (Eds) (1986). *Deception: perspectives on human and nonhuman deceit*. Albany: SUNY Press.

Mitchell, R. W. & Thompson, N. S. (1991). Projects, routines, and enticements in dog-human play. In *Perspectives in ethology*, ed. P. P. G. Bateson & P. H. Klopfer, vol. 9, pp. 189–216. New York: Plenum Press.

Mitchell, R. W., Thompson, N. S. & Miles, H. L. (Eds) (1997). *Anthropomorphism, anecdotes, and animals*. Albany: SUNY Press.

Moore, B. R. (1992). Avian movement imitation and a new form of mimicry: tracing the evolution of a complex form of learning. *Behaviour*, **122**, 231–63.

Moore, T. (1964). Realism and fantasy in children's play. *Journal of Child Psychology and Psychiatry*, **5**, 15–36.

Morgan, C. L. (1900). *Animal behaviour*. London: Edward Arnold.

Morrow, L. M. (1986). Effects of structural guidance in story retelling on children's dictation of original stories. *Journal of Reading Behavior*, **18**, 135–52.

Munroe, R. L. & Munroe, R. H. (1975). *Cross-cultural human development*. Prospect Heights: Waveland.

Musatti, T. (1986). Representational and communicative abilities in early social play. *Human Development*, **29**, 49–60.

Musatti, T. & Mayer, S. (1987). Object substitution: its nature and function in early pretend play. *Human Development*, **30**, 225–35.

Musatti, T., Veneziano, E. & Mayer, S. (1998). Contributions of language to early pretend play. *Cahiers de Psychologie Cognitive/Current Psychology of Cognition*, **17**, 155–81.

Myers, G. (1998). *Children and animals*. Boulder, CO: Westview Press.

Nadel, J. (1994). The development of communication: Wallon's framework and influence. In *Early child development in the French tradition*, ed. A. Vyt, H. Bloch & M. H. Bornstein, pp. 177–89. Hillsdale, NJ: Erlbaum.

Nadel, J. & Baudonnière, P. M. (1982). The social function of reciprocal imitation in 2-year-old peers. *International Journal of Behavioral Development*, **5**, 95–109.

Nadel, J. & Butterworth, G. (Eds) (1999). *Imitation in infancy*. Cambridge: Cambridge University Press.

Nakano, S. & Kanaya, Y. (1993). The effects of mothers' teasing: do Japanese infants read their mothers' play intention in teasing? *Early Development and Parenting*, **2**, 7–17.

Natale, F. (1989). Stage 5 object-concept. In *Cognitive structure and development in nonhuman primates*, ed. F. Antinucci, pp. 89–96. Hillsdale, NJ: Erlbaum.

Natale, F. & Antinucci, F. (1989). Stage 6 object-concept and representation. In *Cognitive structure and development in nonhuman primates*, ed. F. Antinucci, pp. 97–112. Hillsdale, NJ: Erlbaum.

Nelson, K. (1996). *Language in cognitive development: the emergence of the mediated mind*. New York: Cambridge University Press.

Newton, R., Reddy, V. & Bull, R. (2000). Children's everyday deception and performance on false-belief tasks. *British Journal of Developmental Psychology*, **18**, 297–317.

Nichols, S. & Stich, S. (2000). A cognitive theory of pretense. *Cognition*, **74**, 115–47.

Nicolich, L. [McCune-] (1977). Beyond sensorimotor intelligence: assessment of symbolic maturity through analysis of pretend play. *Merrill-Palmer Quarterly*, **23**, 89–101.

Nielsen, M. & Dissanayake, C. (2000). An investigation of pretend play, mental state terms and false belief understanding: in search of a meterepresentational link. *British Journal of Developmental Psychology*, **18**, 609–24.

Nissen, H. W. & Crawford, M. P. (1936). A preliminary study of food-sharing behavior in young chimpanzees. *Journal of Comparative Psychology*, **22**, 383–419.

Noble, W. & Davidson, I. (1991). The evolutionary emergence of modern human behaviour: language and its archaeology. *Man*, **26**, 223–53.

Noble, W. & Davidson, I. (1996). *Human evolution, language, and mind*. Cambridge: Cambridge University Press.

O'Connell, B. & Bretherton, I. (1984). Toddler's play, alone and with mother: the role of

maternal guidance. In *Symbolic play*, ed. I. Bretherton, pp. 337–68. New York: Academic Press.
O'Neill, D. K. (1996). Two-year-old children's sensitivity to a parent's knowledge state when making requests. *Child Development*, **67**, 659–77.
Odum, E. P. (1993). *Ecology*. Sunderland: Sinauer.
Oughourlian, J-M. (1991). *The puppet of desire*. Stanford: Stanford University Press.
Overton, W. F. & Jackson, J. P. (1973). The representation of imagined objects in action sequences: a developmental study. *Child Development*, **44**, 309–14.
Parker, S. T. (1977). Piaget's sensorimotor series in an infant macaque: a model for comparing unstereotyped behavior and intelligence in human and nonhuman primates. In *Primate biosocial development*, ed. S. Chevalier-Skolnikoff & F. E. Poirier, pp. 43–112. New York: Garland.
Parker, S. T. (1984). Playing for keeps: an evolutionary perspective on human games. In *Play in animals and humans*, ed. P. K. Smith, pp. 271–93. Oxford: Blackwell.
Parker, S. T. (1985). A social-technological model for the evolution of language. *Current Anthropology*, **26**, 617–39.
Parker, S. T. (1990). The origins of comparative developmental evolutionary studies of primate mental abilities. In *"Language" and intelligence in monkeys and apes*, ed. S. T. Parker & K. R. Gibson, pp. 3–64. New York: Cambridge University Press.
Parker, S. T. (1991). A developmental approach to the origins of self-recognition in great apes. *Human Evolution*, **6**, 435–49.
Parker, S. T. (1993). Imitation and circular reactions as evolved mechanisms for cognitive construction. *Human Development*, **36**, 309–23.
Parker, S. T. (1996). Using cladistic analysis of comparative data to reconstruct the evolution of cognitive development in hominids. In *Phylogenies and the comparative method in animal behavior*, ed. E. P. Martins, pp. 361–98. Oxford: Oxford University Press.
Parker, S. T. (1999). The development of social roles in the play of an infant gorilla and its relationship to sensorimotor intellectual development. In *The mentalities of gorillas and orangutans*, ed. S. T. Parker, R. W. Mitchell & H. L. Miles, pp. 367–93. Cambridge: Cambridge University Press.
Parker, S. T. (2000). *Homo erectus* infancy and childhood: the turning point in the evolution of behavioral development in hominids. In *Biology, brains, and behavior*, ed. S. T. Parker, J. Langer & M. L. McKinney, pp. 279–318. Sante Fe: School of American Research Press.
Parker, S. T. & Gibson, K. R. (1977). Object manipulation, tool use and sensorimotor intelligence as feeding adaptations in cebus monkeys and great apes. *Journal of Human Evolution*, **6**, 623–41.
Parker, S. T. & Gibson, K. R. (1979). A developmental model for the evolution of language and intelligence in early hominids. *Behavioral and Brain Sciences*, **2**, 367–408.
Parker, S. T. & Gibson, K. R. (Eds) (1990). *"Language" and intelligence in monkeys and apes*. New York: Cambridge University Press.
Parker, S. T. & McKinney, M. L. (1999). *Origins of intelligence*. Baltimore: Johns Hopkins University Press.
Parker, S. T. & Milbrath, C. (1994). Contributions of imitation and role-playing games to the construction of self in primates. In *Self-awareness in animals and humans*, ed. S. T. Parker, R. W. Mitchell & M. L. Boccia, pp. 108–28. New York: Cambridge University Press.

Parker, S. T., Mitchell, R. W. & Boccia, M. L. (Eds) (1994). *Self-awareness in animals and humans*. New York: Cambridge University Press.

Partington, J. T. & Grant, C. (1984). Imaginary companions and other useful fantasies. In *Play in animal and humans*, ed. P. K. Smith, pp. 217–40. Oxford: Blackwell.

Patterson, F. G. P. (1978a). The gestures of a gorilla: language acquisition in another pongid. *Brain and Language*, 5, 72–97.

Patterson, F. G. P. (1978b). Linguistic capabilities of a young lowland gorilla. In *Sign language and language acquisition in man and ape*, ed. F. C. Peng, pp. 161–201. Boulder, CO: Westview Press.

Patterson, F. G. P. (1978c). Conversations with a gorilla. *National Geographic*, 154, 438–65.

Patterson, F. G. P. (1979). *Linguistic capabilities of a lowland gorilla*. Stanford University Ph.D. dissertation. Ann Arbor, MI: University Microfilms International.

Patterson, F. G. P. (1980). Innovative uses of language by a gorilla: a case study. In *Children's language*, ed. K. E. Nelson, vol. 2, pp. 497–561. New York: Gardner Press.

Patterson, F. G. P. (1984). Gorilla language acquisition. *National Geographic Society Research Reports*, Vol. 17, 1976 projects, pp. 677–700. Washington, DC: National Geographic Society.

Patterson, F. G. P. & Cohn, R. H. (1990). Language acquisition by a lowland gorilla: Koko's first 10 years of vocabulary development. *Word*, 41(2), 97–143.

Patterson, F. G. P. & Cohn, R. H. (1994). Self-recognition and self-awareness in lowland gorillas. In *Self-awareness in animals and humans*, ed. S. T. Parker, R. W. Mitchell & M. L. Boccia, pp. 291–300. New York: Cambridge University Press.

Patterson, F. G. P. & Kennedy, M. M. (1987). Fantasy play. *Gorilla*, 11(1), 4.

Patterson, F. G. P. & Linden, E. (1981). *The education of Koko*. New York: Holt, Rinehart & Winston.

Patterson, F. G. P., Tanner, J. & Mayer, N. (1988). Pragmatic analysis of gorilla utterances: early communicative development in the gorilla Koko. *Journal of Pragmatics*, 12, 35–55.

Pederson, D. R., Rook-Green, A. & Elder, J. L. (1981). The role of action in the development of pretend play in young children. *Developmental Psychology*, 17, 756–9.

Pellegrini, A. D. (1985). The relations between symbolic play and literate behavior: a review and critique of the empirical literature. *Review of Educational Research*, 55, 107–21.

Pellegrini, A. D. (1990). Relations between preschool children's symbolic play and literate behavior. In *Play, language and stories*, ed. L. Galda & A. D. Pellegrini, pp. 79–97. Norwood, NJ: Ablex Publishing.

Pepperberg, I. M., Brese, K. J. & Harris, B. J. (1991). Solitary sound play during acquisition of English vocalizations by an African Grey parrot (*Psittacus erithacus*): possible parallels with children's monologue speech. *Applied Psycholinguistics*, 12, 151–78.

Pereira, M. E. & Preisser, M. C. (1998). Do strong primate players 'self-handicap' during competitive social play? *Folia Primatologica*, 69, 177–80.

Perinat, A. & Dalmáu, A. (1989). La comunicación entre pequeños gorilas criados en cautividad y sus cuidadoras [Communication between infant gorillas born in captivity and their caretakers]. *Estudios de Psicología*, 32–34, 11–29.

Perner, J. (1991). *Understanding the representational mind*. Boston: MIT Press.

Perner, J., Baker, S. & Hutton, D. (1994). *Prelief*: the conceptual origins of belief and pretense. In *Children's early understanding of mind*, ed. C. Lewis & P. Mitchell, pp. 261–86. Hillsdale, NJ: Erlbaum.

Perner, J., Leekam, S. R. & Wimmer, H. (1987). Three-year-olds' difficulty with false belief: the case for a conceptual deficit. *British Journal of Developmental Psychology*, 5, 125–37.

Perner, J., Ruffman, T. & Leekam, S. R. (1994). Theory of mind is contagious: you catch it from your sibs. *Child Development*, 65, 1228–38.

Perner, J. & Wilde-Astington, J. (1992). The child's understanding of mental representation. In *Piaget's theory*, ed. H. Beilin & P. B. Pufall, pp. 141–60. Hillsdale, NJ: Erlbaum.

Peskin, J. & Olson, D. (1997). *Children's understanding of misleading appearance in characterization*. Paper presented at Biennial Meeting of the Society for Research in Child Development, Washington, DC.

Pessin, A. & Goldberg, S. (Eds) (1996). *The twin earth chronicles*. London: M. E. Sharpe.

Peters, C. R. (1996). Tempo and mode of change in the evolution of symbolism. In *The handbook of human symbolic evolution*, ed. A. Lock & C. R. Peters, pp. 861–76. Oxford: Clarendon Press.

Peters, H. (1995). *Orangutan reintroduction? Development, use and evaluation of a new method: reintroduction*. Unpublished M.Sc. Thesis, U. Groningen, The Netherlands.

Peterson, C. & McCabe, A. (1996). Parental scaffolding of context in children's narratives. In *Children's language*, ed. C. J. Johnson & J. H. V. Gilbert, vol. 9, pp. 183–96. Mahwah, NJ: Erlbaum.

Piaget, J. (1923). *Le langage et la pensée chez l'enfant* [The language and thought of the child]. Neuchâtel: Delachaux et Niestlé.

Piaget, J. (1936). *La naissance de l'intelligence chez l'enfant* [The origins of intelligence in children]. Neuchâtel-Paris: Delachaux et Niestlé.

Piaget, J. (1937/1954). *The construction of reality in the child* [Trans. M. Cook]. New York: Basic Books. (Original work: J. Piaget (1937). *La construction du réel chez l'enfant*.)

Piaget, J. (1945). *La formation du symbole chez l'enfant*. Neuchâtel: Delachaux et Niestlé.

Piaget, J. (1945/1962). *Play, dreams and imitation in childhood* [Trans. C. Gattegno & F. M. Hodgson]. New York: Norton. (Original work: J. Piaget (1945). *La formation du symbole chez l'enfant*.)

Piaget, J. (1947/1972). *The psychology of intelligence*. Totowa, NJ: Littlefield, Adams & Co.

Pinker, S. (1994). *The language instinct*. New York: Harper Collins.

Pintler, M. H., Phillips, R. & Sears, R. (1946). Sex differences in the projective doll play of preschool children. *Journal of Psychology*, 21, 73–80.

Plooij, F. X. (1978). Some basic traits of language in wild chimpanzees? In *Action, gesture, and symbol*, ed. A. Lock, pp. 111–31. London: Academic Press.

Potts, R. (1996). *Humanity's descent*. New York: Avon Books.

Povinelli, D. J., Bering, J. M. & Giambrone, S. (2000). Toward a science of other minds: escaping the argument by analogy. *Cognitive Science*, 24, 509–41.

Povinelli, D. J. & Cant, J. G. H. (1995). Arboreal clambering and the evolution of self-conception. *Quarterly Review of Biology*, 70, 393–421.

Pratt, M. W. & MacKenzie-Keating, S. (1985). Organizing stories: effects of task difficulty on referential cohesion in narrative. *Developmental Psychology*, 21, 350–6.

Premack, D. (1984). Upgrading a mind. In *Talking minds*, ed. T. G. Bever, J. M. Carroll & L. E. Miller, pp. 181–206. Cambridge, MA: MIT Press.

Premack, D. & Premack, A. J. (1983). *The mind of an ape*. New York: Norton.
Premack, D. & Woodruff, G. (1978). Chimpanzee problem-solving: a test for comprehension. *Science*, 202, 532–5.
Preyer, W. (1887). *L'âme de l'enfant* [The spirit of the child]. Paris: Alcan.
Preyer, W. (1889). *The mind of the child*, Part II: *the development of the intellect*. New York: D. Appleton & Co.
Preyer, W. (1890). *The mind of the child*, Part I: *the senses and the will*. New York: D. Appleton & Co.
Priel, B. & de Schonen, S. (1986). Self recognition: a study of a population without mirrors. *Journal of Experimental Child Psychology*, 41, 237–50.
Quiatt, D. (1984). Devious intentions of monkeys and apes? In *The meaning of primate signals*, ed. R. Harré & V. Reynolds, pp. 9–40. Cambridge: Cambridge University Press.
Quiatt, D. (1997). Silent partners? Observations on some systematic relations among observer perspective, theory, and behavior. In *Anthropomorphism, anecdotes, and animals*, ed. R. W. Mitchell, N. S. Thompson & H. L. Miles, pp. 220–36. Albany: SUNY Press.
Rawlins, R. G. & Kessler, M. J. (Eds) (1986). *The Cayo Santiago macaques*. Albany: SUNY Press.
Reddy, V. (1991). Playing with others' expectations: teasing and mucking about in the first year. In *Natural theories of mind*, ed. A. Whiten, pp. 143–58. Oxford: Blackwell.
Reiss, D. & Marino, L. (2001). Mirror self-recognition in the bottlenose dolphin: a case of cognitive convergence. *Proceedings of the National Academy of Sciences*, 98, 5937–42.
Rensch, B. (1972). Play and art in apes and monkeys. In *Precultural primate behavior*, ed. E. W. Menzel Jr., pp. 102–23. Basel: S. Karger.
Repina, T. A. (1964/1971). Development of imagination. In *The psychology of preschool children*, ed. A. V. Zaporozhets & D. B. Elkonin, pp. 255–77. Cambridge, MA: MIT Press.
Reynolds, P. C. (1976). Play, language, and human evolution. In *Play*, ed. J. Bruner, A. Jolly & K. Sylva, pp. 621–35. Baltimore: Penguin Books.
Reynolds, P. C. (1982). The primate constructional system: the theory and description of instrumental object use in humans and chimpanzees. In *The analysis of action*, ed. M. von Cranach & R. Harré, pp. 185–200. Cambridge: Cambridge University Press.
Reynolds, P. C. (1991). Structural differences in intentional action between humans and chimpanzees–and their implications for theories of handedness and bipedalism. In *Semiotic modeling*, ed. M. Anderson & F. Merrell, pp. 19–46. Berlin: Walter de Gruyter & Co.
Reynolds, P. C. (1993a). Technology–not tool use: cognitive representation of cooperative construction. Paper presented at the American Anthropological Association 92nd Annual Meeting, Washington, DC.
Reynolds, P. C. (1993b). The complementation theory of language and tool use. In *Tools, language, and cognition in human evolution*, ed. K. R. Gibson & T. Ingold, pp. 407–28. New York: Cambridge University Press.
Rheingold, H. L., Cook, K. V. & Kolowitz, V. (1987). Commands activate the behavior and pleasure of 2-year-old children. *Developmental Psychology*, 23, 146–51.
Richert, R. & Lillard, A. S. (2001). 4- to 8-year-olds' understanding of the knowledge prerequisites of drawing and pretending. Unpublished manuscript, University of Virginia
Riguet, C. D., Taylor, N. D., Benaroya, S. & Klein, L. S. (1981). Symbolic play in autistic,

Down's, and normal children of equivalent mental age. *Journal of Autism and Developmental Disorders*, 11, 439–48.

Rivers, A., Bartecku, U., Brown, J. V. & Ettlinger, G. (1983). An unexpected "epidemic" of a rare stereotypy: unidentified stress or imitation? *Laboratory Primate Newsletter*, 22, 5–7.

Roberts, W. P. (2001). The origin of symbolic culture in comparative life-history perspective. In *Play and culture studies*, ed. S. Reifel, vol. 3, pp. 97–108. Norwood, NJ: Ablex.

Roberts, W. P., Krause, M. A. & Fouts, R. S. (1997, April). *Mirror-mediated play in signing chimpanzees as an index of self-awareness*. Paper presented at the 23rd meeting of The Association for the Study of Play, Washington, DC.

Rodman, P. S. & Cant, J. G. H. (Eds) (1984). *Adaptations for foraging in nonhuman primates*. New York: Columbia University Press.

Rogoff, B. (1990). *Apprenticeship in thinking*. Oxford: Oxford University Press.

Romanes, G. J. (1881/1906). *Animal intelligence*, 2nd edn. New York: D. Appleton & Co.

Romanes, G. J. (1889/1975). *Mental evolution in man*. New York: Arno Press.

Roopnarine, J. L., Johnson, J. E. & Hooper, F. H. (Eds) (1994). *Children's play in diverse cultures*. Albany: SUNY Press.

Rorty, A. O. (1995). Understanding others. In *Other intentions*, ed. L. Rosen, pp. 203–23. Sante Fe: School of American Social Research.

Rosen, C. S., Schwebel, D. C. & Singer, J. L. (1997). Preschoolers' attributions of mental states in pretense. *Child Development*, 68, 1133–42.

Rosenberg, A. (1990). Is there an evolutionary biology of play? In *Interpretation and explanation in the study of animal behavior*, ed. M. B. D. Jamieson, vol. 1, pp. 180–96. Boulder, CO: Westview Press.

Rosenblum, L. A. & Cooper, R. W. (Eds) (1968). *The squirrel monkey*. New York: Academic Press.

Rottnek, M. (Ed.) (1999). *Sissies and tomboys*. New York: New York University Press.

Rubin, K. H., Fein, G. G. & Vandenberg, B. (1983). Play. In *Handbook of child psychology*, ed. E. M. Hetherington, vol. 4, pp. 693–774. New York: Wiley.

Ruffman, T., Perner, J., Naito, M., Parkin, L. & Clements, W. A. (1998). Older (but not younger) siblings facilitate false belief understanding. *Developmental Psychology*, 34, 161–74.

Russell, J. (1996). Development and evolution of the symbolic function: the role of working memory. In *Modelling the early human mind*, ed. P. Mellars & K. Gibson, pp. 159–70. Cambridge: McDonald Institute for Archaeological Research.

Russon, A. E. (1996). Imitation in everyday use: matching and rehearsal in the spontaneous imitation of rehabilitant orangutans (*Pongo pygmaeus*). In *Reaching into thought*, ed. A. E. Russon, K. A. Bard & S. T. Parker, pp. 152–76. Cambridge: Cambridge University Press.

Russon, A. E. (1998). The nature and evolution of intelligence in orangutans (*Pongo pygmaeus*). *Primates*, 39, 485–503.

Russon, A. E., Bard, K. A. & Parker, S. T. (Eds) (1996). *Reaching into thought*. Cambridge: Cambridge University Press.

Russon, A. E. & Galdikas, B. M. F. (1993). Imitation in free-ranging rehabilitant orangutans (*Pongo pygmaeus*). *Journal of Comparative Psychology*, 107, 146–61.

Russon, A. E. & Galdikas, B. M. F. (1995). Constraints on great ape imitation: model and action selectivity in rehabilitant orangutan (*Pongo pygmaeus*) imitation. *Journal of Comparative Psychology*, 109, 5–17.

Russon, A. E., Mitchell, R. W., Lefebvre, L. & Abravanel, E. (1998). The comparative evolution of imitation. In *Piaget, evolution, and development*, ed. J. Langer & M. Killen, pp. 103–43. Mahwah, NJ: Erlbaum.

Sabater Pi, J. (1984). *Gorilas y chimpancées del Africa Occidental*. [Gorillas and chimpanzees in western Africa]. Mexico: Fondo de Cultura Económica.

Sade, D. S. (1965). Some aspects of parent-offspring and sibling relations in a group of rhesus monkeys, with a discussion of grooming. *American Journal of Physical Anthropology*, **23**, 1–8.

Saltz, E., Dixon, D. & Johnson, J. (1977). Training disadvantaged preschoolers on various fantasy activities: effects on cognitive functioning and impulse control. *Child Development*, **48**, 367–80.

Salzmann, Z. (1998). *Language, culture, and society*. Boulder, CO: Westview Press.

Sarbin, T. R. (1954). Role theory. In *Handbook of social psychology*, ed. G. Lindzey, vol. 1, pp. 223–58. Cambridge, MA: Addison-Wesley.

Sartre, J. (1948/1966). *The psychology of imagination*. New York: Citadel Press.

Savage-Rumbaugh, E. S. (1984). *Pan paniscus* and *Pan troglodytes*: contrasts in preverbal competence. In *The pygmy chimpanzee*, ed. R. Susman, pp. 395–414. New York: Plenum Press.

Savage-Rumbaugh, E. S. (1986). *Ape language*. New York: Columbia University Press.

Savage-Rumbaugh, [E.] S. & Lewin, R. (1994). *Kanzi: the ape at the brink of the human mind*. New York: John Wiley & Sons.

Savage-Rumbaugh, E. S. & McDonald, K. (1988). Deception and social manipulation in symbol-using apes. In *Machiavellian intelligence*, ed. R. W. Byrne & A. Whiten, pp. 224–37. Oxford: Oxford University Press.

Savage-Rumbaugh, E. S., Shanker, S. G. & Taylor, T. J. (1998). *Apes, language, and the human mind*. Oxford: Oxford University Press.

Scarr, S. & McCartney, K. (1983). How people make their own environments: a theory of genotype-environment effects. *Child Development*, **54**, 424–35.

Schaller, G. B. (1963). *The mountain gorilla*. Chicago: University of Chicago Press.

Schank, R. C. & Abelson, R. P. (1977). *Scripts, plans, goals and understanding*. Hillsdale, NJ: Erlbaum.

Schlesinger, I. M., Keren-Portnoy, T. & Parush, T. (2001). *The structure of arguments*. Amsterdam: John Benjamins.

Schürmann, C. (1982). Mating behaviour of wild orang utans. In *The orang utan*, ed. L. E. M. de Boer, pp. 269–84. The Hague: Dr. W. Junk Publishers.

Schwartzman, H. B. (1978). *Transformations*. New York: Plenum Press.

Schwebel, D. C., Rosen, C. S. & Singer, J. L. (1999). Preschoolers' pretend play and theory of mind: the role of jointly constructed pretence. *British Journal of Developmental Psychology*, **17**, 333–48.

Searle, J. (1975). The logical status of fictional discourse. *New Literary History*, **6**, 319–32.

Searle, J. (1992). *The rediscovery of the mind*. Cambridge, MA: MIT Press.

Sevenster, P. (1977). Displacement activities. In *Grzimek's encyclopedia of ethology*, ed. B. Grzimek, pp. 246–57. New York: Van Nostrand Reinhold.

Shapiro, G. & Galdikas, B. M. F. (1995). Attentiveness in orangutan within the sign learning context. In *The neglected ape*, ed. R. D. Nadler, B. M. F. Galdikas, L. K. Sheeran & N. Rosen, pp. 199–212. New York: Plenum Press.

Shapiro, K. J. (1997). A phenomenological approach to the study of nonhuman animals. In *Anthropomorphism, anecdotes, and animals*, ed. R. W. Mitchell, N. S. Thompson & H. L. Miles, pp. 277–95. Albany: SUNY Press.

Shapiro, L. R. & Hudson, J. A. (1991). Tell me a make-believe story: coherence and cohesion in young children's picture elicited narratives. *Developmental Psychology*, **27**, 960–74.

Shatz, M., Wellman, H. M. & Silbur, S. (1983). The acquisition of mental verbs: a systematic investigation of first references to mental state. *Cognition*, **14**, 301–21.

Sheak, W. H. (1917). Disposition and intelligence of the chimpanzee. *Indiana Academy of Science*, **27**, 301–10.

Shepard, P. (1996). *The others*. Washington, DC: Island Press.

Shore, B. (1996). *Culture in mind*. Oxford: Oxford University Press.

Siegler, R. S. (1996). *Emerging minds*. Oxford: Oxford University Press.

Silk, J. B. (1996). Why do primates reconcile? *Evolutionary Anthropology*, **5**(2), 39–42.

Simpson, M. J. A. (1976). The study of animal play. In *Growing points in ethology*, ed. P. P. G. Bateson & R. A. Hinde, pp. 385–400. New York: Cambridge University Press.

Sinclair, H. (1970). The transition from sensorimotor behavior to symbolic activity. *Interchange*, **1**, 119–20.

Singer, D. G. & Singer, J. L. (1990). *The house of make-believe*. Cambridge, MA: Harvard University Press.

Slade, A. (1987a). A longitudinal study of maternal involvement and symbolic play during the toddler period. *Child Development*, **58**, 367–75.

Slade, A. (1987b). Quality of attachment and early symbolic play. *Developmental Psychology*, **23**, 78–85.

Slaughter, D. & Dombrowski, J. (1989). Cultural continuities and discontinuities: impact on social and pretend play. In *The ecological content of children's play*, ed. M. N. Block & A. D. Pellegrini, pp. 282–310. Norwood, NJ: Ablex.

Small, M. F. (1998). *Our babies, ourselves*. New York: Anchor Books.

Smilansky, S. (1968). *The effects of sociodramatic play on disadvantaged preschool children*. New York: Wiley.

Smilansky, S. & Shefatya, L. (1990). *Facilitating play*. Gaithersburg, MD: Psychosocial and Educational Publications.

Smith, M. (1978). Cognizing the behavior stream: the recognition of intentional action. *Child Development*, **49**, 736–43.

Smith, P. K. (1982). Does play matter? Functional and evolutionary aspects of animal and human play. *Behavioral and Brain Sciences*, **5**, 139–55.

Smith, P. K. (1988). Children's play and its role in early development: a re-evaluation of the "play ethos." In *Psychological bases for early education*, ed. A. D. Pellegrini, pp. 207–26. Chichester: Wiley.

Smith, P. K. (1996). Language and the evolution of mindreading. In *Theories of theories of mind*, ed. P. Carruthers & P. K. Smith, pp. 344–54. New York: Cambridge University Press.

Smith, P. K. (1997). Play fighting and real fighting: perspectives on their relationship. In *New aspects of human ethology*, ed. A. Schmitt, K. Atzwanger, K. Grammer & K. Schäfer, pp. 47–64. London: Plenum Press.

Smith, P. K., Dalgleish, M. & Herzmark, G. (1981). A comparison of the effects of fantasy play tutoring and skills tutoring in nursery classes. *International Journal of Behavioral Development*, **4**, 421–41.

Smith, P. K. & Syddall, S. (1978). Play and nonplay tutoring in preschool children: is it play or tutoring which matters? *British Journal of Educational Psychology*, **48**, 315–25.

Smith, P. K. & Whitney, S. (1987). Play and associative fluency: experimenter effects may be responsible for previous findings. *Developmental Psychology*, **23**, 49–53.

Smuts, B. B. (1985). *Sex and friendship in baboons*. Hawthorne, NY: Aldine.

Smuts, B. B., Cheney, D., Seyfarth, R. M., Wrangham, R. W. & Struhsaker, T. T. (Eds) (1987). *Primate societies*. Chicago: University of Chicago Press.

Smythe, R. H. (1961). *Animal psychology*. Springfield, IL: Charles C. Thomas, Publisher.

Snow, C. E., Dubber, C. & de Blauw, A. (1982). Routines in mother–child interaction. In *The language of children reared in poverty*, ed. L. Feagans & D. C. Farran, pp. 53–72. New York: Academic Press.

Sobel, D. M. & Lillard, A. S. (2001a). The effect of fantasy and action on young children's understanding of pretense. *British Journal of Developmental Psychology*, **19**, 85–98.

Sobel, D. M. & Lillard, A. S. (2001b). Children's understanding of the mind's involvement in pretense: do words bend the truth? *Developmental Science*, **4**, 485–96.

Spencer, H. (1878). *The principles of psychology*, vol. 2. New York: D. Appleton & Co.

Sperber, D. (Ed.) (2000). *Metarepresentations*. Oxford: Oxford University Press.

Stahl, S. M. (1998). *Essential psychopharmacology*. Cambridge: Cambridge University Press.

Stambak, M. & Sinclair, H. (1990). Introduction. In *Les jeux de fiction entre enfants de 3 ans*, ed. M. Stambak & H. Sinclair, pp. vii–xviii. Paris: Presses Universitaires de France. (English translation: (1993). Introduction. *Pretend play among 3-year-olds*. Hillsdale, NJ: Erlbaum.)

Stein, D. M. (1981). *The nature and function of social interactions between infant and adult male yellow baboons* (Papio cynocephalus). Ph.D. Thesis, University of Chicago.

Stern, C. & Stern, W. (1909/1999). *Recollection, testimony and lying in early childhood*. Washington, DC: APA.

Stern, D. (1985). *The interpersonal world of the infant*. New York: Basic Books.

Stern, W. (1914/1924). *Psychology of early childhood*. New York: Henry Holt.

Stringer, C. & Gamble, C. (1993). *In search of the Neanderthals*. New York: Thames and Hudson.

Strawson, P. F. (1959/1963). *Individuals*. Garden City, NJ: Anchor Books.

Strum, S. C. (1987). *Almost human*. New York: Norton.

Subbotsky, E. V. (1993). *Foundations of the mind*. Cambridge, MA: Harvard University Press.

Suits, B. (1977). Words on play. *Journal of the Philosophy of Sport*, **4**, 117–31.

Sully, J. (1896). *Studies of childhood*. London: Longmans, Green & Co.

Super, C. M. & Harkness, S. (1986). The developmental niche: a conceptualization of the interface of child and culture. *International Journal of Behavioral Development*, **9**, 545–70.

Sutton-Smith, B. (1986). The spirit of play. In *The young child at play: reviews of research*, ed. G. G. Fein & M. Rivkin, vol. 4, pp. 3–16. Washington, DC: Naeyc.

Sutton-Smith, B. (1997). *The ambiguity of play*. Cambridge, MA: Harvard University Press.

Sutton-Smith, B., Botvin, G. & Mahoney, D. (1976). Developmental structures in fantasy narratives. *Human Development*, **19**, 1–20.

Svendsen, M. (1934). Children's imaginary companions. *Archives of Neurology and Psychiatry*, **2**, 985–99.

Swartz, K. B., Sarauw, D. & Evans, S. (1999). Comparative aspects of mirror self-recognition in great apes. In *The mentalities of gorillas and orangutans*, ed. S. T.

Parker, R. W. Mitchell & H. L. Miles, pp. 283–94. Cambridge: Cambridge University Press.
Symons, D. (1978). *Play and aggression*. New York: Columbia University Press.
Tanner, J. (1985, December). Koko and Michael, gorilla gourmets. *Gorilla*, **9**(1), 3.
Tart, C. T. (Ed.) (1972). *Altered states of consciousness*. New York: Doubleday Anchor.
Tayler, C. K. & Saayman, G. S. (1973). Imitative behaviour by Indian Ocean bottlenose dolphins (*Tursiops aduncus*) in captivity. *Behaviour*, **44**, 286–98.
Taylor, A. R. (1898). *The study of the child*. New York: D. Appleton & Co.
Taylor, M. (1999). *Imaginary companions and the children who create them*. Oxford: Oxford University Press.
Taylor, M. & Carlson, S. M. (1997). The relation between individual differences in fantasy and theory of mind. *Child Development*, **68**, 436–55.
Taylor, M. & Carlson, S. M. (1999). Retrospective reports of imaginary companions. Unpublished data.
Taylor, M. & Carlson, S. M. (2000). The influence of religious beliefs on parental attitudes about children's fantasy behavior. In *Imagining the impossible*, ed. K. Rosengren, C. Johnson & P. Harris, pp. 247–68. Cambridge: Cambridge University Press.
Taylor, M., Carlson, S. M., Gerow, L. & Charlie, C. (1999). A longitudinal follow-up of children with imaginary companions. (Unpublished data.)
Taylor, M., Cartwright, B. S. & Carlson, S. M. (1993). A developmental investigation of children's imaginary companions. *Developmental Psychology*, **29**(2), 276–85.
Taylor, M. & Dishion, T. (1999). Imaginary companions in an ethnically diverse sample of 12-year-old children. (Unpublished data.)
Taylor, M., Hodges, S. & Kohanyi, A. (1999). Minds of their own: the relationship between fiction writers and the characters they create. (Unpublished data.)
Taylor, M. & Luu, V. (1999). Parental interpretations and reactions to children's pretend play behaviors. (Unpublished data.)
Teleki, G. (1973). *The predatory behavior of wild chimpanzees*. Lewisburg, PA: Bucknell University Press.
Terr, L. (1990). *Too scared to cry*. New York: Harper and Row.
Thierry, B. (1984). Descriptive and contextual analysis of eye-covering behavior in captive rhesus macaques (*Macaca mulatta*). *Primates*, **25**, 62–77.
Thompson, R. K. R., Oden, D. L. & Boysen, S. T. (1997). Language-naive chimpanzees (*Pan troglodytes*) judge relations-between-relations in an abstract matching task. *Journal of Experimental Psychology: Animal Behavior Processes*, **23**, 31–43.
Thompson-Handler, N., Malenky, R. K. & Badrian, N. (1984). Sexual behavior of *Pan paniscus* under natural conditions in the Lomako Forest, Equateur, Zaire. In *The pygmy chimpanzee*, ed. R. L. Susman, pp. 347–66. New York: Plenum Press.
Thorpe, W. H. (1956/1963). *Learning and instinct in animals*, 2nd edn. Cambridge, MA: Harvard University Press.
Tomasello, M. & Call, J. (1997). *Primate cognition*. Oxford: Oxford University Press.
Tomasello, M., George, B. L., Kruger, A. C., Farrar, M. J. & Evans, A. (1985). The development of gestural communication in young chimpanzees. *Journal of Human Evolution*, **14**, 175–86.
Tomasello, M., Savage-Rumbaugh, S. & Kruger, A. C. (1993). Imitative learning of actions on objects by children, chimpanzees, and enculturated chimpanzees. *Child Development*, **64**, 1688–705.

Trabasso, T., Stein, N. L. & Johnson, L. R. (1981). Children's knowledge of events: a causal analysis of story structure. In *Learning and motivation*, ed. G. Bower, vol. 15, pp. 237–82. San Diego: Academic Press.

Trinkaus, E. & Shipman, P. (1993). *The Neandertals*. New York: Knopf.

Turner, V. (1974). *Dramas, fields, and metaphors*. Ithaca, NY: Cornell University Press.

Tylor, E. B. (1871). *Primitive culture*. London: Murray.

Ujhelyi, M., Buk, P., Merker, B. & Geissman, T. (2000). Observations on the behavior of gibbons (*Hylobates leucogenys, H. gariellae*, and *H. lar*) in the presence of mirrors. *Journal of Comparative Psychology*, 114, 253–62.

Ungerer, J., Zelazo, P. R., Kearsley, R. B. & O'Leary, K. (1981). Developmental changes in the representation of objects in symbolic play from 18 to 34 months of age. *Child Development*, 52, 186–95.

Užgiris, I. Č. (1981). Experience in the social context: imitation and play. In *Early language*, ed. R. L. Schiefelbusch & D. D. Brisher, pp. 139–67. Baltimore: University Park Press.

Užgiris, I. Č. (1999). Imitation as activity: its developmental aspects. In *Imitation in infancy*, ed. J. Nadel & G. Butterworth, pp. 186–206. Cambridge: Cambridge University Press.

Valentine, C. W. (1942/1950). *The psychology of early childhood*. London: Methuen & Co.

Valsiner, J. (1988). Ontogeny of co-construction of culture within socially organized environmental settings. In *Child development within culturally structured environment*, ed. J. Valsiner, vol. 2, pp. 283–97. Norwood, NJ: Ablex Publishing.

van Hooff, J. A. R. A. M. (1967). The facial displays of the catarrhine monkeys and apes. In *Primate ethology*, ed. D. Morris, pp. 7–68. London: Weidenfeld and Nicolson.

van Hooff, J. A. R. A. M. (1972). A comparative approach to the phylogeny of laughter and smiling. In *Non-verbal communication*, ed. R. A. Hinde, pp. 209–41. New York: Cambridge University Press.

van Hooff, J. A. R. A. M. (1973). A structural analysis of the social behavior of a semi-captive group of chimpanzees. In *Social communication and movement*, ed. M. von Cranach & I. Vine, pp. 75–162. London: Academic Press.

Vasey, P. L. (1998). Female choice and inter-sexual competition for female sexual partners in Japanese macaques. *Behaviour*, 135, 579–97.

Veneziano, E. (1981). Early language and nonverbal representation: a reassessment. *Journal of Child Language*, 8, 541–63.

Veneziano, E. (1990). *Functional relationships between social pretend play and language use in young children*. Poster presented at the Fourth European Conference on Developmental Psychology, Stirling, Scotland.

Veneziano, E. (2001). Interactional processes in the origins of the explaining capacity. In *Children's language*, ed. K. Nelson, A. Aksu-Koc & C. Johnson, vol. 10, pp. 113–41. Mahwah, NJ: Erlbaum.

Veneziano, E. & Sinclair, H. (1995). Functional changes in early child language: the appearance of references to the past and of explanations. *Journal of Child Language*, 22, 557–81.

Veneziano, E., Sinclair, H. & Berthoud, I. (1990). From one word to two words: repetition patterns on the way to structured speech. *Journal of Child Language*, 17, 633–50.

Verba, M. (1990). Construction et partage de significations dans les jeux de fiction entre enfants. In *Les jeux de fiction entre enfants de 3 ans*, ed. M. Stambak & H. Sinclair, pp.

1–29. Paris: Presses Universitaires de France. (English translation: (1993). Construction and sharing of meanings in pretend play. *Pretend play among 3-year-olds*. Hillsdale, NJ: Erlbaum.)

Visalberghi, E., Fragaszy, D. M. & Savage-Rumbaugh, S. (1995). Performance in a tool-using task by common chimpanzees (*Pan troglodytes*), bonobos (*Pan paniscus*), an orangutan (*Pongo pygmaeus*), and capuchin monkeys (*Cebus apella*). *Journal of Comparative Psychology*, **109**, 52–60.

Visalberghi, E. & Limongelli, L. (1994). Lack of comprehension of cause–effect relationships in tool-using capuchin monkeys (*Cebus apella*). *Journal of Comparative Psychology*, **108**, 15–22.

Vygotsky, L. S. (1930–1966/1978). *Mind in society* [Trans. & Ed. M. Cole, V. John-Steiner, S. Scribner & E. Souberman]. Cambridge, MA: Harvard University Press.

Vygotsky, L. S. (1933/1978). The role of play in development. In *Mind in society*, L. S. Vygotsky, pp. 92–104. Cambridge, MA: Harvard University Press.

Vygotsky, L. S. (1935/1978). Interaction between learning and development. In *Mind in society*, L. S. Vygotsky, pp. 79–91. Cambridge, MA: Harvard University Press.

Wachs, T. D. (1993). Multidimensional correlates of individual variability in play and exploration. In *The role of play in the development of thought*, ed. M. H. Bornstein & A. W. O'Reilly, pp. 43–53. San Francisco: Jossey-Bass.

Walker, A. & Leakey, R. E. (Eds) (1993). *The Nariokotome Homo erectus skeleton*. Cambridge, MA: Harvard University Press.

Wallon, H. (1942/1970). *De l'acte à la pensée* [From act to thought]. Paris: Flammarion.

Wallon, H. (1965/1984). The psychological development of the child. In *The world of Henri Wallon*, ed. G. Voyat, pp. 133–46. New York: Aronson.

Walton, K. L. (1990). *Mimesis as make-believe*. Cambridge, MA: Harvard University Press.

Warkentin, V., Baudonnière, P. M. & Margules, S. (1995). Emergence de la fonction symbolique et développement cognitif chez l'enfant entre 17 et 20 mois [Emergence of the symbolic function and cognitive development in children between 17 and 20 months]. *Enfance*, **2**, 179–186.

Watson, J. S. (1870). *The reasoning power in animals*, 2nd edn. London: L. Reeve & Co.

Watson, M. W. & Fischer, K. W. (1977). A developmental sequence of agent use in late infancy. *Child Development*, **48**, 828–36.

Watterson, B. (1990). *The authoritative Calvin and Hobbes*. Kansas City: Andrews and McMeel.

Watts, D. P. & Pusey, A. E. (1993). Behavior of juvenile and adolescent great apes. In *Juvenile primates*, ed. M. E. Pereira & L. A. Fairbanks, pp. 148–72. Oxford: Oxford University Press.

Weir, R. H. (1962). *Language in the crib*. The Hague: Mouton & Co.

Wellman, H. M., Cross, D. & Watson, D. K. (2001). A meta-analysis of theory of mind development: the truth about false belief. *Child Development*, **72**, 655–84.

Wellman, H. M. & Estes, D. (1986). Early understanding of mental entities: a reexamination of childhood realism. *Child Development*, **57**, 910–23.

Wellman, H. M., Hollander, M. & Schult, C. A. (1996). Young children's understanding of thought bubbles and of thoughts. *Child Development*, **67**, 768–88.

Wellman, H. M. & Woolley, J. D. (1990). From simple desires to ordinary beliefs: the early development of everyday psychology. *Cognition*, **35**, 245–75.

Werner, H. & Kaplan, B. (1963). *Symbol formation*. New York: Wiley.

Whiten, A. (1996). Imitation, pretense, and mindreading: secondary representation in comparative primatology and developmental psychology? In *Reaching into thought*, ed. A. Russon, K. Bard & S. T. Parker, pp. 300–24. New York: Cambridge University Press.

Whiten, A. (2000). Chimpanzee cognition and the question of mental re-representation. In *Metarepresentations*, ed. D. Sperber, pp. 139–67. Oxford: Oxford University Press.

Whiten, A. & Byrne, R. W. (1988). Tactical deception in primates. *Behavioral and Brain Sciences*, 11, 233–73.

Whiten, A. & Byrne, R. W. (1991). The emergence of metarepresentation in human ontogeny and primate phylogeny. In *Natural theories of mind*, ed. A. Whiten, pp. 267–81. Oxford: Blackwell.

Whiten, A. & Byrne, R. W. (Eds) (1997). *Machiavellian intelligence II*. Cambridge: Cambridge University Press.

Whiten, A., Goodall, J., McGrew, W. C., Nishida, T., Reynolds, V., Sugiyama, Y., Tutin, C. E., Wrangham, R. W. & Boesch, C. (1999). Chimpanzee cultures. *Nature*, 399, 682–5.

Whiten, A. & Ham, R. (1992). On the nature and evolution of imitation in the animal kingdom: reappraisal of a century of research. *Advances in the Study of Behavior*, 21, 239–83.

Wilson, F. R. (1998). *The hand*. New York: Pantheon Books.

Wittgenstein, L. (1921/1974). *Tractatus logico-philosophicus*. London: Routledge & Kegan Paul.

Wittgenstein, L. (1949/1992). *Last writings on the philosophy of psychology*, vol. 2, *The inner and the outer*. Oxford: Blackwell.

Wolf, D. & Grollman, S. (1982). Ways of playing: individual differences in imaginative play. In *The play of children*, ed. K. Rubin & D. Pepler, pp. 46–63. New York: Karger.

Wolf, D. P. (1984). Repertoire, style, and format: notions worth borrowing from children's play. In *Play in animals and humans*, ed. P. K. Smith, pp. 175–93. New York: Blackwell.

Wolf, D. P., Rygh, J. & Altshuler, J. (1984). Agency and experience: action and states in play narratives. In *Symbolic play*, ed. I. Bretherton, pp. 195–217. London: Academic Press.

Wolfberg, P. J. (1999). *Play and imagination in children with autism*. New York: Teachers College Press.

Woolley, J. D. (1995a). The fictional mind: young children's understanding of imagination, pretense, and dreams. *Developmental Review*, 15, 172–211.

Woolley, J. D. (1995b). Young children's understanding of fictional versus epistemic mental representations: imagination and belief. *Child Development*, 66, 1011–21.

Woolley, J. D. & Boerger, E. (2002). The development of beliefs about the origins and controllability of dreams. *Developmental Psychology*, 38. (In press).

Woolley, J. D. & Bruell, M. (1996). Children's awareness of the origins of their mental representations. *Developmental Psychology*, 32, 335–46.

Woolley, J. D. & Phelps, K. E. (1994). Young children's practical reasoning about imagination. *British Journal of Developmental Psychology*, 12, 53–67.

Woolley, J. D. & Wellman, H. M. (1990). Young children's understanding of realities, nonrealities, and appearances. *Child Development*, 61, 946–61.

Woolley, J. D. & Wellman, H. M. (1992). Children's conceptions of dreams. *Cognitive Development*, 7, 365–80.

Woolley, J. D. & Wellman, H. M. (1993). Origin and truth: young children's understanding of imaginary mental representations. *Child Development*, **64**, 1–17.

Wrangham, R. W. & Peterson, D. (1996). *Demonic males*. Boston: Houghton Mifflin.

Wright, J. L. (1992). *Correlates of young children's narrative competence: maternal behaviors and home literacy experiences*. Unpublished Ph.D. dissertation, University of Maryland, College Park.

Wulff, S. B. (1985). The symbolic and object play of children with autism: a review. *Journal of Autism and Developmental Disorders*, **15**, 139–48.

Wundt, W. (1884/1907). *Lectures on human and animal psychology*, 2nd edn. London: Swan Sonnenschein & Co.

Yerkes, R. M. & Yerkes, A. W. (1929). *The great apes*. New Haven: Yale University Press.

Youngblade, L. M. & Dunn, J. (1995). Individual differences in young children's pretend play with mother and sibling: links to relationships and understanding of other people's feelings and beliefs. *Child Development*, **66**, 1472–92.

Zazzo, R. (1975). Des jumeaux devant le miroir: questions de méthode [Twins in front of the mirror: methodological questions]. *Journal de Psychologie*, **4**, 389–413.

Zazzo, R. (1977). Image spéculaire, conscience de soi [Specular image, self-awareness]. In *Psychologie expérimentale et comparée* [Experimental and comparative psychology], ed. G. Oléron, pp. 325–38. Paris: Presses Universitaires de France.

Zazzo, R. (1981). Miroirs, images, espaces [Mirrors, images, spaces]. In *La reconnaissance de son image chez l'enfant et l'animal* [Self-recognition in children and animals], ed. P. Mounoud & A. Vinter, pp. 77–110. Neuchâtel-Paris: Delachaux et Niestlé.

Zazzo, R. (1985). Conscience de soi et méthode des doubles perceptifs [Self-awareness and double perception method]. In *Etudier l'enfant de la naissance à 3 ans* [Study of the child from birth to 3 years old], ed. P. M. Baudonnière, pp. 199–211. Paris: CNRS.

Zeller, A. (1992). Grooming interactions over infants in four species of primate. *Visual Anthropology*, **5**, 63–86.

Zeller, A. (1994). Evidence of structure in macaque communication. In *The ethological roots of culture*, ed. R. A. Gardner, B. Gardner & B. Chiarelli, pp. 15–39. The Netherlands: Kluwer Academic Publishers.

Zeller, A. (1996). The interplay of kinship organization and facial communication in the macaques. In *Evolution and ecology of macaque societies*, ed. J. E. Fa & P. G. Lindburg, pp. 527–50. New York: Cambridge University Press.

Zeller, A. (1998). *Out of awareness: into perception*. Paper presented at American Anthropological Association Meetings, Philadelphia.

Zentall, T. R. (1996). An analysis of imitative learning in animals. In *Social learning in animals*, ed. C. M. Heyes & B. G. Galef, pp. 221–43. San Diego, CA: Academic Press.

Zukow, P. (1989). Siblings as effective socializing agents: evidence from central Mexico. In *Sibling interaction across cultures*, ed. P. Zukow, pp. 79–105. New York: Springer-Verlag.

Author index

Abelson, R. P., 15
Abravanel, E., 252
Acredolo, L. P., 311
Adang, O., 309
Agayoff, J., 5–12, 14–15, 17, 21, 30, 39, 41, 43, 229, 235, 286, 291, 303, 308, 314–15
Aldis, O., 39, 41, 167, 178–9, 186
Allen, C., 167, 178
Altmann, J., 198
Altmann, S. A., 198–9
Altshuler, J., 59, 145
Ames, L. B., 144
Amsterdam, B., 75
Anderson, J. R., 74, 194
Anscombe, G. E. M., 104
Antinucci, F., 52, 183
Apfel, N., 5, 8, 13, 15, 17, 258, 308, 311
Applebee, A. N., 143–4
Arlitt, A. H., 28
Aronson, J. N., 109
Asendorpf, J. B., 73, 75, 80, 90, 314
Astington, J. W., 112, 117, 137, 138
Atran, S., 270
Austin, J. L., 104, 115

Badrian, A., 198
Badrian, N., 198, 208
Baker, S., 111
Baldwin, J. M., 25–6, 28, 31
Bard, K. A., 243, 315
Barkow, J. H., 270
Bartecku, U., 192
Bartsch, K., 106, 108, 128, 143
Bates, E., 5, 9, 13, 34, 40, 93, 286, 311
Bateson, G., xiv, 3, 6, 38–9, 102, 167, 186, 196, 201, 269, 309
Baudonnière, P.-M., 12, 73–6, 80, 89–90, 112, 114, 310, 312–14
Bauer, G. B., 6

Beck, B. B., 183
Beeghly, M., 106
Beeghly-Smith, M., 70–1, 311
Behrend, D. A., 105
Bekoff, M., 167, 178
Belkhenchir, S., 12, 73, 112, 114, 310, 312–14
Bellugi, U., 40
Benaroya, S., 314
Benigni, L., 5, 9, 13, 40, 286
Benninga, J., 311
Bergen, D., 197
Bering, J. M., 168
Berlin, B., 270
Berntson, G. G., 219, 223
Berridge, D. M., 137
Bertenthal, B. I., 75, 89
Berthoud, I., 72
Bertrand, M., 41
Béteille, A., 199
Biben, M., 39
Bickerton, D., 269, 279, 286
Black, J., 91–2, 98–100
Blades, M., 224
Blake, J., 229, 239
Bloom, F. E., 197
Blurton-Jones, N. G., 276
Boccia, M. L., xvii, 269, 315
Boerger, E., 119
Boesch, C., 6, 15, 41, 52, 270, 272
Boesch, H., 52
Bogin, B., 274
Bonen, S., 131
Bonvillian, J. D., 14, 41, 287
Bornstein, M. H., 129, 178
Botvin, G. J., 143, 144, 146–7, 151
Boucher, J., 4, 13, 134, 257, 312–13
Boulanger-Balleyguier, G., 74
Bourchier, A., 19
Bouvier, J., 315

Bowlby, J., 198
Boyatzis, C. J., 310
Boysen, S. T., 8, 11, 16, 52, 210, 213, 219, 223–5, 227
Bråten, S., 162
Braunwald, S. R., 259
Brese, K. J., 29
Bretherton, I., 5, 12–13, 15, 21, 32, 34, 40–1, 70–1, 91–3, 102, 106, 132, 178, 281, 286, 308, 310–11, 316
Breuer, J., 25
Breuggeman, J. A., 16, 39, 41, 187–8, 192, 263, 307, 312
Bridge, C., 311
Brislin, R. W., 259
Bronfenbrenner, U., 271
Bronowski, J., 40
Brooks-Gunn, J., 4, 13, 73
Brown, E., 113
Brown, J. R., 137
Brown, J. V., 192
Bruell, M. J., 110, 112, 118, 121, 124–5
Bruner, J. S., 30, 144, 147, 196, 257, 269, 279
Bruskin, C., 47
Budwig, N., 71
Bühler, C., 32
Bühler, K., 15, 27–8, 32–3, 35
Buk, P., 315
Bull, R., 11, 308
Burton, F. D., 186, 192
Butterworth, G., 205
Byers, J. A., 178
Byrne, R. W., 3, 10, 20, 45, 53, 183, 186, 188–90, 194, 201, 229, 231–2, 235–7, 240, 242–3, 252–3, 263, 281, 309, 313–14

Cachel, S., 277
Cacioppo, J. T., 6, 223
Call, J., 14, 183, 210, 272, 281
Camaioni, L., 5, 9, 13, 40, 286
Candland, D. K., 278
Cant, J. G. H., 198, 252
Carey, S., 105
Carlson, S. M., 4, 5, 7, 15, 17–18, 20–1, 30, 98, 112–13, 129, 138, 167, 170, 179, 286, 314
Carn, G., 12, 73, 112, 114, 310, 312–14
Carpendale, J. I., 143, 152
Carpenter, C. R., 198
Carroll, L., 102
Carruthers, P., 4, 134, 257, 312–13
Cartwright, B. S., 170
Casby, M. W., 13
Case, R., 252
Cerbone, M. J., 91–2, 94, 96–7
Chance, M. R. A., 39, 199
Chandler, M. J., 71, 143, 152
Chapman, M., 75–6, 89

Charlie, C., 170
Charman, T., 136
Cheney, D. L. 183, 188, 197
Chevalier-Skolnikoff, S., 39, 41, 44, 52, 233, 315
Chompsky, N., *xv*
Christie, J. F., 132
Christopherson, H., 28, 132
Claparède, E., 27–8
Clark, E., 169
Clements, W. A., 137
Coates, S. W., 4
Cocking, R. R., 310
Cohn, R. H., 40, 287, 304
Colburn, D., 115
Cole, D., 286
Cole, M., 91
Colmenares, F., 190
Cook, K. V., 13
Cooper, J. R., 197
Cooper, R. W., 198
Copple, C. E., 310
Cords, M., 235
Cornelius, C. B., 131
Corrigan, R., 75, 310
Corsaro, W. A., 156, 163
Cosmides, L., 270
Crawford, M. P., 26, 34, 205, 309
Cross, D., 110
Crum, R. A., 311
Csikszentmihalyi, M., 197
Cunningham, A., 26, 241
Curcio, F., 311, 314
Curenton, S., 105
Currie, G., 21, 315
Custance, D. M., 315
Custer, W. L., 110–11

Dale, N., 91–2, 94
Dalgleish, M., 132
Dalmáu, A., 264
Dansky, J. L., 131
Darling, L. D., *xv*, 5, 13, 16, 19, 28, 142
Darwin, C., 23–4, 42, 167, 197
Davidson, I., 8, 44, 270
Davis, A., 19
Davis, D. L., 110, 121, 124
Davis, W., 3, 8–9, 11
Dawkins, R., 270
Deacon, T. W., 239, 270
de Blauw, A., 15
de Gramont, P., 165
Delery, D. B., 47
Delfour, F., 315
DeLoache, J. S., 8, 11, 211–14, 217, 219, 222–4, 227–8
de Lorimier, S., 113, 137

de Rivera, J., 18, 279
de Schonen, S., 75
Desmond, A., 40
De Stefano, L., 310
de Veer, M. W., 74, 203
DeVore, I., 197
de Waal, F. B. M., 168, 186, 198, 201, 208, 235, 241, 280–1, 314–15
Dias, M. G., 114, 139, 141
Dickson, L., 92
DiLalla, L. F., 299
Dishion, T., 170
Dissanayake, C., 311
Dixon, D., 145
Dockett, S., 140–1
Dolgin, K. G., 105
Dolhinow, P., 197
Dombrowski, J., 129–30
Donald, M., 44, 279
Donelan-McCall, N., 137
Dowson, T. A., 8
Doyle, A.-B., 113, 137
Draper, D., 293
Drucker, J., 32
Dubber, C., 15
Dunn, J., 70–1, 91–2, 94, 112, 137, 139, 258, 303

Eibl-Eibesfeldt, I., 36–7, 129, 314
Einstein, A., 115
El'konin, D. B., 8, 30, 32
Elder, J. L., 8, 286, 310–11
Ellison, C., 20, 304
Emory, G. R., 39
Estes, D., 116
Ettlinger, G., 192
Evans, A., 6
Evans, S., 315
Ewer, R. F., 37

Fagan, R., 39, 42, 256
Fagot, J., 8
Farrar, M. J., 6
Farver, J. A. M., 91–2, 94, 98, 179
Fedigan, L. M., 183–5
Fegley, S., 73, 75, 314
Fein, G. G., xv, 4–6, 8–9, 11, 13–17, 19, 22, 28, 60, 91, 102, 142, 145, 147, 150, 258, 273, 278, 286, 308, 310–12, 316
Feitelson, D., 98
Feldman, C. F., 144
Fellner, W., 6
Fenson, L., 9–10, 13, 60, 93, 310–11
Ferenczi, S., 163
Fernandez, J. W., 163
Field, R. M., 270
Field, T., 310

Fiese, B. H., 92, 95, 97
First, E., 15
Fischer, K. W., 75, 89, 93
Flaum, M., 51
Flavell, E. R., 111, 303
Flavell, J. H., 103, 106, 111, 113, 285–6, 303
Fleagle, J. G., 275
Folven, R. J., 13, 41
Fonagy, P., 136
Forguson, L., 113
Formanek-Brunell, M., 28
Forys, S., 51
Fouts, D. H., 271–2
Fouts, R. S., 14, 258, 271–2, 286–8, 297, 314
Fragaszy, D. M., 52, 194
Frank, R., 20, 304
Freeman, N. H., 137
Freud, S., 25, 28
Friedman, L., 131
Fromberg, D. P., 197
Fuentes, A., 197
Fung, H., 99
Furuichi, T., 280

Galda, L., 273, 278
Galdikas, B. M. F., 50, 229–30, 240, 251, 315
Galef, B. G., 270
Gallup, G. G., Jr., 14, 74–5
Gamble, C., 275–7
Ganea, P. A., 108
Gardner, B. T., 16, 40, 51, 168, 210, 271–2, 314
Gardner, H. W., 271
Gardner, R. A., 16, 40, 51, 168, 210, 271–2, 314
Gargett, R. H., 275
Garvey, C., 60, 92, 179, 278, 286
Gaskins, S., 92, 98
Gauthier, C., 5, 19, 27, 186, 241, 266, 312–13
Gautier-Hion, A. C., 186, 241–2, 249, 254
Geertz, C., 269
Geissman, T., 315
Gelman, R., 105
Gelman, S., 166
Gentner, D., 227–8
George, B. L., 6
Gerow, L. E., 110, 170
Gesell, A., 74
Giambrone, S., 168
Gibson, J. J., 271
Gibson, K. R., xiv, 41, 54, 208, 243, 269–71, 278, 314–15
Giffin, H., 60
Glaubman, R., 145
Gogate, L., 47
Goldberg, S., 114
Goldman, L. R., 4, 269, 273, 278
Golinkoff, R., 70–1

Golomb, C., 109, 131, 180
Gómez, J. C., *xv*, 4–5, 9, 17, 27, 34, 254–5, 259, 267, 286, 308, 311, 313, 315
Göncü, A., 98, 179
Goodall, J. [van Lawick-], 9, 15–16, 18, 39–40, 51, 198, 209, 210, 241, 272, 277, 308, 311–12, 315
Goodwyn, S. W., 311
Gopnik, A., 113, 117, 304
Gordon, D. E., 4, 132
Gordon, W., 293–4
Gottfried, G. M., 110, 112, 134, 167
Gottlieb, G., 271
Gouin-Décarie, T., 74
Gould, C. G., 22
Gould, J. L., 22
Gould, S. J., 269, 273
Gouzoules, H., 183
Gouzoules, S., 183
Gowing, E. D., 131
Grant, C., 178
Green, F. L., 111, 303
Greenfield, P. M., 280
Gregory, R. L., 204
Grice, H. P., 6, 34, 38
Grize, J.-B., 62
Grollman, S., 310–11
Groos, K., 3, 5, 16, 23–7, 36–7, 39, 42, 309, 311
Groth, L. A., *xv*, 5, 13, 16, 19, 28, 142
Guillaume, P., *xvii*, 3, 5, 12, 15, 21, 29–31, 33, 35, 40, 73, 308, 311, 314
Guinet, C., 315
Guthrie, S. E., 8, 17–18

Hahn, E., 241–2
Haight, W. L., 91–2, 95–100
Hall, G. S., 16, 26–8
Hall, W. S., 20, 304
Ham, R., 194
Hamm, M., 20
Hannan, M. B., 223
Hanson, K., 164
Han-tih, H., 92, 99–100
Harkness, S., 271
Harlow, H. F., 202
Harmer, S., 113
Harper, S. E., 9, 13–14, 40, 229, 233, 243, 271–2, 281, 288, 308–9, 314–15
Harris, B. J., 29
Harris, P. L., 3–4, 6–7, 9, 13, 19, 21, 61, 92, 100, 112–14, 130, 135, 139, 141, 157, 235, 237, 242, 303, 313
Harrisson, B., 241
Hart, D., 73, 75, 314
Harvey, P. H., 315
Hashimoto, C., 280, 284

Hatfield, E., 6
Hawkes, K., 276
Hayaki, H., 36, 272, 314
Hayes, C., *xvii*, 13–14, 17, 34–5, 50–1, 54, 168, 210, 243, 271–2, 285, 287, 314
Hayes, K. J., 34–5, 54
Haynes, O. M., 129
Hecht, B. F., 169
Heckeby, E., 92
Herzmark, G., 132
Heyes, C. M., 74, 304
Hickerson, P., 244, 248–9, 254
Hickling, A. K., 110, 112, 134, 167
Higgins, E. T., 258, 286
Hiller, B., 288, 297–8, 302
Hillix, W. A., 9, 14–18, 40, 258, 272, 281, 285, 312, 314
Hinde, R. A., 37
Hirsch, A. D., 272
Hirschfeld, L., 166
Hodges, S., 170
Hoff, M. P., 260
Hollander, M., 123
Hooper, F. H., 4, 129, 179
Hoppe-Graff, S., 12, 15
Hornaday, W. T., 23
Howe, N., 113, 137
Howes, C., 92, 98, 135
Hoyle, H. W., 286
Hoyt, A. M., 34, 241, 309
Huber, P., 167
Hudson, J. A., 145–6, 149
Huffman, M. A., 311
Humphrey, N. K., 163
Hutt, C., 28, 132
Hutt, S. J., 28, 132
Huttenlocher, J., 258, 286
Hutton, D., 111
Huxley, T. H., 197

Ingmanson, E. J., 20–1, 280–4
Ingold, T., 270
Inhelder, B., 59

Jackowitz, E. R., 8, 310–11
Jackson, J. P., 112, 310
Jain, J., 98
James, W., 24, 28, 36, 162
Jarrold, C., 4, 13, 134, 257, 312–13
Jenkins, J. M., 112, 137–8
Jensvold, M. L. A., 14, 258, 271–2, 286–8, 297, 314
Jézéquel, J. L., 74
Joffre, K., 107, 109, 111
Johnsen, E. P., 132
Johnson, C. N., 19, 106

Johnson, D., 98
Johnson, J. E., 4, 129, 145, 179
Johnson, L. R., 144
Johnson, N. B., 269
Jolly, A., 39, 41, 186, 269, 308
Jolly, C. J., 199, 275
Jones, W., 24
Joseph, R. M., 107, 109–11, 134, 168

Kagan, J., 73, 311
Kahn, P. H., 166
Kanaya, Y., 309
Kano, T., 280
Kaplan, B., 13, 31–3, 41, 44, 50, 91, 237, 242
Katcher, A., 163
Kaugars, A. S., 105
Kavanaugh, R. D., 3–4, 6–7, 9, 14–15, 19, 61, 91–2, 94, 96–7, 100, 103, 112, 135, 156, 170, 235, 237, 242, 303, 313
Kaye, K., 14
Kearsley, R. B., 112, 310–11
Keller, H., 34
Kellogg, L. A., 33, 35
Kellogg, W. N., 33, 35
Kendon, A., 44
Kennedy, M. M., 285
Keren-Portnoy, T., 62
Kessler, M. J., 200
Killen, M., 235
King, B. J., 183
King, C. E., 13, 142
Klein, L. S., 314
Koewler, J. H. III, 310
Kohanyi, A., 170
Köhler, W., 9, 26–7, 309
Kolowitz, V., 13
Konner, M., 129
Kosmitzki, C., 271
Kotosky, L., 228
Kramer, T., 179
Krause, M. A., *xv*, 13–15, 208, 269, 272, 315
Kruger, A. C., 6, 240, 303
Kuczaj, S. A., 114
Kuersten, R., 180
Kuhlmeier, V. A., 8, 11, 16, 210, 213, 219, 224–5, 227
Kummer, H., 198
Kupfermann, K., 164
Kuroda, S., 200, 280
Kyriakidou, C., 137

La Voie, J. C., 286
Labrell, F., 309
Ladygina-Kots, N. N., 33, 35
Lang, E. M., 241
Langer, J., 243, 252, 312

Lanting, F., 198, 208, 280
Lawson, J., 52
Leakey, R. E., 275
Leekam, S. R., 117, 137
Leevers, H. J., 139
Lefebvre, L., 252
Leslie, A. M., *xv*, 3, 14, 19, 43, 47, 60, 104, 133, 212, 235–7, 242–3, 253, 255–7, 262, 268, 286, 303, 312–13
Lethmate, J., 74
Levin, G. R., 179
Lewin, R., 18, 35, 280–1
Lewis, C., 137
Lewis, M., 4, 13, 73, 165, 314
Lewis, V., 13
Leyhausen, P., 37
Lezine, I., 54, 59, 75, 311
Lillard, A. S., *xvii*, 3–5, 7–8, 13, 17, 19–20, 59–60, 71, 102–14, 120–2, 124, 127–8, 133–4, 138, 143, 167–8, 242, 303, 313
Limongelli, L., 52
Lindburg, D. G., 201
Linden, E., 168, 258, 271–2, 287, 314
Lindsay, W. L., 24, 309
Liss, M. B., 28
Lock, A., 41, 269
Lorenz, K., 8, 18, 26, 33, 36–8, 308
Lovell, L., 286
Lowe, M., 5, 13, 15–16, 18–19
Lucariello, J. 144–5, 147
Luquet, G. H., 8
Luu, V., 170
Lyons, B. G., 311
Lyons, J., 72
Lyytinen, P., 9, 310

Machiavelli, N., 204
MacKenzie-Keating, S., 145
MacKinnon, J., 251
MacWhinney, B., 115
Mahoney, D., 144, 151
Main, M., 94
Malenky, R. K., 208
Malfait, N., 114
Mandler, J. M., 43, 143
Manzig, L. A., 270
Maple, T. L., 260
Margules, S., 12, 73, 75–6, 89, 112, 114, 310, 312–14
Maridaki-Kassotaki, K., 137
Marino, L., 315
Markman, A., 228
Markovits, H., 114
Marler, P., 183
Marriott, C., 113
Marten, K., 315

Martín-Andrade, B., xv, 4–5, 9, 17, 27, 34, 255, 267, 286, 308, 311, 313, 315
Martini, M., 129
Masiello, T., 92
Matevia, M. L., 9, 14–18, 40, 258, 272, 281, 312, 314
Matheson, C. C., 135
Matsuzawa, T., 17, 243, 314
Matthews, W. S., 286, 310
Mauro, J., 169
Mayer, N., 287
Mayer, S., 8–9, 60–1, 63, 310, 315
McCabe, A., 14
McCartney, K., 271
McCune, L., 5–7, 9–12, 14–15, 17, 21, 30, 39, 41, 43–9, 54, 229, 235, 237–9, 243, 286, 291, 303, 308, 314–15
McCune-Nicolich, L., 44, 46–7, 51, 54, 59, 91, 286, 303
McDonald, K., 14, 40, 51, 168, 240, 243, 281, 309, 314
McGrew, W. C., 15
McKinney, M. L., 21, 39, 44, 46, 54, 183, 243, 269, 272–3, 316
McNamara, K., 273
McNew, S., 34, 70–1, 311
Mead, G. H., 17, 30, 164
Meck, E., 105
Meins, E., 136
Mellars, 275
Meltzoff, A. N., 75, 166, 314
Menzel, E. W., Jr., 40, 52, 74
Merker, B., 315
Meyer-Holzapfel, M., 37
Michel, J., 73
Middleton, J., 277
Mignault, C., 39, 50, 272, 311
Milbrath, C., 20, 186
Miles, H. L., xvii, 3, 9, 11, 13–14, 16–18, 40, 51, 168, 229, 233, 243, 271–2, 281, 285, 287–8, 291, 308–9, 314–15
Millar, S., 4, 38
Miller, K. F., 212
Miller, P. H., 103
Miller, P. J., 91–2, 94–7, 99, 141
Mills, E., 18, 26
Mintz, J., 92, 99–100
Mitchell, P., 111, 133, 177, 311
Mitchell, R. W., xiv–xv, xvii, 3, 5–16, 19–21, 23, 26, 29, 34, 38–41, 44–6, 53–4, 74–6, 89, 110, 113, 127–8, 157, 161, 163, 166, 167–8, 178–9, 186, 192, 194, 196, 204, 208, 229, 233, 235–7, 242–4, 249, 252, 256, 260, 263, 268, 269, 271–2, 281, 288, 299, 303, 307–10, 312–15
Moore, B. R., 315
Moore, M. K., 166
Moore, M. S., 4
Moore, T., 18
Moorin, E. R., 4, 6, 8–9, 11, 13–14, 22, 308, 312, 316
Morgan, C. L., 26
Morrow, L. M., 144, 147
Moses, L. J., 110
Mukobi, K. L., 213, 219, 224–5, 227
Munn, P., 70–1
Munroe, R. H., 277
Munroe, R. L., 277
Murphy, R. C., 309
Musatti, T., 8–9, 59–61, 63, 310, 315
Mutter, J. D., 271
Myers, O. E. [G.], Jr., 17, 154, 156, 165, 298

Nadel, J., 30, 74, 205
Naito, M., 137
Nakamo, S., 309
Natale, F., 52
Neal, M., 8
Nelson, K., xv
Newton, R., 11, 308
Nichols, S., 104
Nicolich, L., 10, 44, 46–9, 59, 91, 308
Nielsen, M., 311
Nishida, T., 15
Nissen, C., 35
Nissen, H. W., 26, 309
Noble, W., 8, 44, 270

O'Brien, C., 71
O'Connell, B., 92–3
O'Connell, J. F., 276
O'Leary, K., 112, 310
O'Neill, D. K., 70
O'Reilly, A. W., 129, 178
Oden, D. L., 219
Odum, E. P., 271
Olson, D., 112
Orlansky, M. D., 14, 41
Oughourlian, J.-M., 32
Overton, W. F., 112, 310

Pagel, M. D., 315
Painter, K. M., 129
Parke, R., 91–2, 98–100
Parker, S. T., xiv, xvii, 15, 17, 20–1, 39, 41, 44, 46, 54, 183, 186, 208, 243, 269–70, 272–3, 278, 308, 314–16
Parkin, L., 137
Partington, J. T., 178
Parush, T., 62
Patterson, F. G. P., 9, 14–18, 40, 51, 168, 258, 271–2, 281, 285, 287, 289–91, 294–6, 303–4, 312, 314
Payne, R. G., 39

Pederson, D. R., 8, 286, 310–11
Pellegrini, A. D., 60, 145
Pèpe, F., 12, 73, 112, 114, 310, 312–14
Pepperberg, I. M., 29
Pereira, M. E., 39
Perinat, A., 264
Perner, J., xv, 60–1, 71, 111, 117, 137, 257, 262, 268, 313
Perrine, R., 39
Peskin, J., 112
Pessin, A., 114
Peters, C. R., 269, 277
Peters, H., 238
Petersen, A. C., 208, 271
Peterson, C., 14
Peterson, D., 17, 314
Petrakos, H., 113, 137
Phelps, K. E., 117
Phillips, M., 291, 300–1
Phillips, R., 28
Piaget, J., xv, xvii, 3–5, 9, 11–13, 19, 27, 29–35, 39–41, 43–9, 52, 59, 62, 69, 71, 73, 91, 237, 242, 255–8, 270, 278, 286, 308, 311–14
Pinker, S., xv
Pintler, M. H., 28
Pion, N., 114
Piserchia, E. A., 311, 314
Plooij, F. X., 40, 311
Potts, R., 276
Poulin-Dubois, D., 74
Pouliot, T., 74
Povinelli, D. J., 74, 168, 252
Pratt, M. W., 145
Preisser, M. C., 39
Premack, A. J., 210, 213, 222
Premack, D., 52, 210, 213, 222, 267, 304
Preyer, W., 27, 32, 74
Priel, B., 75
Propp, V., 146–7
Psarakos, S., 315
Pusey, A. E., 277

Quiatt, D., 16, 21, 185, 308, 311

Ramsay, D., 93, 165, 314
Rapson, R. L., 6
Rattermann, M. J., 228
Rawlins, R. G., 200
Reddy, V., 10–11, 308–9
Redfern, S., 136
Reiss, D., 315
Renderer, B., 144
Rensch, B., 27, 241
Repina, T. A., 32
Reynolds, P. C., 3, 7, 12, 17, 38–9, 196, 204–5, 209

Reynolds, V., 15
Rheingold, H. L., 13
Riguet, C. D., 314
Rinaldi, C. M., 113, 137
Rivers, A., 192
Roberts, W. P., xv, 13–15, 208, 269, 272, 277–8, 315
Rodman, P. S., 198
Rogoff, B., 91
Romanes, G. J., 23–4
Rook-Green, A., 8, 286, 310
Roopnarine, J. L., 4, 129, 179
Rorty, A. O., 21
Rosen, C. S., 112, 134, 138, 168
Rosenberg, A., 167
Rosenblum, L. A., 198
Rosengren, K. S., 212
Roth, R., 197
Rottnek, M., 28
Roug-Hellichius, L., 47
Rubin, K. H., 310–11
Ruffman, T., 137
Russell, J., 313
Russon, A. E., xvii, 3, 5–6, 9–11, 15–16, 19–20, 27, 40, 45, 50, 53, 186, 194, 229–35, 238, 241, 243, 251–2, 266, 308, 311–13, 315
Rygh, J., 59, 145

Saayman, G. S., 17, 40, 315
Sabater Pi, J., 262
Sade, D. S., 189
Saltz, E., 145
Salzmann, Z., 274, 278
Sarauw, D., 315
Sarbin, T. R., 18, 25, 279
Sartre, J., 43
Savage-Rumbaugh, E. S., 6, 14, 16, 18, 35, 40, 51–2, 168, 205, 240, 243, 280–1, 285, 303, 309, 314
Scarr, S., 271
Schaller, G. B., 198, 235
Schank, R. C., 15
Schlesinger, I. M., 62
Schult, C. A., 123
Schürmann, C., 235
Schwartzman, H. B., 4, 129, 269
Schwebel, D. C., 112, 134, 138, 168
Searle, J., 45, 104
Sears, R., 28
Sevenster, P., 37
Seyfarth, R. M., 183, 188, 197
Shanker, S. G., 16, 205, 285
Shapiro, G., 230
Shapiro, K. J., 161
Shapiro, L. R., 145–6, 149
Shatz, M., 128

Sheak, W. H., 33
Shefatya, L., 131
Shepard, P., 163, 166
Shin, Y. L., 179
Shipman, P., 275
Shore, B., 271
Shore, C., 34, 93, 311
Sidall, M. Q., 286
Siegler, R. S., 128
Silbur, S., 128
Silk, J. B., 235
Simpson, M. J. A., 39
Sinclair, H., 54, 59–60, 66, 71–2, 75
Singer, D. G., 169
Singer, J. L., 112, 134, 138, 168–9
Slade, A., 51, 92–5, 97, 100
Slaughter, D., 129–30
Small, M. F., 271
Smilansky, S., 129, 131
Smith, M., 107
Smith, P. K., 4, 13, 15, 19–20, 39, 42, 54, 113–14, 129–41, 178, 257, 286, 303–4, 312–13
Smuts, B. B., 183–4, 186–7, 197, 201, 308
Smythe, R. H., 36
Snow, C. E., 15, 115
Sobel, D. M., 106, 111
Spelke, E. S., 105
Spencer, C., 224
Spencer, H., 23
Sperber, D., 312
Spitzer, S., 144
Stahl, S. M., 197
Stambak, M., 59–60
Stein, D. M., 183
Stein, N. L., 144
Stein, S., 71
Stern, C., 28
Stern, D., 162
Stern, W., 24, 27–9, 35
Stich, S., 104
Strawson, P. F., 21
Stringer, C., 275–7
Struhsaker, T. T., 197
Strum, S. C., 183, 308
Suarez, S. D., 74
Subbotsky, E. V., 117
Sugiyama, Y., 15
Suits, B., 242
Sully, J., 5, 24–5, 27–9, 31–2, 316
Suomi, S. J., 39
Super, C. M., 271
Sutton-Smith, B., 143–4, 146–7, 150–1, 177–8
Svendsen, M., 169
Swartz, K. B., 315
Syddall, S., 132

Sylva, K., 269
Symons, D., 41, 179, 202

Tanner, J., 51, 287–9, 298–9
Tart, C. T., 197
Tayler, C. K., 17, 40, 315
Taylor, A. R., 27
Taylor, M., 4–5, 7, 15, 17–18, 20–1, 30, 98, 110, 112–13, 129, 138, 167, 170, 172, 179, 286, 314
Taylor, N. D., 314
Taylor, T. J., 16, 205, 285
Teleki, G., 200
Terr, L., 4
Tessier, O., 113, 137
Thierry, B., 241–2, 249
Thompson, N. S., xvii, 3, 39, 74, 309
Thompson, R. K. R., 219
Thompson-Handler, N., 208
Thornburg, K., 311
Thorpe, W. H., 36
Tomasello, M., 6, 14, 183, 210, 240, 270, 272, 281, 303
Tooby, J., 270
Trabasso, T., 144
Trinkaus, E., 275
Tuermer, U., 98
Turkheimer, E., 108
Turner, V., 269
Tutin, C. E., 15
Tyler, S., 28, 132
Tylor, E. B., 270

Ujhelyi, M., 315
Ungerer, J., 112, 310
Užgiris, I. Č., v, xvii, 14, 310, 314

Vadeboncoeur, I., 114
Valentine, C. W., 27–8, 39, 308–9, 311
Valsiner, J., 269
Vandenberg, B., 310–11
van den Bos, R., 74, 203
van Hooff, J. A. R. A. M., 258
van Lawick, H., 199
Vasey, P. L., 5, 19, 27, 186, 241–3, 251, 266, 312–13
Venet, M., 114
Veneziano, E., xv, 5–6, 9, 13–14, 19, 21, 32, 59–61, 63, 66, 71–2, 308
Verba, M., 60
Vihman, M. M., 47
Visalberghi, E., 52, 183, 194
Volterra, V., 5, 9, 13, 40, 286
Vygotsky, L. S., 3, 5, 31–2, 69, 91, 94, 267, 273

Wachs, T. D., 271
Walker, A., 275
Wallon, H., 29–30, 74

Author index

Walton, K. L., 3-4, 8-9, 142, 151, 279
Wang, X., 92, 99-100
Warkentin, V., 12, 73, 75-6, 80, 89-90, 112, 114, 310, 312-14
Watson, D. K., 110
Watson, J. S., 24
Watson, M. W., 8, 93, 299, 310-11
Watterson, B., 170
Watts, D. P., 277
Weir, R. H., 29
Wellman, H. M., 106, 108, 110, 112, 115-18, 128, 134, 167
Wells, O., 51
Werner, H., 13, 31-3, 41, 44, 50, 91, 237, 242
White, R., 275
Whiten, A., 3, 10, 15, 20, 183, 186, 188-90, 194, 201, 229, 231-2, 235-7, 240, 242-3, 253, 263, 309, 313-15
Whitney, S., 140
Whittall, S., 113
Whittington, S., 91-2, 94, 96-7
Wilde-Astington, J., 71
Wilkins, G., 163
Williams, K., 92, 99-100

Williamson, C., 311
Wilson, F. R., 209
Wimmer, H., 117
Wittgenstein, L., 3, 11, 19-20, 32, 38
Wolf, D. P., 59, 145, 179, 310-11
Wolfberg, P. J., 13
Wooding, C., 92, 258
Woodruff, G., 52, 304
Woolley, J. D., 4, 7-8, 20, 103, 106, 110-12, 115-28, 168, 299
Wrangham, R. W., 15, 17, 197, 314
Wright, J. L., 145
Wulff, S. B., 4
Wundt, W., 23-4

Yerkes, A. W., 28, 287
Yerkes, R. M., 28, 287
Youngblade, L. M., 112, 139, 303

Zazzo, R., 73-4
Zelazo, P. R., 112, 310-11
Zeljo, A., 105
Zeller, A., 7, 9, 17, 20, 41, 183-95, 308, 311, 315
Zukow, P., 98

Subject index

accommodation, 313
action, *xv*, 120–1, 125, 145, 150
    repertoire of, 308
adaptation, 42, 162–3, 198, 209, 270–1, 274, 276, 279, 315
adornment, 275, 277, 279
affect, *see* emotion
affection, 94–5
affordances, 271, 279
agency, 152
agonism, 198–203, 206, 315
alarm calling, 193–4
alligator, 16, 146–7, 149, 150–1, 156, 289, 291, 294–5, 299–301
amusement, 244
animal, 3–9, 11–18, 20–6, 29, 32–4, 36–39, 41–2, 45, 50–1, 113, 167–9, 176–9, 203, 205, 270, 277, 299, 307–9, 311–16
    as object of pretense, 105, 108, 121–4, 150, 153, 170–1, 174–5, 285, 288, 298, 304
    see also animal-role pretend play
animal-role
    imitation, 154, 159–61
    pretend play, 16–17, 154–66, 297–9
animation, 59, 103, 105–6, 108, 112, 142–3, 150–3, 156, 162, 165, 258, 271, 285–6, 288–293, 298, 301, 307, 313
animism, 16, 27–9, 33
ant, 23, 167–8
ant-fishing, 210, 272, 314–15
anthropomorphism, *xiv*, *xvii*, 42, 169
ape, great, *xiv*, 8, 10, 12–17, 20–1, 23–4, 32–5, 42, 44, 51–4, 129, 133, 156, 168–9, 197, 203–8, 210, 222, 227–9, 240–1, 255, 258, 262, 266–8, 269–73, 275, 278, 281, 285–8, 297, 303–4, 309, 313–15
    see also bonobo, chimpanzee, gorilla, orangutan
ape, lesser, 198–9, 242, 315

appearance-reality, 138
apprenticeship, 27, 137, 315
art, 3, 8, 26–7, 38, 100, 197, 270, 275, 277, 279
as if, acting, 3–4, 7, 17, 167, 234–5, 241, 255, 260, 266, 268, 272, 275, 302, 309–10, 314
assimilation, 313
attachment, 94, 100, 132, 136
attention, 198–9, 201, 270, 283, 309, 315–16
attribution
    of function, 286
    insubstantial material, 285–6, 300–1
    insubstantial situation, 286
    see also character attribution, mental state attribution
atypicality, 230–2, 234–8
audience, 156, 258
autism, 4, 13, 133, 311, 313–14

baboon, 184, 190, 198, 201, 208, 242
babysitting, 15, 274
badger, 36, 314
bear, 26
beaver, 18, 156, 158
bee, 22
belief, 103, 106, 117–18, 120, 133, 143, 270, 273, 279, 304
    false, 60, 104, 110, 133, 136–8, 143, 256, 303
bodily translation/matching, 154, 160–2, 165, 315
    see also kinesthetic-visual matching
body part substitution, 310
body supported pretense, 310–11
bonobo, *xiv*, 6, 16, 18, 40, 51–2, 54, 198, 208, 242, 280–4
brain, 44, 105–6, 163, 196–7, 202–3, 208–9, 270–1, 278
    development, 54–5, 197, 207–8, 271, 274–5, 279
    see also neurophysiology

Subject index

branch dragging, 280
breach, 144–51
budgerigar, 270
burial, 275, 278

canalization, 13, 16, 270
canonical, 144–51
caregiving, 54, 91–101, 123, 135, 137, 142,
    183–4, 187–9, 191–5, 196, 198–9, 201–2,
    204–7, 209, 259–61, 272, 274, 277, 284
    see also grandmother
cartoon, 158
cat, 24, 156, 161, 164, 168, 170, 301–2
catharsis, 4, 22
causality, understanding of, 205
cetacean, 39, 42, 315
    see also dolphin, killer whale
character, 111, 118–19, 142–4, 146–53, 160,
    162, 287, 300, 304
    attribution, 286, 297–9
    see also impersonation
cheating, 245, 247–8, 250, 252
chest-beating, 266
child, *xiv*, 3–21, 23, 26–9, 32–5, 39, 42, 45–51,
    53–5, 59–72, 73–90, 91–101, 102–114,
    115–28, 129–30, 132–41, 142–53, 154–66,
    167–80, 196, 198–200, 204–5, 207–9,
    211–14, 217, 222–4, 227–8, 233, 235, 239,
    249, 255–8, 262, 266–8, 269–79, 281,
    285–8, 291, 298–9, 303–4, 307–16
    autistic, 4, 13, 133, 311, 313–14
    see also caregiving
CHILDES database, 115–16
childhood, 23–5, 27, 208, 269–70, 273–8
chimpanzee, *xiv, xvii*, 6, 16, 18, 26, 33–6,
    39–41, 45, 50–2, 54, 75, 168–9, 198–200,
    204–5, 208–9, 210, 213–28, 235, 272,
    274, 277, 281, 285, 287, 300, 304, 311–12,
    314–15
    pygmy, *see* bonobo
classification, 163, 165–6, 270
cognition, *xv–xvi*, 10, 26, 41, 46, 78, 87, 89–90,
    106, 109, 112, 114, 129–32, 163, 167–8, 179,
    196, 208–9, 219, 229–30, 235, 237, 241–3,
    245, 248, 251–4, 255–6, 259, 261–2,
    267–8, 269–70, 274, 276, 278, 284, 311–12
    see also cognitive, representation
cognitive
    body, 196, 203–7, 209
    development, 39, 131, 208
command, 13
commentary, 92–7
communication, *xiv–xv*, 14, 34, 38–40, 46,
    60–2, 67, 70–1, 73–4, 157, 166, 178–9,
    183–5, 191, 196, 204, 237, 259, 267, 279,
    280–1, 284, 287, 307, 309, 311

    see also conversation, language
comparative developmental
    evolutionary approach, *xiv*, 307
    method, *xv*, 43, 45, 166, 307–8, 311
complementarity, 94–6, 135, 205–6
concealment, 230–3, 236
conflict, 278, 280
consciousness, 145, 150, 165, 197
    divided, 25–6
    see also self-awareness
constructivism, *xv–xvi*
contemplation, 143
context, 11, 91, 93, 203, 208, 307–12,
    315–16
    substitution, 259, 262
    see also decontextualization
contingency testing, 309
continuity, 15, 165, 197
convention, 5, 13, 28, 159–60, 270, 273, 278
conversation, 116, 136–7
cooperative task/work group, 205–6
costume, 104
creativity, 101, 131, 169, 177–8, 197, 308
critical/sensitive period, 90, 278
cross-fostering, 10, 33–5, 40, 42, 50–1, 271–2,
    275, 281, 287, 303, 314
    see also enculturation
culture, 4, 8, 13–18, 42, 53, 55, 91–2, 97–101,
    129, 141, 142, 144–5, 160, 163, 165, 179,
    240, 269–71, 273, 278
    see also enculturation

daydream, 27–8, 115
decentering, 9
deception, *xiv, xvii*, 5, 7, 9–11, 23–6, 28, 33,
    39–40, 45, 143, 168, 185–8, 195, 229,
    235–40, 263, 278, 281, 307–10, 313
    see also self-deception
decontextualization, *xv*, 9, 11, 144, 256, 258,
    262–3, 267–8, 310
decoupling, 47, 133, 267, 312–13
design, 166, 260
desire, 106, 108, 143
development, *xiv–xv*, 13, 15, 19–21, 23, 25–33,
    36, 38–41, 43–54, 59–62, 67, 73–9, 81,
    84–5, 89, 92, 94, 98–101, 119, 122, 127–8,
    130–1, 133, 135–40, 143–53, 154, 159,
    163–4, 166, 197, 203, 207–9, 211, 235,
    256–9, 269–71, 273, 276–8, 286, 303,
    307–15
    see also brain, cognitive, language,
        sensorimotor, zone of proximal
developmental
    psychology, 23, 42, 166, 313
    rates/timing, 163, 208, 273, 275
    see also comparative developmental

discontinuity, 167, 179
displacement, 60, 266
　activity, 37
dissociation, 25–6
distancing, 32–3
distraction, 187–9, 232, 236, 262, 297
dog, 14, 22–4, 26, 36, 39, 156–7, 161–2, 167–8, 170, 175, 309–10
doll, 7–8, 12, 16–19, 21, 26–8, 31, 33, 40, 47–51, 53, 61–5, 67, 78, 93, 97, 102–4, 106–110, 121, 142, 159, 161, 168–71, 177, 210, 259, 271–2, 285–99, 303, 307, 313–14
　see also Moe task, toy
dolphin, 6, 17, 40–1, 170, 175, 314–15
domestication, 163
dominance, 183, 185, 189, 191, 196, 198–202, 277, 282–3
double-take, 231
drawing, see image making
dream, 28, 102, 197
　understanding of, 115–16, 118–20
dress up, 13, 51, 168, 272
drive state, 37

eating, 315
　mock, 285, 311, 314
　ritualized, 234–5, 237–9
ecology, 269–71, 276, 278
embarrassment, 35, 287
emotion, 36, 39, 44, 132, 142, 144, 147–50, 161–3, 165, 178, 180, 204, 281, 283, 300, 303, 307
　see also empathy, fear, sympathy
empathy, 161, 163, 280–4
enculturation, 10, 16, 163, 195, 229–30, 240, 259, 272, 287, 303–4, 314
　see also cross-fostering, culture, sign language
epigenesis, 270
ethogram, 307
ethology, 23, 36–9, 42, 203
evocation, 9–10, 12, 31, 38, 43, 46, 59, 72, 90, 161, 263–4, 307
evolution, xiv–xvi, 3, 6, 8–9, 15, 23, 38, 41–2, 44, 130, 141, 163, 196–7, 208–9, 236, 255–7, 260, 262, 267–8, 269–71, 273–6, 279, 304, 307, 315
　social, 163, 196, 269, 270–1, 273–6, 279, 304
　see also comparative developmental, zone of proximal
exaggeration, 6, 46, 59, 187, 230, 238, 295, 300, 308
exaptation, 130
experience, 4–5, 8–12, 15–16, 18, 20–2, 28, 43–4, 46–7, 52, 110, 117–18, 129, 132, 137, 140–1, 143–4, 149–151, 162, 164, 168, 176, 179, 201, 219, 228, 250, 252–3, 260, 299, 312, 315–16
　as, 8
exploration, 19, 94, 96, 99–100, 114, 132, 160, 164–5, 193, 245, 265–6, 298, 309
externalization, 104, 278
extractive foraging, 41, 315
eye-closing/covering, 5, 19, 27, 186, 241–54, 266, 313

face-to-face interaction, 198, 204–5
facial
　expression, 261, 287
　funny, expression, 300
　recognition, 207–8
faking, 296, 298–9, 303, 309–10
fantasy, 3, 27–8, 36, 98–101, 111, 115, 119, 129, 138–9, 145, 158, 167, 169, 176, 180, 248
fear, 299
　pretend, 18, 168, 285, 299
feedback loop, 271
feeding, mock, 265–6
feigning, 6–7, 23–6, 41, 168, 186–7, 230–2, 235, 238–9, 307, 309
females, post-reproductive, 276–7
fiction, 10, 12, 30, 32, 100, 115–28, 144, 146, 156, 158
flow, 197
focal animal sampling, 307
frame, 102–3, 111–12, 154, 158, 161, 188, 205, 279
free time, 15, 55, 62, 278
function, 14, 20, 23, 36, 49–50, 54, 59–61, 71–2, 73–4, 76, 80, 84, 89, 91, 98–9, 130–41, 142–3, 153, 162, 165, 172, 178–9, 206–9, 256, 268, 278, 286
　see also attribution, semiotic, symbolic

game, 5, 9, 15, 19, 22, 26–7, 32–3, 39, 41, 157, 164, 186, 188, 241–54, 261, 266, 281, 299, 309, 313
　see also play
gaze, 63, 183, 187, 199, 204, 283–4
genes, 273
gesture, 32, 34, 41, 92, 185–7, 191, 231, 239, 280
gibbon, see ape, lesser
gorilla, xiv–xv, 14–18, 26, 34, 39–40, 51–4, 198–9, 235, 241, 255, 258–68, 274, 285–304, 308, 311, 315
grandmother, 276–7
grooming, 183, 186, 194, 199, 205–7, 209, 236, 280, 283, 311
　of leaf, 16, 39

groping, 245–8, 250–2
gullibility, 279

habitat, 183, 198
hallucination, 29
hazing, 38
heterochrony, 273
hide and seek, 153
hiding, 33–4, 183, 185, 188–90, 210–12, 214–20, 222, 224–5, 227–8, 232, 281, 298
  knowledge, 302–3
history, *xv*, 23–42, 130–4, 279, 316
  life, 274–6
hominid, *xiv–xvi*, 8, 209, 267
  grade, 206–7, 209
Homo erectus, 275
homology, 203
honesty, 236, 238
human, *xiv*, 4, 6, 8, 10, 14–19, 21, 23–4, 28, 33–8, 40–2, 43–6, 50–1, 53, 113, 167–9, 176–7, 191–3, 197, 204–5, 207, 229–33, 235–6, 239–40, 255–65, 267–8, 269–79, 281, 285–8, 290–1, 298, 303–4, 309–10, 314–15
  ancestor, 255, 267, 275–7
  uniqueness, *xvii*, 115, 129, 168–9, 178–9, 286, 314
  see also child, hominid
hunting, 163, 277
hypermorphosis, 273
hysteria, 38

identification, 293
identity, 135, 163
image
  false, 231, 236–7
  making, 3, 8, 27, 262
  mental, 31, 210, 248–50, 252–4, 312–13
  see also representation
imaginary
  companion, 5, 17–18, 21, 28, 30, 113, 138, 167–80, 272, 285, 288–95, 298–9, 307, 314
  creature, 18–19, 142, 255, 281, 314
  object, 12, 23–5, 30, 32, 35–6, 40, 51, 80, 104, 112, 168, 189, 204, 237, 243, 263, 271–2, 281, 285, 302, 307, 310–12, 314
  properties, 259, 263–5, 278, 285–6, 300–1
  situation, 31, 168, 189, 242, 286
imagination, *xiv*, *xvii*, 3–6, 10, 13, 16, 18–22, 23–4, 26–33, 41, 43, 61, 67, 98, 100–2, 104, 106, 112, 135, 139, 145, 150, 156, 160–1, 177–9, 186, 205, 210, 237, 241, 248, 250, 254, 261, 263–5, 281, 286, 315–16
  understanding of, 116–18, 120

imaginings
  believed-in, 18, 279
  mandated, 142
imitation, *xiv–xvii*, 5, 7–9, 11–13, 15, 18, 21, 23–4, 26–7, 29–37, 39–42, 44–6, 50, 53, 73–6, 129, 134, 154, 159–61, 166, 179, 184, 186, 190–5, 196, 203–5, 208, 229, 242, 256, 264, 292, 300, 312–15
  see also kinesthetic-visual matching, simulation
immature activity, 36–7
impersonation, 24, 33–4, 138, 258, 307, 313–15
incomplete action, 258, 262, 302
infant, *see* child
inhibition, 23, 81, 132, 219, 236, 238, 307
innovation, 5
insect, 312
  see also ant, bee
insight, 52
instead of, acting/using, 17, 266, 268
instinct, 17, 24, 36–7, 260, 264
intelligence, 30, 36, 41, 71, 138, 206
  Machiavellian, 200, 204, 207
  sensorimotor, 30–1, 41, 46, 52, 71, 96
  social, 163
intention, 4, 6, 9, 12, 16, 18, 20–1, 24–5, 27–8, 45–7, 53, 60–1, 67, 70–2, 100, 104, 106–14, 125, 133, 154, 156–8, 162–6, 167, 178, 183–4, 187, 201, 203, 235, 237, 239, 244, 251, 256–8, 260–1, 265–6, 280–1, 286, 300, 304, 311, 315
internal state behavior, 150
internal state terms, *see* mental state attribution
internalization, 44, 47, 273, 278
interpretive diversity, 138
intersubjectivity, 154
intimacy, 95
intrusion, maternal, 95, 97
invention, 6, 11, 18, 24, 28, 30, 35, 144

justification/explanation, 14, 62, 65–9, 70, 160

keepaway, 22, 24, 39, 41, 309
killer whale, 315
kinesthetic
  empathy, 161
  memory, 209
kinesthetic–visual matching, 12–13, 21, 29–30, 75–6, 161, 163, 203–4, 207, 314–15
knowledge, 5, 9–11, 17–21, 23, 25, 28–9, 37–8, 42, 60, 66–7, 70–1, 103, 105, 108–11, 115, 117, 120–2, 124, 127–8, 133–5, 143–4, 146–7, 149, 151–3, 158, 165, 168, 179–80, 185, 191, 197, 203, 215, 270, 273, 279, 284, 302, 304, 307–9, 315–16

SUBJECT INDEX

language, *xvi*, *xvii*, 5–7, 10, 13–14, 16, 18, 21, 25, 29, 32, 40–1, 44, 46–7, 50, 54, 63, 70, 73–5, 89, 92, 95–7, 113, 115, 127–8, 135, 142, 154, 157–9, 162–6, 178–9, 197, 256–9, 269, 272, 275, 277–9, 286, 303, 308, 310, 314
   comprehension, 35, 40, 272, 287, 314
   development and acquisition, *xv*, 41, 46–7, 49–50, 59–62, 65–6, 68–9, 72, 74–5, 131, 278, 285–6
   see also sign language, syntax
   learning, 8, 11, 15–16, 21, 37, 45, 47, 50, 52–4, 74, 76–77, 83, 89–90, 100, 102, 113, 131, 163, 183, 185, 190–1, 194, 197, 201, 203–4, 207–9, 222, 233, 242, 249, 265, 267, 269–70, 273, 286–7, 302, 312
   social, 163, 209, 270, 281, 315
   see also imitation
lexigram, 54, 280
literacy, 273, 278
literal interpretation, *see* reality
luring, 233

macaque, *xiv*, 7, 16, 18, 39, 41, 52, 184–95, 198–203, 208, 241–54, 312–13, 315–16
machine, shrinking/enlarging, 212
make-believe, *see* pretense
map, mental, 22
mapping/matching, perceptual, 15, 44, 228, 255–6, 314
   intersimian, 203–4
   see also imitation, kinesthetic-visual matching, simulation
mark test, 75
mask, 281
maternal behavior, *see* caregiving
maturation, 55, 77, 82, 89–90, 274
me, 49, 74, 293, 297–8
meaning, 59–61, 63–4, 67, 70, 72, 153, 159–60, 165, 279, 287
   non-natural, 6, 38
meat consumption, 277
meme, 270
memory, 245, 250, 252–4, 267
   false, 304
mental representation, *see* representation
mental state attribution, 19–21, 28–9, 36, 45, 59–60, 67, 70–1, 102, 104–10, 112–13, 115–28, 134–8, 140–1, 142–53, 154, 156–7, 170–7, 210, 237, 272, 277, 281, 285, 287, 297, 300–1, 304, 311
   see also perspective taking, theory of mind
mental state terms, *see* mental state attribution

metacognition, 167–8, 270, 312
metacommunication, *xiv–xv*, 6, 38–9, 135–6, 179–80, 201, 257
metaphor, 163, 165
metarepresentation, *xiv–xv*, 10, 129–30, 133–6, 140, 229, 237, 239, 242–3, 249, 251–2, 256, 312–13
methodological problems, 132, 139–41
migration, 276
mimesis, 44, 278
miming, 310–11
mind, *xiv*, 111, 269
mind-mindedness, 136
mindreading, 256
mirror, 266
   image reversal, 204
   reflection, 176, 203, 208, 266, 272
   understanding, 74–5
   see also self-recognition
misrepresentation, 188
misunderstanding, 280
mock activities, 24, 265–6, 285, 311, 314
modeling, 93, 98
modulation, 287
Moe task, 7–8, 17, 19, 108–11, 121–2, 128, 138, 143
mongoose, 27
monkey, *xiv–xv*, 7, 20, 26, 39, 41, 44, 52–3, 156–7, 160, 162, 170, 175, 183–4, 186, 190–2, 194–5, 198–9, 202–3, 208, 242, 255–6, 287, 290, 315
   grade, 206–8
   see also baboon, macaque
monster, 18, 142, 281
morality, 99, 270
mother-child, *see* caregiving
motivation, 196, 201, 203, 206, 281, 300
   see also intention
Mousterian, 278
myth, 269

narrative, 143–53, 287
   thought, 14, 144–5, 149
natural selection, 23, 281, 315
Neandertal, 275–6
negation, 201
negotiation, 135, 137, 154, 158, 277–8
neoteny, 273
nest construction, 262–3, 293
neurophysiology, 196–7, 203, 206
   see also brain
niche, 271, 277–8
norm, 5–6, 12, 15–16, 18, 177, 313
novelty, 45, 311

object
  manipulation, 167, 193, 261–3, 267
  permanence understanding, 12–13, 46, 52, 73–6, 78–82, 87–90
  types of, 310–11
  see also imaginary object, prop, substitution
ontogenetic observation, 307–8, 312, 316
operations, 313
orangutan, *xiv*, 18, 24, 40, 50–4, 168, 198, 229–40, 241, 243–5, 247–54, 274, 277, 285, 291, 315
orientation, dual, 211–12, 218–19
ornamentation, 277

paleoanthropology, 269–70
Paleolithic, 275–8
pantomime, 300, 311
parental investment, 274
parrot, African grey, 14, 29, 315
peeking, 242, 244–8, 250–2
peer interaction, 136–7, 273, 277
perceptual modalities, coordination of, 207, 209, 245, 248–9, 252, 266, 313
  see also exploration, kinesthetic–visual matching
performance, 129, 135
personality, 156, 159, 169, 314
personification, 27–8, 143, 169
perspective taking, 21, 59–60, 66–7, 70–1, 102, 152–3, 164, 195, 281, 284, 303–4
  see also mental state attribution
pet ownership, 155
photograph, 11, 52, 119, 210–12, 217–19, 222
phylogeny, 206
picture, 35, 62, 116, 145
pinniped, 40, 315
pivot, 31
plan, 10–12, 15, 28, 41, 45, 47, 49, 94, 120, 135, 207, 237, 249, 251, 275, 277, 286, 315
plasticity, 279
play, 5–6, 8, 14–15, 18–19, 23–39, 41–42, 43–55, 59–62, 64–5, 91–100, 140, 154–66, 167, 169, 178, 186, 188, 190, 193, 195, 196–7, 203, 205–6, 210, 232, 235, 241–54, 258, 261, 263, 266, 269, 278, 285–6, 307–9, 311–12, 314–16
  approach-avoidance, 201–2, 205
  autotelic, 242, 249
  chasing, 6, 18, 202, 206
  ethos, 131
  face, *xiv*, 258, 260–1, 264–5
  fighting, 23, 25, 34, 38–9, 41, 167, 179, 201–2, 256

  invitation/solicitation, 184, 262–3, 314
  mothering, 16, 192–5, 259–61
  object, 8, 210, 261–2, 311
  role, 17, 20, 113, 129, 131, 139, 194, 286
  rough-and-tumble, 201–2, 206, 256, 262
  schematic, 256, 260
  signal, 258, 265, 280
  social, 207, 209, 263
  solitary, 91–2, 95
  stranding, 315
  see also doll, pretend play
pleasure, 4, 25, 30, 36–7
pointing, 127, 270
political, 200–1
polypod, 207
pongid grade, 206–8
practice, 24–5, 36, 41, 140, 202
pre-pretend, 268
pre-representational skill, 52
precursor, 4, 9, 34, 255–7, 262–3, 267, 307, 312
prediction, 163, 252
presymbolic, *xiv*, 30–1, 48, 235, 237, 243, 268
pretend/pretense, *xiv–xv*, *xvii*, 3–10, 12–22, 23–5, 27–30, 32–42, 43–55, 59–61, 63–67, 70–2, 91–101, 102–14, 115–16, 118–21, 123–8, 132–5, 138–40, 142–3, 154–66, 167–9, 177–80, 183–90, 192–5, 196, 201, 206, 209, 210, 213, 222, 229, 231, 235–40, 241–3, 245, 248–9, 253–4, 255, 257–60, 263, 265–7, 269, 271–3, 277–9, 281, 285–8, 291, 297–300, 302–4, 307–16
  as word, 111, 257, 304
  awareness of, 103–4, 112, 133–4, 308
  communicative, 34
  definition of, 4–6, 103–4
  display, 18
  fear, 18, 168, 285, 299
  signal, 300
  sociodramatic, 129, 131, 145
  understanding of, 3, 7, 27, 102–14, 115, 119–28, 168, 297–8, 316
  see also self-pretense, symbolic
pretend play, *xv*, 4, 11, 13–16, 21, 30, 42, 43–9, 51–5, 59–67, 70, 73–6, 78–80, 83, 85–90, 91–101, 102–3, 114, 115, 129–41, 142, 154–5, 157–60, 162–6, 167–8, 178–9, 204–5, 210, 212, 229, 255–7, 267, 269, 271, 273, 278–9, 285–6, 288, 303, 309, 316
  animal-role, 16–17, 154–66, 297–9
  caregiver-child, 91–101, 135, 137
  social, 257–8, 291, 303
  see also doll, toy

SUBJECT INDEX

pretend scenarios, 61, 112
   creating, 15
   elaborating, 97, 99–100, 286
   initiating, 95–8, 100, 135
   maintaining, 95, 96–8, 135
pretender, types of, 310
primate, 19, 27, 39, 42, 196–209, 258, 274
   nonhuman, 44, 46, 50, 53, 196, 203, 205, 241–2, 286
   societies, 196–9
   see also ape, baboon, human, macaque, monkey
primatology, 196–7
problem solving, 28, 52, 140, 183, 267, 315
projection, 16, 104, 143, 168, 242
prop, 4, 8–10, 12, 16, 46, 142, 145–50, 156–7, 159, 168, 279, 310
   see also doll, object, scale model, toy
proper conduct, 99, 158–60
protoculture, 278
protogrammatical linguistic system, 278–9
psychoanalysis, 163–4

re-representation, 313
reading, 123, 287
reality, 7–14, 16–22, 23–8, 30–6, 39, 46–7, 50, 52–4, 59–61, 67, 70–2, 103–4, 106–8, 111–14, 115–20, 125–6, 133–4, 138, 142, 156, 158, 168, 178–80, 187, 210–12, 236–8, 242, 248–9, 253, 255–6, 258, 260, 264, 278–9, 299, 303, 307, 309–13, 316
reasoning, 110, 122–6, 139, 167
recapitulation, 26
reciprocity, 205–6
reconciliation, 280
redirected activity, 37, 188
reenactment, 308
reference, 47, 63, 71, 267, 300
   specificity of, 212–13
   to past, 14, 62, 65–70
reinvention, 269, 273, 278
relational shift, 227–8
religion, 163, 179, 279
replication, 308, 314–16
representation, *xiv–xv*, 6, 8–13, 15–17, 29–34, 42, 43–55, 59–61, 67, 70–2, 73, 90, 103–4, 107–9, 111–13, 117–18, 120, 124–6, 129, 133–5, 142–6, 150–3, 155–6, 161–2, 165, 167–8, 180, 192, 203–4, 210–13, 217–20, 222–3, 229, 235–9, 242–3, 245, 249–54, 256, 260–2, 267–8, 277, 281, 303, 310–14
   cortical/neural, 203, 206, 209
   double/dual, 59, 70, 211
   multiple, 237, 239
   object-supported, 310

   primary/first-order, 90, 133–4, 243, 249–54, 313
   secondary/second-order, 90, 133, 243, 252–3, 312–13
   see also image, metarepresentation, misrepresentation
representational
   diversity, 124–6
   intelligence, 46
   play, 44–7, 50
   see also pretend play, symbolic play
resemblance, 6, 8–9, 16, 24–5, 36, 112, 259
rigidity, 219, 222
ritual, 163, 269
role, 12, 17, 20, 24, 26, 30–2, 39, 42, 94, 135–6, 138, 154, 157, 159–60, 163, 179, 194–5, 205–6, 239, 241, 278, 304
   enactment, 138, 255, 259, 265–6, 273, 314
   learning, 185
   play/taking, 113, 129, 131, 139, 194, 286
   reversal, 185, 206
   see also perspective taking
routine, 311
rule, 31–2, 42, 159, 164, 245

scaffolding, cultural, 13–16
scale model, comprehension of, 11, 210–28
schema, 5, 15–19, 30–2, 41, 144, 209, 256, 264–5, 267–8, 311
   see also script
schematization, 262, 266, 268
school success, 101
script, 10–11, 13–18, 32, 41, 96, 129, 134–5, 146, 159–60, 233, 236, 263–5, 273, 313, 316
   violation, 16, 39
scrounging, 233
search pattern, 215, 218–222
seeing as, 8
self, 24, 30, 32, 155, 162, 164–6, 251, 254
self-awareness, *xiv, xvii*, 40, 155, 164–5, 287
self-concept, 252
self-deception, 25–6
self-handicap, 39
self-illusion, 25–6
self-pretense, 9, 41, 50, 185–7, 195, 229, 237, 307, 315
self-recognition, *xvii*, 12–13, 21, 29, 35, 40, 42, 51, 73–6, 78–80, 83–90, 113, 203, 208, 272, 297–8, 312–15
self-stimulation, 242
self/other comparison, 7, 11, 15, 21, 60, 71, 110–11, 128, 138, 154, 162, 165, 204, 281
semiotic function, 91, 256, 268, 278

Subject index

sensorimotor, 207–8, 243
  development, 11–12, 30–1, 43–52, 59, 71, 73, 96
  intelligence, *xv*, 41, 46, 52, 71
serious interpretation, *see* reality
sex differences, 24, 26, 28, 39, 92, 100, 147–8, 155, 219, 221, 244, 246
sexual expression, 196, 207–9, 280
sharing, 200, 207, 209
siblings, 91–2, 94, 98, 136–7, 155, 277
sign language, 13–15, 18, 40, 54, 169, 229–30, 240, 243, 271–2, 278, 285–8, 296–7, 304, 314
  see also language, symbol
signification, 10, 30–2, 59, 61, 73, 91
similarity, 7–9, 210, 224, 228
  see also resemblance
simulation, *xiv–xv*, 3, 5–11, 23–4, 26, 30, 33, 36–40, 42, 45–6, 59, 113, 135, 161, 166, 184–6, 188, 190–5, 196–7, 201–9, 237–9, 244, 249, 264–5, 281, 308–9, 311
simulative modality, 196–7, 201–9
skill training hypothesis, 178
social
  centrality, 198–200
  cognition, 91, 113–14, 229, 280–2, 284
  organization, 30, 196, 198, 207–9, 275, 277, 280
  theory, 199–201, 206
  see also learning
sociality, 91–101, 113, 135, 138, 144, 156, 188, 196, 198, 203–4, 206, 277–8, 280–1, 284, 291, 315
socialization, 99, 156, 159–60, 163, 165
sociolinguistic understanding, 287
sofa construction, 262–3
speaking for, 142, 145, 151–3, 291
speech, *see* language, sign language
speechlessness, 25, 154, 158–9
status, *see* dominance
stereotypy, 18, 27
stimulation seeking, 249
story, 14, 123, 135, 142–53, 287, 300–1
strange situation, 94
strategy, 41, 92, 98, 236, 252, 263
stubbornness, 302
style, 100, 277
subjectivity, 112, 134, 161
substitute activity, 37
substitution
  body part, 310
  context, 259, 262
  object, 17, 23–4, 30–2, 36–9, 41, 47, 51, 61, 67, 70, 96, 113, 133, 212, 237, 255, 258–62, 268, 271–2, 286, 291, 293–7, 300–1, 307, 310–12

symbolic, 71, 256
  see also grooming of leaf, imaginary object
surreptitiousness, 232–3
symbol, *xiv*, *xvi*, 11–13, 27, 29, 31, 34, 44, 60, 73, 90, 196, 201–3, 206, 208, 212–13, 222, 235, 237–40, 242, 256–8, 260, 263, 267–8, 269–73, 275–9, 281, 287
  use, 40, 240, 255, 258, 267–8, 280, 286
symbolic
  function, 73, 76, 84, 89, 268, 313
  play/pretense, *xiv–xv*, 4, 11–12, 27, 30, 32, 41, 91, 93, 129, 210, 213, 229, 237–40, 255–6, 260, 267, 269, 271
  see also substitution
symbolism, *xv*, 6, 42, 94, 163, 166, 286, 303
symbolization, 30–1, 47
sympathy, 20, 29, 163, 281
synchromimesis, 6, 205
syntax, 47, 50, 54, 66
systems approach, 271, 276–8

teaching, 6, 41, 90, 99–100, 193, 229, 240, 268, 270, 272–3, 303, 309, 315
  see also cross-fostering, enculturation
teasing, 5, 23, 26, 38, 41, 298–9, 307–9
technology, 197, 205–6, 273, 277
temperament, 230
tension, 280, 282–3
theater, 115
theft, 231–2, 236
theory of mind, *xiv*, 19–21, 59–60, 71, 116, 129–30, 132–41, 142–3, 153, 268, 303–4
  see also mental state attribution, perspective taking
thinking, 14, 17, 19, 32, 43–4, 102–3, 105–6, 108–9, 111–13, 115, 121–6, 134, 142, 144–5, 149–50, 163, 166, 304
thought bubble, 107, 112, 121, 123–7
threat, 38, 183, 193–4, 200
  veiled, 39, 187
threshold effect, 130
tonic lithogrip, 207, 209
tool, 90
  making/modification, 40, 270, 312
  social, 14, 145, 236, 267, 270
  use, 33–4, 40–1, 46, 51–3, 183, 204–5, 210, 270, 275, 278, 312, 315
totemism, 166
toy, *xvii*, 13, 16, 18, 27–9, 32–3, 35, 47–9, 51, 59, 62–5, 80, 109, 115–16, 142–53, 157, 170–1, 204, 211–12, 219, 221, 223, 229, 243, 271–2, 285, 288, 290–1, 293, 295–6, 298–9, 301–4, 307, 310
  see also doll, prop, scale model
tracing, *see* image making

transaction, 271, 278
transitional object, 170
trauma, 4, 22, 164
trial and error, 52
trying, 106–112
Twin Earth, 114

universality, 129, 133, 198, 255, 273

vacuum activities, 18, 37
verbalization, 59–72
　see also language, sign language

vigilance, 262
vocalization, 183, 187–9, 193–4, 258, 260, 264
voluntary control, 207–9, 238, 242

what if, 5, 27

zone of proximal
　development, 94, 273, 278
　evolution, 267

Printed in the United Kingdom
by Lightning Source UK Ltd.
102480UKS00002B/182